CONSTRUCTION SUPERVISION

CONSTRUCTION SUPERVISION

JERALD ROUNDS and ROBERT SEGNER

WILEY

John Wiley & Sons, Inc.

Library of Congress Cataloging-in-Publication Data:

Rounds, Jerald L.
    Construction supervision/Jerald L. Rounds, Robert O. Segner.
        p. cm.
    Includes index.
    ISBN 978-0-470-61496-9 (hardback); 978-0-470-95038-8 (ebk); 978-0-470-95062-3 (ebk); 978-1-118-00987-1 (ebk); 978-1-118-00988-8 (ebk); 978-1-118-00989-5 (ebk)
1. Building–Superintendence.   I. Segner, Robert O.   II. Title.
    TH438.R675 2011
    690.068–dc22

                                                                2010042178

Printed in the United States of America

SKY10067435_021624

DEDICATION

This book is dedicated to a stronger construction industry through better-prepared supervisors.

CONTENTS

PREFACE

THE ROLE OF THE SUPERVISOR

Supervision is a topic of fundamental importance to the construction industry. The construction supervisor is the person who plans, directs, and coordinates onsite activities that result in turning drawings and specifications into reality. The supervisor is in the most influential position of anyone on or off the job site to make or break a construction project, to make the project profitable or unprofitable, to make the job safe or unsafe, and to develop satisfied customers or lose dissatisfied ones.

The supervisor's job is highly complex and requires extensive knowledge and skills. The supervisor must be familiar with the craft skills required to execute the project. He must be highly skilled in communication, both written and oral. He must understand how to work with and engage people at all levels through a variety of legal and organizational relationships. The supervisor must be able to read, interpret, and execute construction contracts and must be aware of human resource law, as well. She must be able to plan, schedule, and coordinate the work of the project. She must understand construction costs and the interaction of cost, schedule, production, and quality, all within the context of maintaining a safe work environment. According to Don Kawal, retired CEO of Klinger Constructors, Inc., Albuquerque, New Mexico:

> A field supervisor is a valuable and key member of the project team transforming the project documents into a physical facility. It takes a variety of skills to make everything happen in the field. Projects have technical expectations which are growing in complexity such as new materials and assemblies, higher quality requirements, green accounting, and safety commitments. Effectively motivating craft performance whether company or subcontractor employees requires planning, creativity and understanding. Making the best from limited resources whether labor, tools, or equipment is typical. Each day is different with crew composition changes, unexpected deliveries, weather changes and revised scopes of work. The supervisor manages risk and uncertainty with plans with built in alternatives for surprises. The supervisor is always looking for better methods that are safer and more productive. The tough and barking

supervisor does not get the job done. It is a nimble, clever and people oriented supervisor that experiences the best results. Personal satisfaction with the quality and quantity results experienced every work day is the primary motivator.

David Long, Senior Vice President of Miller Electric Company, Jacksonville Florida also recognizes the importance of the supervisor.

We know the field supervisor is a key and unique team member within our organization. Our field supervisor is expected to know the field installation aspect of our projects such as efficiency, safety and productivity as well as project documentation, contracts, regulatory compliance and accounting. On all projects the field supervisor will have a great impact upon the successfulness of the project both financially and relational. Education for the field supervisor is essential for any successful electrical contractor.

The first-line supervisor in construction, the foreman, has moved from the ranks of craft labor onto the first rung of management. This makes supervision a radically different position from that of craft worker, which they came from. As management, they now represent the company and make decisions that can have far-reaching consequences for it. However, foremen often move back and forth between supervision and craft work. As a consequence, they must handle their workers skillfully, because the craft worker they manage this week could be the foreman for whom they work next week.

Unfortunately, there are few books that focus on construction supervision. There are many fine books on project management and specific aspects of project management, such as project scheduling and project controls. There are many books on construction materials and methods and on managing production, focused either broadly across the industry or on specific types of construction. There are books that cover various aspects of construction cost estimating, bidding, and cost control. There is, however, a severe lack of books on supervising field operations.

Ten years ago the authors embarked upon a project to develop an extensive training program for construction supervisors. They built upon their experience in the industry, together with background in both practicing and teaching project management, to develop a nine-day field supervision training program. In doing so, they were struck by the fundamental importance of the supervisor's role and the lack of books on field supervision. The authors feel that this book will begin to fill the gap in texts available to support field supervisors.

INTENDED AUDIENCE

The book is intended to meet the needs of a diverse group. First, it is intended to be beneficial for supervisors at all levels, including foreman, general foremen, superintendents, and general superintendents. There is basic material in each chapter that will be useful to the new foreman who has recently entered the supervisory track as well as to the experienced supervisor who has forgotten some of the elements of basic foremanship. There is also more advanced material in each chapter that can expand the horizons of the less experienced supervisor, while expanding the knowledge base and skill set of the experienced supervisor.

The book has been centered around supervision in the commercial sector of the construction industry. It will, however, have much to say to those engaged in residential construction, as well as those in the industrial and heavy civil sectors. With extensive background in general construction, and more than a decade of training in the specialty areas, the authors believe that the book is applicable broadly across construction specialties, as well as for general construction.

The book is also designed to form the basis of a semester college or university course in Construction Supervision. Over the many years the authors have spent in higher education, they have observed few courses on construction supervision. They hope, therefore, that the existence of the text will cause more construction educators to think about adding a course of study in field supervision, since it plays such a key role in the construction process.

THE BOOK'S STRUCTURE AND ORGANIZATION

The book has been divided into four sections with a total of 20 chapters. The intent of *Section I: Setting the Stage* is to provide an overview of the construction industry and to provide a basic understanding of the characteristics of the supervisor and the job of supervision in the construction industry.

Section II: Soft Skills focuses on the nontechnical areas that are of fundamental importance to successfully carrying out the duties of a supervisor. Chapters in this section cover communication, both oral and written, team building, managing employees to embrace diversity and eliminate discrimination, and managing the human resources through leadership and motivation. The final chapter provides practical guidance in solving problems, since a significant portion of the supervisor's day is taken up with recognizing and dealing with problems as they crop up.

Section III: Technical Skills addresses the more technical areas of supervision. It begins by establishing the importance of a safe work environment and then provides guidance in how to secure that safe work environment. Next, the contract is introduced as a very powerful tool to help manage field operations. The next chapter addresses managing the physical resources on the job, including the materials and installed equipment that go into the project, as well as the tools and construction equipment required for that installation. The next two chapters focus on developing an accurate projection of job costs and then managing the actual costs to maintain them within the projection. The last two chapters in this section address control of time and the management of the production processes.

Section IV: Project Supervision considers the overall project, first presenting project organization and then moving successively through mobilization, supervising throughout the project, and closeout of field operations. The intent of the last chapter is to provide guidance to the supervisor in how to develop professional skills necessary to successfully supervise projects well into the future. Because construction is a dynamic industry that is rapidly evolving, the book closes with suggestions for how supervisors can prepare themselves to maintain relevancy in an industry that is continually changing and prepare to move up the supervisory ladder.

MAKING THE MOST OF THE BOOK

Since the book is designed to meet the needs of two distinct groups, each will be dealt with separately.

For the supervisor, the book can be read in sequence, starting at the beginning and moving through successive chapters as time permits. The supervisor should reflect, as she reads the book, on how the information can be useful in providing a better understanding of such things as the supervisor's roles and responsibilities, and why things happen the way they do on the job. It should also be useful in providing tools that can be used to better control field operations within the sphere of responsibility of the supervisor.

The supervisor can also use the book as a reference for specific topics when certain situations arise. Each chapter is designed to stand alone so that, when a supervisor is confronted by a situation or when a question arises, the supervisor should be able to look at the relevant chapter for guidance.

In an academic environment, the book is generally structured to enable a chapter or two to be covered each week of a semester. The flow of the information should be appropriate to introduce supervision to college students without a great deal of construction industry experience or background and to lead them to an understanding of the critical role the field supervisor plays, what supervision is, and how supervision is best accomplished. A very practical time to use such a course in an academic environment would be toward the end of the program of studies, where the course demonstrates to students the use of tools they have learned in other courses, such as scheduling, cost control, and construction contracts, all focused on efficiently executing the construction methods introduced earlier in the course of studies.

ACKNOWLEDGEMENTS

The authors would like to thank all the people who have contributed to this book, but especially some who have made notable contributions. First, we would like to recognize the contribution of Mr. Albert Wendt, whose vision for supervision, supported by a very generous contribution to the Electrical Contracting Foundation, enabled the authors to develop the Electrical Project Supervision course from which the inspiration for this book came.

We would also like to recognize the patience and unwavering support of our wives, Beverly and Suzanne, in making this book possible.

SECTION I

SETTING
THE STAGE

CHAPTER 1

OVERVIEW OF THE CONSTRUCTION INDUSTRY

INTRODUCTION

The construction supervisor functions in a business environment that is fraught with challenges and filled with opportunities. In this environment, the work is demanding, both physically and mentally. The environment is permeated by risk and uncertainty. There is endless variability in the type of work to be done on construction projects and among the people who manage and perform the work, and among the contract systems and project delivery methods being employed. Seemingly, the industry grows more complex and more demanding every day.

Yet the construction industry can also be tremendously rewarding. For those who learn how to manage the elements of challenge presented by the industry, the benefits are both numerous and long-lasting. Financially, construction can be very rewarding to those who are successful in the practice. The numerous variables that present risk and uncertainty also render the work endlessly challenging and interesting. And certainly, few other professions offer the very tangible fulfillment and sense of accomplishment that the construction industry provides. Almost without exception, those who have had a hand in the building of a construction project are able to view the completed project and, with a sense of enormous pride and satisfaction, to say, "I built that."

This book will begin by making note of some characteristics and truisms pertaining to the construction industry. It will also be noted that, for the construction supervisor, the construction industry is filled with enormous opportunity, especially for those who are willing to learn how to manage within the industry, and who are willing to continue their learning so as to meet the continuing challenges and complexities of an evolving industry.

CONSTRUCTION VOLUME AND IMPACT ON SOCIETY

The construction industry has long been, and certainly continues to be, a major force in the economy of the United States. By any standard of measure—fraction of U.S. Gross Domestic Product comprising construction, amount of direct and indirect employment of the U.S. workforce comprising the construction industry, number of workers employed in the industry, or percentage of the total U.S. workforce employed in the industry—the construction industry is a huge component of business in the United States. Additionally, construction is increasingly becoming a globally integrated industry, as more and more U.S. construction firms work internationally, and as more and more firms from around the globe perform work in the United States.

In addition to being very large, the construction industry is also very diverse. Work is performed in many different industry segments, including commercial, institutional, and residential buildings. These buildings are typically designed under the leadership of architects and are frequently referred to as "architectural construction."

Additionally, the industry includes what is referred to as "engineered construction facilities." Examples include industrial facilities, such as refineries, processing plants, fresh water and wastewater treatment plants, and manufacturing facilities, as well as utilities, pipelines, transmission lines, roadways, airports, bridges, dams, and so forth. These types of facilities are typically designed by engineers; hence the term "engineered construction projects."

A great deal of construction work is characterized as new construction—the construction of a new facility on a vacant site. Additionally, a large segment of construction work consists of remodeling, restoration, renovation, and adaptive reuse of existing facilities.

Construction work is performed by companies, which may be large or small, in terms of volume of work performed and number of people employed. Some of these firms, known as specialty contractors, specialize in a particular market segment, while others, known as general contractors, choose to encompass a broad scope of types of work performed.

Construction work is performed by a number of different project delivery methods, including design-bid-build, also known as linear construction; design-build; design-procure-construct (DPC); phased construction, also known as fast track; job order contracting; and others. Additionally, a number of different types of contractual arrangements are employed in the performance of the work, including single contract contracting; multiple prime contracts, also known as separate contracts contracting; construction management agency and construction management at-risk; and others. Construction contracts may be defined as public or private, and may take the forms of lump sum, also known as hard money, unit price, cost reimbursable, and others. In addition, construction contracts are awarded by a number of different methods, including competitive bid, negotiated, competitive sealed proposals, and others.

For the construction supervisor, this tremendous volume and diversity in the construction industry, equates to opportunity. The business community, and society in general, place a huge reliance upon the construction industry. This means that, for the person skilled in the performance and management of construction work, opportunity abounds. One of the primary purposes of this book is to add to the knowledge base of those who supervise construction projects, so as to enhance their opportunities for continuing success in the construction industry, as well as to enhance the chances for a successful project for the owner.

PROFIT, PROFITABILITY, AND
THE SUPERVISOR'S IMPACT

In a free enterprise economy, the basic reason for a construction firm to be in business is to earn a profit from its performance of construction contracts. This fundamental premise is central to the operation and continuance of the business enterprise.

Therefore, the construction work that a company undertakes must be performed in compliance with the requirements of the contract documents for each project, and also must be performed at a cost equal to or less than the contracted cost of completing the work. This implies, in turn, that the work must be performed and managed with cost consciousness and budget consciousness in mind, and in such a way that the company will earn a profit from the performance of the work.

The construction supervisor plays a huge role in determining the profitability of the construction work that a construction firm performs. As the management person closest to the workface, that is, to where the work is actually performed by skilled construction craft workers, the supervisor continually makes decisions and takes actions that directly affect the cost of the work, as well as the duration of the project and the quality and the safety of the work.

While many others in the construction firm also have a role in ensuring the profitability of construction projects, it is the supervisor who plays a central role. Therefore, it is incumbent upon the supervisor to be knowledgeable of the environment in which the work is performed, of the work itself, and of the best way to perform the work, and also to have the knowledge to perform the work in such a way as to fulfill all of the objectives for each project, including profitability.

COMPETITION, RISK, AND CONSTRUCTION COMPANY FAILURE

In addition to the other truisms regarding the construction industry that have been noted, certainly there are three additional important aspects of the construction industry that are also worthy of note. First, the construction industry has been for many years, and certainly continues to be, one on the most highly competitive of all industries. Second, risk permeates the industry in general and construction operations in particular. Third, the failure rate for construction companies is among the highest of any business.

Competition is at the heart of most contract awards in the construction industry. For many years in recent U.S. history, owners (who decide what the contract award system will be for a construction project) have employed lump-sum competitive bidding or unit-price competitive bidding as the primary method for awarding contracts to prime contractors. In these contract award systems, a series of documents, called the contract documents for the project, describes in detail the work to be done and sets forth the owner's requirements in the work. With this complete set of contract documents in place, contractors who are interested in the project will prepare proposals, or bids, wherein they set forth their proposed prices on a lump sum or unit price basis. These are the prices for which they would be willing, if selected by the owner, to enter into a contract to fulfill all of the requirements of the contract documents in constructing the project for the owner.

Contractors' proposals are submitted to the owner on a specified date and time, and in a designated location. Typically, the contractor who submits the lowest bid, or the lowest valid bid, is selected by the owner to be the contract recipient. So contractors are in competition with one another for the contract award, based on the lowest price for which they are willing to enter a contract to fulfill the contract requirements as established by the owner.

The rationale on the part of the owners in their use of this contracting method is that if the contract documents completely describe all aspects of the work to be performed, and if all of the bidding contractors prepare their proposal prices based on this same information, then the owner will receive the benefit of all of the contractors competing with one another for the contract award. The owners can then make a decision based on the price submitted by each contractor. Thus, the owner will know what he or she will receive, that is, what the contract deliverables will be, as described in the contract documents. Additionally, the owner should be able to have the work performed at the best possible price, based on the competition by contractors for the contract award.

While other methods of contract award are frequently employed today, competitive bidding is still very commonplace. Even when methods of contract award other than competitive bidding are utilized by owners today, competition among contractors for the award of the contract from the owner remains central to the project delivery method of choice. The competition may be based on many different criteria, such as contractors' record of successful projects completed in the past, the quality of work performed, quality assurance programs, safety records, qualifications and credentials of the contractors' personnel, and so forth.

Competition among contractors for the contract from the owner remains at the heart of owners' seeking to obtain maximum value for their construction contract expenditures. Additionally, in the same fashion in which owners place contractors into competition for the award of the prime contract (the contract between the owner and the prime contractor), prime contractors typically employ a competitive methodology with subcontractors for the award of the subcontracts to the specialty contractors who will be selected to perform work on the project.

From the foregoing, it can be clearly seen that competition among contractors is, historically and today, very deeply embedded in the culture of construction contracting. Every indication is that this fact will remain a constant in the industry. Construction contractors continually are in quest of some means of achieving a competitive advantage, so as to maximize their prospects for the award of construction contracts.

Like competition, risk is an element that is constantly present in the construction industry. The risks that a construction contractor must overcome, in a company and on a project, are both numerous and significant. A great deal of the work on a construction project is inherently hazardous in many of its aspects. Construction sites, and the environment in which the work on a construction project is performed, vary greatly with regard to numerous factors, which translate into risk for the construction contractor. Heavy and cumbersome materials and equipment must be installed under a variety of conditions, and often at considerable heights or depths, or in confined spaces. Many of the materials and systems to be installed in a construction project are themselves inherently dangerous. The productivity of the skilled construction labor force is subject to many variables, and thus production rates become undependable and difficult to predict. Many construction operations are sensitive to weather, and the variability and unpredictability of the weather can wreak great havoc on a construction project. The dependability of suppliers, subcontractors, and the skilled construction craft labor force, varies widely. Much of the work involves the use of machinery and equipment, and its use introduces other elements of risk into the process. Many construction components, and systems and subsystems, must be installed with great precision—even the smallest error, or the smallest deviation from a standard, can render a system inoperative or dangerous. Many construction projects consist of a very large quantity of materials and products, each of which must be procured, managed, and properly installed; their sheer number provides a management challenge and introduces risk into the construction process. The dollar amounts—in the construction contract amount, subcontract amounts, materials, equipment, and labor prices—are huge. By their very nature, the financial commitments on a construction project are a source of risk. Contract award methods that place contractors and subcontractors into competition for the award of construction contracts introduce risk. Successful contractors and successful construction supervisors are those who come to terms with the risk inherent in the business, and who learn methods to recognize, mitigate, and deal effectively with the numerous risks that the work in the industry entails.

In an environment of intense competition and enormous risk, many construction companies do not survive. In fact, for prime contractors and subcontractors alike, the failure rate of construction businesses is among the highest of any business classification. Those companies that endure and are successful are those that recognize and successfully come to terms with the competition, the uncertainty, and the risk that are inherent in the construction business. Supervisors who can recognize these facts, and who can learn to manage in this uncompromising environment, are those who are much more likely to succeed in the future.

DEFINITIONS AND ROLES OF CONSTRUCTION TEAM MEMBERS

A number of different people compose the team that is typically formed for the construction of a project. It is important for the supervisor to recognize the names, typical functions of, and typical relationships between these people. These are set forth in the paragraphs that follow, and are graphically depicted in Figure 1.1.

While there may be some amount of variance in who these people are, and in the roles they play on the construction team, according to the type of project and by the project delivery method being employed, these parties and their functions and relationships are typical of those that are utilized on a building construction project, with a single contract system in use, and with a lump-sum competitive bid contract award method in use.

Owner

The owner initiates the entire design and construction process, and all of the events that follow, when he or she perceives a need for a new facility, for additional space, or for renovated space.

The owner analyzes finances, determines budget, determines equity and borrowed capital necessary. The owner seeks the services of a professional designer, architect or engineer, to produce a design and to lead the design process. He or she enters a contract with the architect or engineer of choice.

The owner's basic expectations of designer are that he or she will:

- Produce a design that will satisfy the needs of the owner, within the constraints of the owner's budget

- Assist the owner in forming a contract with the construction contractor

- Oversee the construction of the project by the construction contractor, to protect the owner's interests

- Assist the owner during the contract warranty period

The owner enters into a contract with the contractor, as well as with the designer, and throughout the project coordinates the work of the contractor and designer, funds the project as it proceeds, works with the design team to resolve problems, and accepts the project when it has been completed.

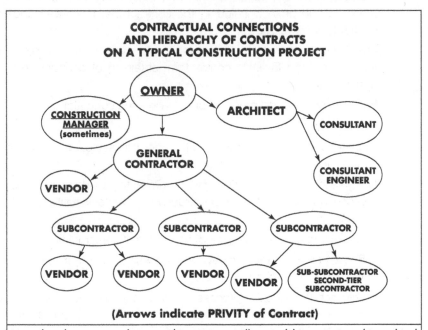

CONTRACTUAL CONNECTIONS AND HIERARCHY OF CONTRACTS ON A TYPICAL CONSTRUCTION PROJECT

(Arrows indicate PRIVITY of Contract)

Note that the contracts between the parties as illustrated here are in a hierarchical arrangement, with the owner at the top of the hierarchy.

The solid lines (arrows) between parties represent the existence of privity of contract, that is, the existence of an actual contractual connection between the parties. In the method of project delivery depicted here, there is privity of contract between the owner and the architect, and there is privity between the owner and the prime contractor; there is no privity of contract between the architect and the prime contractor or the owner and the subcontractor.

Figure 1.1   Contractual Connections and Hierarchy of Contracts on a Typical Building Construction Project

Architect or Engineer

Architect—for building construction projects.

Engineer—for industrial facilities, fresh water and wastewater treatment plants, and manufacturing facilities, as well as pipelines, transmission lines, utilities, roadways, airports, bridges, dams, and so on.

The architect or engineer is referred to as the primary designer, or as the designer of record.

The architect or engineer enters a contract with the owner to provide the following basic services:

- Assist the owner in development of owner's program of requirements

- Produce a design which will satisfy the needs of the owner, which can be built for the amount of money in the owner's budget

- Produce drawings, specifications, and other contract and bid documents for the project, including the contract that the owner and the contractor will execute
- Assist the owner with making contractors aware of the existence of the project, and facilitate contractors' obtaining of contract and bid documents
- Facilitate contractors' proposal preparation and submittal
- Conduct the bid opening
- Counsel and assist the owner with regard to selection of, and final contract formation with, the contractor
- Observe the construction during its progress, to provide reasonable assurance to the owner that the construction contractor is fulfilling contract requirements
- Administer change orders during the course of construction
- Approve prime contractor's payment requests and authorize the owner to make payment to the contractor
- Administer the punchlist process, as well as the project closeout process, issue the Certificate of Substantial Completion, and administer the contractor's Request for Final Payment
- Assist the owner during the warranty period

Consulting Engineer

The consulting engineer is an engineer who has professional expertise in a system or component to be included into the design by the primary designer, whether the primary designer is an architect or an engineer. The consulting engineer is usually retained and paid by the primary designer. Typical examples include civil engineer, structural engineer, and mechanical engineer, among others.

Consultant

The consultant is a person who may not be an engineer or an architect, but who is an expert with regard to a product or system that is to be incorporated into the project. The consultant is utilized by the primary designer, architect, or engineer, to provide assistance with some aspect of the design. Usually, this person or firm is retained by the primary designer and is paid directly by the primary designer for his or her services on a consulting basis.

Construction Manager

A construction manager is sometimes utilized on construction projects and sometimes not, at the election of the owner. Construction managers function in different capacities, and their responsibilities vary considerably, depending upon the terms of their contract with the owner.

A general definition, which will serve us well in the context of considering team members on a construction project, is: the construction manager is one who enters a contract with the owner, and by the terms of that contract, represents the interests of the owner in his contracts with the architect (or engineer), and with the prime contractor.

Construction managers may function in an "agency" capacity, wherein they are legally bound to act in the best interests of the owner, and they provide consulting, counsel, and assistance to the owner; the owner then acts upon that counsel or not, at his discretion.

Additionally, construction managers may function in an "at risk" capacity, wherein their contract with the owner no longer recognizes them as an agent of the owner, but makes them financially responsible for delivering the project to the owner within specified limits of time and money. This form of construction management contract is often referred to as CMAR, construction manager at risk.

Prime Contractor

The prime contractor enters into a contract with the owner to fulfill all of the requirements set forth in the contract documents. The prime contractor usually provides all materials, labor, equipment, support staff, and other resources that are necessary to do so.

A person is defined as a prime contractor inasmuch as he or she enters into a contract with the owner. Usually contractors who function in this capacity refer to themselves as general contractors, building contractors, and similar titles.

Subcontractor

A subcontractor enters into a contract with the prime contractor to perform a defined segment of the work on a project. Traditional subcontractor or specialty contractor trades include: plumbing; electrical; heating, ventilating, and air conditioning; masonry; roofing; drywall; tile setting; glazing; and the like. Today, subcontractor specialists are available to perform almost any task or scope of work on a construction project.

Sub-Subcontractor

The sub-subcontractor enters into a contract with a subcontractor to perform some specified aspect of the work for a project for the subcontractor. This person is also referred to as a second-tier subcontractor.

Vendor or Materials Supplier

The vendor or materials supplier enters into a contract with the prime contractor or with a subcontractor, to provide a material or products specified for the project. This person provides no labor for installation. The contract for materials purchase is often referred to as a purchase order, or purchase agreement, or purchase order agreement.

BECOMING AN EFFECTIVE SUPERVISOR

It is within this complex and demanding domain, where many different people have important roles to play, where numerous different contracting methods are employed, and where risk, variability, and uncertainty are constant companions, that the construction supervisor functions. Supervising and managing effectively in this environment can be a daunting task. As has been noted, however, it is also a most fulfilling and rewarding opportunity.

Those who can meet the challenges, who can master the complexities, who can be effective managers and decision makers, and who can consistently fulfill project objectives are those who have assured themselves a bright and rewarding future in management in the construction industry. Providing guidance and assistance to the person who aspires to these goals is one of the primary objectives of this book.

SUMMARY

Key points of learning in this chapter include the following:

- The construction industry is huge, and diverse, and has a tremendous impact upon the economies of the United States and other companies around the world;
- Opportunities for success and satisfaction abound for those who are successful in construction;
- Construction companies are in business to earn a profit, and those who supervise construction work have an enormous impact on the profitability of construction work;
- Competition is intense and risk is high in the construction industry; but there are tremendous rewards for those who can manage in such a way as to be successful;
- Construction is performed by a number of different contracting and contract award methods;
- There are a variety of people who comprise the construction team on projects: owner, architect, engineer, consultant, construction manager, prime contractor, subcontractor, sub-subcontractor, materials suppliers;
- Effective supervisors, those who can consistently fulfill project objectives in the complex and demanding environment of the construction industry, have a bright and rewarding future in the industry.

Learning Activities

Obtain a copy of Engineering News-Record magazine, and the issue containing the "ENR Top 400 Contractors" feature, published every year in May.

Additionally, you may wish to read the "Top 600 Specialty Contractors" feature, published in the magazine every year in October.

The ENR magazine may be subscribed to by someone in your company, or you can obtain a copy from your local library.

You can also view these feature articles of the magazine online at: http://enr.construction.com/toplists/

These features, and the other information in the magazine as well, will help you not only to see who the largest firms are, but also will enable you to see the tremendous impact the construction industry has on all aspects of the U.S. and worldwide economy.

You can broaden and enhance your own learning with this information. In addition, you may wish to share some of this information with others in your company in your "toolbox talks."

CHAPTER 2

SUPERVISION AND THE SUPERVISOR

INTRODUCTION

The supervisor, and the function and domain of construction supervision, can be defined in a number of different ways. Company policy and company organizational plan certainly will influence the definition within a construction company. Collective bargaining agreements frequently contain their own elements of definition. A variety of different management references likewise contain various definitions of construction supervision.

DEFINITION OF SUPERVISOR

For our purposes in this book, we will base our definition of construction supervision in the federal statutes. The federal government has, in two federal statutes, provided a legal definition of a supervisor.

The Fair Labor Standards Act of 1938 (which is also referred to as the "Minimum Wage Law") defines a supervisor as follows:

An executive whose primary duty consists of the management of a customarily recognized department or subdivision; who customarily directs the work of two or more employees; who has the authority to hire or fire other employees, or whose suggestions and recommendations as to the hiring and firing, and as to the advancement and promotion or any other change of status, will be given particular weight; who

15

customarily and regularly exercises discretionary powers; and who does not devote more than twenty percent of his work to activities which are not closely related to the work described above.

Additionally, the Taft-Hartley Act of 1941 includes the following provision:

[A supervisor is] ... any individual having authority, in the interest of the employer, to hire, transfer, suspend, lay off, recall, promote, discharge, assign, reward, or discipline other employees, or responsibility to direct them, or to adjust their grievances, or to effectively recommend such action if, in connection with the foregoing, the exercise of such authority is not merely of a routine or clerical nature, but requires the use of independent judgment.

According to the federal statutes, then, the supervisor is defined as a member of management in the workplace. This is significant in construction, inasmuch as most people who perform construction supervision advance to that stature and function after having been construction craft workers. With a background and training in performing skilled construction craft work, individuals who excel and who also demonstrate leadership and management potential are often promoted to the position of supervisor. When this occurs, these individuals become members of management on a construction project and in a construction company. While these individuals clearly retain their affinity for, and certainly their identity with, their craft, and perhaps with their craft trade union as well, when they become supervisors, they have, according to the definition provided in these federal statutes, become members of management.

THE SUPERVISOR'S FUNCTIONAL ROLE IN A CONSTRUCTION COMPANY

In a functional sense, supervisors provide the operational link between the construction craft workers and the management team, both on a construction project and within a construction company. Figure 2.1 depicts schematically the typical functional organization of a construction company and of a construction project team, from the executive level in the company, through the project manager and superintendent, to the construction supervisor, to the construction craft workers.

The supervisor, then, can be said to be the link between those who actually perform the construction work on a project (the construction craft workers) and those who have responsibility for the management of the project (the superintendent and the project manager), and in turn with those responsible for management and operation of the construction company (the support and executive levels of the company). The management functions performed by the supervisor are critically important to the success of every construction project, and in like fashion, are vital to the profitability and to the continuance of the construction company.

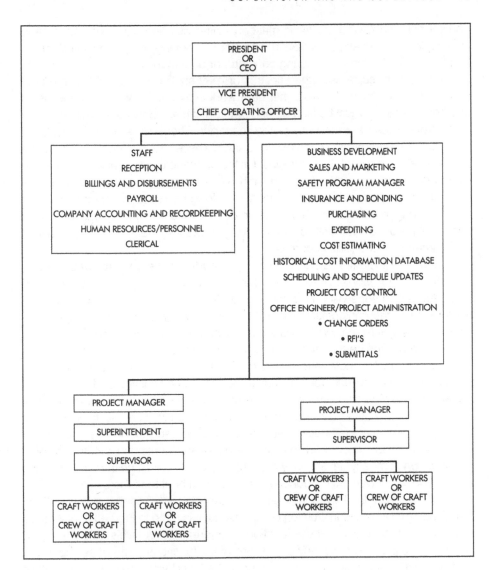

Figure 2.1 Construction Company Functional Organization

THE SUPERVISOR AS MANAGER

Although he or she may well continue to perform some craft work on construction projects, the supervisor is decidedly a member of the management team, both for the project and for the construction company. As a craft worker, the construction worker utilizes a set of skills referred to as craft skills or technical skills. These are the skills that are directly related to the performance of the construction work on a project, such as building concrete formwork; installing reinforcing steel; assembling structural steel components; installing conduit, junction boxes, wiring, transformers, and switches; laying masonry units; installing pipe, connectors, and fixtures; installing heating, ventilating, and air conditioning components and equipment; placing, finishing, and curing concrete; and so forth.

In the role of supervisor, the constructor may continue to exercise some craft skills in the performance of his or her daily work activities. However, in supervision another set of skills must be learned and applied, if the supervisor is to be effective in this role. These are the skills of the manager. These skills are referred to as management skills or as human relations skills. These skills are decidedly different from craft skills. It is important to realize that the constructor's success in supervision will depend upon the success with which he or she masters and applies these new and different skills. In addition, it is true that the longer the supervisor remains in the role of manager, and the further the constructor advances in management, the less important his or her craft skills become and the more important these management and human relations skills become.

Figure 2.2 illustrates some of the differences that have been referred to here. Construction craft workers have different characteristics and different attitudes than the construction supervisor, and they use different tools and apply different skills from those of the construction supervisor. Similarly, the construction supervisor, while he or she is typically well grounded in craft skills, must develop different skills and attitudes in the performance of the supervisory function.

It is important to note, and it is critically important for the reader as well as for the supervisor to understand, that while the characteristics and skills of the supervisor are *different* from those of the craft workers, this does not in any way imply that they are *better*. It is emphasized that both craft workers with their skills and management workers with their skills, are *absolutely essential* for the performance of a construction project and for the operation and continued profitability of a construction company. In fact, they are mutually dependent—neither craft workers nor management workers can function effectively without the other.

In similar fashion, if we again turn our attention to the depiction of functions performed in a typical construction company, as illustrated in Figure 2.1, it is to be emphasized that no one of these functions, departments, or levels within the company is more important than any other. The "office functions" are just as important, and are just as valuable, as the "field functions," and vice versa. In fact, neither would have relevance, and neither would even exist, without the other. Rather, all of the functional components and all of the individual members must function together as a *team*. Further emphasis regarding the importance of the concept of *teams*, and of team building and team maintenance, will be provided in subsequent chapters of this book.

TRANSITION TO MANAGEMENT

As we have noted, when construction craft workers become construction supervisors, they become part of management of both the construction project, and also of the construction company. In this supervisory capacity, they perform management functions rather than construction craft worker functions.

CHARACTERISTICS OF CONSTRUCTION CRAFT WORKERS	CHARACTERISTICS OF CONSTRUCTION MANAGEMENT WORKERS
Usually are paid (compensated) on an hourly or on a piece-work basis.	Usually are compensated on a salary basis, or often on a "guaranteed hours per week" basis.
Take great pride in their skills and abilities.	While retaining their affinity for their craft, and their understanding of craft work, increasingly apply the skills of management.
Take pride in knowing how to utilize the tools and equipment associated with construction work, and with the performance of their craft.	Think in terms of fulfilling project objectives, and learn that their success is measured in terms of achieving project objectives through their management of the work of the craft workers.
Think in terms of fulfilling assignments and performing tasks.	
Are grounded in the philosophy that their responsibility is to provide an hour of skilled labor for an hourly rate of pay.	Think in terms of planning for the fulfillment of project objectives, through assigning work tasks to the craft workers, and to managing the successful completion of those work activities.
Operate on the basis of "I get paid for what I do; if I am not going to be paid for it, I will not do it."	Think in terms of doing what is necessary for the success of the work, and for the success of the project. They are willing to perform a great deal of work, and to give a great deal of thought, outside the 8 AM to 5 PM, 5 days per week project work time, often working in the evenings and on weekends.
Interact on a day-to-day basis primarily with the members of the craft team—the supervisor and the other craft workers.	
Tend not to think a great deal in terms of "big picture" concepts. They are not greatly concerned about project planning, schedules, resource management, cost reports, documentation, invoices, change orders, RFIs, etc.	Take part in hiring and dismissal decisions regarding craft workers, and in evaluations of craft worker performance.
Operate from a perspective that it is management's responsibility to assign the tasks to be done, and to have available for the use of the craft workers the proper materials, tools, equipment, support elements (water, sanitary facilities, etc.), and safety gear as necessary for the proper performance of the assigned work.	Deal with discipline issues regarding enforcement of company policies, project policies, regulations, and legislation.
	Increasingly, the tools with which they perform their work are the project contract documents, and estimates, schedules, cost reports, change orders, RFIs, purchase orders, elements of documentation, written communication, verbal communications, etc.
	Develop a much broader vision regarding the craft work, and how it fits in as a component of overall project success.
	Frequently interact with other members of the project team, both those within their company and those of the general contractor, and subcontractors, the superintendent, the project manager, and sometimes with members of the office staff.
	Often interact with the owner, and architect and engineers for the project.
	Frequently interact with other consistencies beyond the members of the project team— building inspectors, safety inspectors, etc.

Figure 2.2   Characteristics of Construction Craft Workers and Management Workers

There are five basic functions that have been referred to as the "fundamental functions of management." They are: planning, organizing, directing, controlling, and staffing. Each of these functions will be defined, and its relevance to the construction supervisor will be set forth, in this section.

Planning

Planning can be defined as setting goals and objectives, and determining specific elements to be accomplished, in order to ensure their fulfillment.

Planning involves determining how to get from where we are now to where we want to be.

Planning is done at all levels of the construction company organization and by all members of the project management team.

Planning is done in terms of both long-term and short-term planning.

Planning is best accomplished by deliberate thought, followed by writing down the plan. When the plan is written down, it becomes part of project documentation, and importantly, it then also becomes a communication tool for assessing the plan, and for communicating it to others. (The importance of documentation and communication will be further emphasized in subsequent chapters.)

It has been observed that planning is the management function that many managers do not perform as well as they should (even though they frequently believe that they are doing so).

Failure to plan properly, and/or failure to properly monitor the plan, is the source of innumerable problems and difficulties of all kinds on construction projects.

Planning avoids many of the problems and difficulties that occur when planning is absent; additionally, proper planning enables the finding of effective solutions to problems that do occur.

Planning is the management function that the supervisor does well to focus on, and to which he or she should devote a great deal of time and attention.

Organizing

Organizing can be defined as lining up and obtaining all of the necessary resources to fulfill the plan, and giving them the structure to implement both the long-term plan and the short-term plan.

The resources necessary for the fulfillment of the plan include: materials, equipment, tools, people, documents, and time, and so forth.

These resources do not simply appear. Rather, management effort, in the performance of the organizing function, is necessary to obtain each of them and to place them into a proper array for the fulfillment of the plan.

Directing

Directing consists of communicating the plan and energizing the human resources to accomplish the plan. Directing entails setting objectives, assigning tasks, giving instructions, and communicating and enforcing policies.

Controlling

Controlling can be defined as monitoring the plan and its execution. Controlling entails measuring results, comparing results with expectations, evaluating the significance of differences, and doing what is necessary to make corrections when there are deviations from expectations.

Controlling also involves making the determination as to whether to "stay the course" with the existing plan, to modify the plan, or to formulate a new plan.

Staffing

Staffing is defined as locating, hiring, training, and developing the people who are necessary to implement the plan, and to fulfill the objectives.

Staffing includes recruiting, hiring, evaluating, promoting, reprimanding, dismissing, compensating, and rewarding the people who are performing the project.

It is important for the supervisor to understand that the staffing function also includes training and professional development of the people in a construction company. Making it possible for the people in a construction company to expand their skills and to enhance their knowledge base is a basic function of management. This investment in people through training and professional development is an investment that will pay dividends over a long period of time.

These are the fundamental management functions that the supervisor will engage in during the course of performing his or her role in management. In other chapters of this book, further reference will be made to these management functions and to the manner in which they apply to construction supervision.

ATTRIBUTES OF SUCCESSFUL SUPERVISORS

As can be gathered from the foregoing, the construction supervisor must become skilled in a variety of competencies as he or she functions in a supervisory capacity. If the supervisor were planning, and setting a goal to be the very best he or she could be, that person might do very well to seek to learn as much as possible regarding effective supervision. Providing some of that learning is one of the intents of this book.

One of the most effective methods of developing the competencies needed to be an effective supervisor may well be to emulate accomplished supervisors

ATTRIBUTES OF THE BEST LEADERS AND MANAGERS
AS DETERMINED BY PRACTICING CONSTRUCTORS

Good People Skills	Teacher
Good Communicator—Verbal and Written	Dedicated
Respectful—Treats People with Respect	Develops People
Charismatic	Consistent
Organized	Problem Solver
Knowledgeable	Goal Oriented, Goal Setter
Leads by Example	Good Planner
Earns the Respect of Others	Disciplined
Open-Minded	Optimistic
Confident	Calm
Honest	Fair
Good Decision Maker	Motivator
Delegator	Humble
Sincere	Provides Recognition
Follows Through	Trustworthy
Visionary	Accessible, Approachable
Willingness to Share Information	Team Builder
Crisis Management—Can Be Counted on in a Crisis	Sets High Expectations
Good Listener	Personable

Figure 2.3   Attributes of the Best Leaders and Managers

who have demonstrated their supervisory and leadership effectiveness over time. A synopsis of the attributes of such supervisors is shown in Figure 2.3.

The list of attributes in Figure 2.3 was derived from queries made to the participants in construction supervision and construction project management seminars. These seminars were presented in all parts of the United States, over a span of more than 10 years. The responses are those of thousands of participants in these programs.

During the conduct of these seminars, the facilitator provided the following guidance to the participant group. "Think of the best, and most effective, construction leader or supervisor you have ever worked for; one whom you intuitively recognized as a good leader, one whom you admired and respected more than any other, one whose effectiveness was unrivaled, one who was universally regarded as one of the very best at what he or she does. Now think about, and then say aloud, the attributes that this person possessed, that caused him or her to be as effective as he was in supervision and leadership, and/or that caused you to admire him as much as you do."

The facilitator wrote the participants' responses on a whiteboard or flip chart for all to see, as the participants voiced their thoughts. The exercise continued until the responses from the group slowed significantly or stopped. While there was some small amount of variance in the responses, it is interesting to note that those attributes summarized in Figure 2.3 were repeatedly provided by the different groups of participants, with a very high degree of consistency.

It should also be noted that no survey validation criteria were applied, and no statistical analysis was performed with regard to the responses that were provided. The responses shown in Figure 2.3 are simply a compilation of the most frequently repeated responses to the query, made to a large sample size of constructors, in various regions of the country, over a long period of time. It is noteworthy to recall that these responses were provided by constructors who were describing the attributes of the very best and most effective supervisor and leader they could envision.

It is suggested here, as it was to each group of participants in the seminars, that this set of characteristics may well provide the best possible specification or profile of attributes that a person seeking to become a better construction supervisor could aspire to. It is recommended that the supervisor who is setting a goal of being the best he or she can be might do well to make note of these attributes, and to make frequent reference to them, and then seek to cultivate these qualities, if he or she wishes to be considered in the same manner, as the best of the best, by those whom he or she supervises.

LEVELS OF SUPERVISION—CAREER PATHS

Many construction supervisors spend their entire careers working in a supervisory capacity. People who function effectively in this capacity can have extremely rewarding and successful careers.

For those who are successful in construction supervision and who wish to further broaden their horizons, to continue learning and developing, and to assume further responsibility, numerous additional opportunities await. Some of those additional opportunities are discussed in this section.

Some supervisors may aspire to become a project superintendent, a position that is referred to in some companies as a general superintendent. The superintendent is usually the primary manager at a construction site and typically has overall responsibility for the management, scheduling, and coordination of a construction project. This management position involves coordinating all of the trades who will perform work on the project. Additionally, it entails managing all of the day-to-day operations and making the day-by-day decisions relative to all aspects of the project throughout the time of construction.

Others may aspire to become project managers. Project managers are best described as the link between field operations and office operations in a construction company. A project manager might perform management functions for one project at a time, assisting a foreman or superintendent in managing the daily operations on that project, and providing the interface between the foreman or superintendent and the company's office management. Alternately, depending on company policy and also on the size and complexity of the projects being performed, a project manager might perform these functions on a portfolio of several construction projects at a time.

Additionally, in some companies the estimating and project management functions are combined. A person might prepare the cost estimate for a project, culminating in the submittal of a proposal. If the proposal is accepted and a construction contract is awarded, the person who prepared the estimate becomes the project manager, who assists in managing and coordinating the construction of the project.

Whether the estimating and project management functions are combined or separate is a matter of policy determined by the construction company's management. Some companies prefer to keep the estimating function discrete from project management functions, while other companies have found advantage in having the same person prepare the estimate and then serve as project manager for constructing the project.

Additional opportunities may also await the supervisor in the performance of any of a variety of different functions in the construction company office. For example, becoming an estimator may be an attractive option. Estimators have the responsibility for analyzing the construction documents for a forthcoming project and determining what the costs of construction for that project are predicted to be. The estimated costs that result from the estimator's determinations are used to prepare proposals, with the hope that the contractor's proposal will be accepted and that a construction contract will result. Further discussion of construction cost estimates and the process by which estimators make their determinations in an estimate, is provided in Chapter 12. The supervisor should know that good estimators are among a contractor's most valued assets. If a person has an interest in and a talent for estimating, he or she can enjoy a very rewarding career performing this function.

Additional opportunities are also available in construction scheduling (discussed in Chapter 14), and in schedule analysis, management, and updating. In addition, there are opportunities in construction project cost reporting, cost analysis, and cost control. Additional insight into the cost control process is provided in Chapter 13, as well as in several other sections of this book.

Similarly, the important matter of expediting, coordinating, and controlling the purchase of the materials and equipment to be installed in construction projects that the company is performing, may be appealing. The purchasing and expediting of materials and installed equipment are discussed in various sections of this book, so that the supervisor can form an idea of what these functions entail.

SUMMARY

For the person who is in construction supervision, or who is preparing to enter the world of supervision and management, and who is aspiring to be the best and most effective supervisor he or she can be, a challenging, and satisfying, and rewarding opportunity awaits. If he or she enjoys supervision and is effective in the many facets of construction supervision, a person can have a fulfilling and bountiful career performing these functions. If he or she has ambitions of performing additional tasks, the world of management of construction offers numerous additional opportunities.

Learning Activities

1. Make a copy of, or write out, Figure 2.2, "Characteristics of Construction Craft Workers and Management Workers" from this chapter, and then think of items you could add to the list, both for craft workers and for management workers.

   Think of yourself and your work in both categories.

   Reflect on how different your work as a supervisor is from how it was as a craft worker, and think of the different level of responsibility you have now as a member of management.

2. Make an organization chart for your company and its office management and staff personnel. Include the names and contact information for all of the people. List the project manager and superintendent and general foreman (with contact information), as appropriate, who interface the management of your project with the company office.

3. Look again at Figure 2.3, "Attributes of the Best Leaders and Managers."

   Add to the list with any additional qualities or characteristics you may have thought of.

   Write out the list on a tablet or sheet of paper, including your own additions. Underline or highlight those qualities you most admire, and those you would most like to emulate and to cultivate in yourself, in order to be thought of by those with whom you work as the best of the best.

4. Do some long-range planning and goal-setting, and think in terms of where you would like to see your career evolve, perhaps in supervision or in some other role in the company.

SECTION II

SOFT SKILLS

CHAPTER **3**

ORAL COMMUNICATION

INTRODUCTION

Communication is of critical importance in construction. Poor communication is the root cause of many problems in the industry. Poor communication can impact safety, production, quality, cost, schedule, and all other aspects of a job. Thus, it is important to work to continuously improve communication skills, both oral and written.

This chapter will begin by defining communication and exploring its critical nature. It will then focus on oral communication, considering barriers to oral communication, and the means to overcome the barriers. Some techniques will be provided to improve oral communication. Since listening plays a major part in oral communication, active listening will be discussed. Finally, two topics heavily dependent upon oral communication will be addressed: customer relations and negotiating.

DEFINING COMMUNICATION

Communication can be defined as:

Communication: The process of passing information and understanding from one person to another.

It is well understood that the *process* of communication is passing information from one person to another. More importantly, the *purpose* of communication is

29

to pass understanding. Without understanding, the communication process is incomplete. Incorrectly understanding the message is also a cause of breakdown of communication. The focus of the definition of communication is passing understanding, not just information.

CHARACTERISTICS OF COMMUNICATION IN CONSTRUCTION

Communication is an essential part of construction. Three characteristics of construction communication help in understanding its importance. Communication in construction is *critical*, *time-consuming*, and *pervasive*. Each of these is explored below.

Critical

It has been determined that two-thirds of all errors at the construction job site are due either directly or indirectly to communication. Take a few minutes to think of instances from your experience when a communication breakdown has caused an error. These errors take many forms. Installation might not be correct, because of the fact that directions were not given correctly, not heard correctly, or not understood correctly. Accidents often happen on the construction site because of garbled communication. Production is impeded because a message was not received, resulting in lack of required resources: tools, materials, or workers at the workface. Claims or back charges are assessed because proper notification was not provided. These are just a few examples of how communications problems can affect a job. They illustrate that breakdowns in communication often cause minor irritations, but also can result in major ramifications throughout any construction project.

Time-Consuming

Supervisors generally spend 60% to 90% of their time communicating. As stated in Chapter 2, a major difference between the craft worker and the supervisor is that supervisors work primarily with people rather than the tools of their craft to accomplish their tasks. Instead of producing work directly, with their own hands, supervisors direct the work of others and accomplish their work objectives through others. They coordinate the contributions to the project of many different entities, including subcontractors, suppliers, designers, and even the owner. They design, implement, and monitor construction processes. All of this is accomplished through communication. On the job site, in the context of supervision, most of this communication is oral.

   Proper communication requires time. Ideas must be organized and then appropriate words selected to effectively convey the ideas. It takes time to choose the proper words to communicate the technical message required, within the context of the other party's knowledge, culture, and language skills.

Advances in technology have both helped and hindered communication. Communication devices are ubiquitous. If something is needed right now, it can be summoned by telephone (generally a cell phone that is immediately available), by walkie-talkie, by texting, or by many other means. This is convenient but has led to inefficiencies. It tends to minimize the emphasis on pre-planning since, if something is overlooked, it can be solicited immediately. It also tends to be disruptive to the person on the other end of the communication. When I contact someone at my convenience, it is highly likely that they will be distracted from the activity in which they are engaged.

Pervasive

Communication plays a part in virtually everything the supervisor does. Whether the supervisor is directing the work of a crew, evaluating the performance of workers, developing documentation, asking for clarifications, or providing answers to questions, it involves communication.

One of the key functions of the supervisor is to serve as the link between the worker and the rest of the world. Information the worker gets comes primarily through the supervisor. This information could be from the designer, another contractor, or the company office. Information the rest of the world needs about the work is communicated through the supervisor. Such information may be needed by the company office, other contractors involved in the project, material suppliers, or the project owner. All of this information, whether going to the crew or coming from the workface, passes through the supervisor.

Because of the fundamental importance of communication to the work of the construction supervisor, it can be concluded that time and effort spent in honing communication skills is well invested. The rest of this chapter focuses on oral communication. The next chapter deals with written communication and documentation.

IMPROVING ONE-ON-ONE ORAL COMMUNICATION SKILLS

Oral communication, at best, is not very effective. Studies have shown that when people are attentive, they absorb about 50% of what they hear. After a week, people remember only about 20% of what they initially absorbed. So after a week, 90% of what a person hears is lost.

However, oral communication is essential. It is quicker than written communication. In addition, since many people do not have a high level of reading comprehension or a high level of writing skill, oral communication is more efficient. It is simply impractical to write everything out.

Oral communication is a skill, and as a skill, it can be improved with training and practice. One way to improve oral communication is to recognize barriers to effective oral communication and to develop the means to overcome those barriers. The barriers can be overcome by removing them or by mitigating their effects so that they do not negatively impact the communication.

Barriers to Effective Oral Communication

It is the responsibility of the sender of a message to identify any barriers and to deal with those barriers. The sender knows when a message is being sent. The sender knows the intent of the message. So, only the sender can recognize when a message has not been properly received. Thus, when communicating, the sender must check to make sure that the message was received and that the message received was the message intended to be received.

When there is a barrier that could potentially obstruct the communication, the sender must identify the barrier, then evaluate the barrier to determine if the communication can take place. If it is determined that the communication may be impaired, action must be taken to remove the barrier or mitigate its effects. After the communication is complete, the sender must ensure that the correct message was received and understood.

Three types of barriers to effective communication will be addressed:

- Physical barriers
- Sender barriers
- Listener barriers

Physical Barriers

Physical barriers occur when something in the physical environment obstructs the message. There are many ways the environment can keep the message from getting through. Three examples are:

- Noise
- A hearing impairment
- Distractions

Construction, by nature, is noisy. There are many tools that cause noise. There is often noisy equipment operating in the vicinity. Construction activities, such as hammering or shoveling, cause noise. If these are going on in the vicinity of the communication, the first remedy might be to mitigate the problem by speaking more loudly and clearly. The next remedy might be to relocate to a quieter place, thus removing noise as a barrier. A third alternative is to write out the message. Although it is effective, writing is time-consuming and cumbersome and may not be an appropriate remedy.

A consequence of the noisy nature of construction is that hearing in construction workers is often impaired. This is especially true of older workers who began their careers at a time when noise abatement was not practiced in construction and whose hearing tends to be deteriorating with age. If a hearing impairment is identified, the mitigations identified above for a noisy environment might be considered. A quiet place might be sought for the communication or a written message may be used.

A longer-term mitigation for a person who is hard of hearing might be to procure a hearing aid. Hearing aids are very effective, but carry with them their own set of problems, such as amplifying background noise.

For those with hearing impairments, it is particularly important to speak directly to the listener. Comprehension of an oral message is often enhanced if the listener can see the speaker's lips, facial expressions, and gestures. People with a severe hearing impairment might have developed lip-reading skills, and whereas a noisy environment might cause a problem for an oral communicator with good hearing, it may cause no problem at all for a skilled lip reader.

Distractions are abundant on construction sites. Most construction workers enjoy being on a work site and can be distracted by interesting operations taking place nearby. Distractions could also come from off the job site, such as when an interesting person walks by or a fancy car passes by. Simple distractions can often be overcome by maintaining eye contact, which will generally draw a similar response from the listener. Another simple mitigation might be to reorient the position of the speaker and the listener so that the distraction is behind the listener. If this is not enough to keep the listener's attention, then perhaps the conversation needs to be relocated into a less distracting environment.

Sender Barriers

With a sender barrier, either the content of the message is flawed or the means of transmission is flawed. Content problems could be the result of missing content or inappropriate language. Transmission barriers could be the result of physical impairments or the sender's level of capability in oral communication.

An incomplete message may be the result of the speaker assuming more knowledge of the subject than the listener has, or it may be the result of careless preparation of the message, resulting in something being left out. This problem of missing information can often be eliminated by better preparation, more thoroughly thinking about the message prior to sending it, and formulating a more complete message.

The problem of inappropriate language might be the result of the words that are employed being unfamiliar. This could be a language problem arising between communicators from different regions, in which case, the speaker or the listener might need to learn some of the other's familiar language structure and words. It could also be a problem with the use of technical language or jargon. This is not often a problem when people from the same technical discipline are talking, but as the technical background of those engaged in the conversation diverges, it becomes more of a problem. The speaker should always watch the listener to try to identify when terminology may be puzzling. If the speaker anticipates that a problem in technical comprehension might arise, the speaker could request that the listener ask about any terms not understood. Speakers should also work to avoid technical jargon, especially when talking with someone from outside the technical area.

Problems with words might also arise with the use of culturally inappropriate or offensive language. Supervisors should always conduct themselves in a professional manner, working to avoid inflammatory or inappropriate language. Much more will be said of this in Chapter 6, but suffice it to say that there are appropriate ways to get a strong message across and inappropriate ways. Inappropriate ways are counterproductive and should be avoided. Sometimes inappropriate language creeps into a conversation. When that occurs, the speaker should apologize, move on, and learn from the experience, and the listener should accept the apology and also move on.

Problems with the means of transmission could result from the speaker having a physical impediment or simply never learning how to speak effectively. Physical impediments can often be treated either medically or with speech therapy. Not knowing how to speak effectively can be overcome through training and practice. Effective speaking techniques, such as proper formulation of a message and the use of gestures, facial expressions, and voice inflections, can be learned through classes. Since oral communication plays such a key part in the supervisor's role, improvements in speaking are always to be encouraged.

Sender barriers are often difficult to detect, since they originate with the speaker. A speaker should always scrutinize his or her own oral communications to try to identify when a problem arises and what that problem is. Once identified, sender barriers are generally not difficult to correct because the problem originates from the speaker, and the speaker controls the situation. The key to eliminating sender barriers is timely detection of a problem.

Listener Barriers

Listener barriers are often the most difficult to identify and to correct. With a listener barrier, the listener's ability to receive the message is impaired because of something that resides within him or her.

The listener may be inattentive. This may be the result of a distraction for which the remedies are similar to those described for a noisy environment. The inattention may come from a lack of interest, in which case it is important that the sender determine how to regain or enhance the interest of the listener. The listener may have a physical or emotional problem. Inattention may be the result of not sleeping at night or the listener may have problems outside the context of the work site. Such difficulties as family problems or financial problems can often distract workers while at their job. If the supervisor detects such physical or emotional problems in a worker for whom he or she has responsibility, the worker should be referred to an appropriate counselor. These types of problems imperil important areas such as safety, quality, and productivity and need to be addressed by a professional.

On the other hand, the listener's mind may just be wandering back to an entertaining event the previous evening, or to the approach of lunch, or any of a myriad of other distractions. When this situation is detected, the speaker could simply pause until the listener's attention returns, or ask the listener to be more attentive.

Another type of listener barrier occurs when the listener tends to jump to conclusions. Often the conclusion to which the listener jumps is not the correct one. If this is the barrier, the listener has to be asked to withhold conclusions until the communication is complete.

Sometimes listeners simply reject the message as contrary to experience or their beliefs. The listener's response might be, "That is not the way we did it at my last job," or, "That is not the way I was taught to do it." Supervisors should encourage workers to think creatively and come up with alternatives, where appropriate, but they also need to be able to stop the discussion at some point, make a decision, and move forward with the support of the workers.

Listener barriers are often the most difficult to deal with. They are usually easier to detect than sender barriers but more tricky for the supervisor to mitigate or remove because they often become personal. The objective is to overcome listener barriers without offending or alienating the worker. Initiative and creativity are very important, and every effort must be made not to discourage these valuable contributions.

Other Barriers

There are many other barriers to oral communication. With an increasingly diverse construction workforce, speakers with different native tongues or accents can often present a problem. In this case, the speaker must speak more slowly and clearly, must use simple, common words, and perhaps learn key words and expressions in the listener's language.

Culture may cause distractions. The physical distance between speaker and listener varies greatly from culture to culture. How forceful or how direct the message can be varies significantly in different cultures. The amount of eye contact allowable can vary from culture to culture, and even touching may be acceptable in some cultures and totally wrong in other cultures. When workers from a different culture become the responsibility of a supervisor, it is wise for that supervisor to learn a bit about the culture, especially in terms of acceptable and unacceptable characteristics of communication.

TECHNIQUES FOR IMPROVING THE EFFECTIVENESS OF ORAL COMMUNICATION

Four techniques have been demonstrated to be particularly useful in improving oral communication. These techniques are:

- Repetition
- Tell-back
- Feedback
- Follow-up

Repetition

Repetition differs a bit depending on whether the speaker is speaking to a group in a formal setting or speaking to an individual or a small group in an informal situation. For a formal presentation, a speech should be organized with an introduction that tells the audience what is going to be presented, then the body of the presentation in which the speaker relates the message with as much detail as appropriate, and finally a conclusion with a summary that reviews the key points of what was presented. This process prepares the audience, then provides the information, then repeats important points to help improve retention.

In a one-on-one situation, repetition is used to reinforce the message in such a way that the important elements are less likely to be forgotten. The speaker should present the message. He or she should then repeat it, if possible in a somewhat different form. The speaker should highlight important points. He or she might ask questions to determine how well the message was received, and might repeat points again to correct, clarify, or emphasize the message even more.

Tell-Back

The purpose of tell-back is to ensure that the message was received as intended. With this technique, the message is delivered, then the speaker asks the listener to repeat the message back to the sender. The sender can then correct erroneous parts or repeat forgotten parts and ask for the listener to tell it back again. This cycle is repeated until the response is the same as the message, thus ensuring that the message was received and understood as intended.

Tell-back can be used not only to ensure that the correct message has been received but also to reinforce the message so that it will not easily be forgotten.

Feedback

Feedback has a different focus than the other techniques. The purpose of feedback is to find out what the sender or receiver thinks about the message. From the standpoint of the receiver, is the message accepted, embraced, or rejected? From the standpoint of the sender, is the message strongly felt, is it debatable, or is it inflexible?

When giving feedback, focus needs to be placed on specific facts. The participants should be as objective as possible and avoid getting personal. The responder should make a clear statement in a positive and instructive manner about how he or she feels about the message. Responders should limit their feedback to key issues instead of trying to cover every point. The feedback is finished with an invitation to the other party to respond to the feedback.

Feedback has many uses. In addition to the stated objective of determining how the message is perceived, it is useful for drawing a person into a discussion. It is also useful for gaining the attention of an inattentive listener. The feedback process can be used as a means to solicit creative ideas and also to get buy-in from the listener.

The feedback process can help a supervisor move from a dictatorial leadership style to a more interactive and positive style. Leadership styles are addressed in detail in Chapter 7.

Follow-Up

Since oral communication is very ineffective and fleeting, it is important to follow up oral messages in one or more ways.

Follow-up should be carried out any time oral instructions are given to ensure that they are followed and to ensure that the action resulting from the instructions produced the anticipated outcome. After giving instructions, the listener should be informed that verification will follow. If verification is indicated, then it needs to be done. If the message is given that verification will be done, but it is not done, doubt is cast on future communications and particularly on follow-up to future communications.

Important oral communications need to be followed up in writing. If the supervisor is the recipient of an oral notification or instruction, they should ask for follow-up in writing. If written verification is not received from the sender, the supervisor should follow up by writing out the message they received orally and sending it to the person who gave the oral message to verify that the message was correctly understood. It also documents that the message was given, and it puts the message into a form that will not change over time. Follow-up in writing can eliminate the possibility of the message being incorrectly received. It can also prove invaluable later should a dispute over the oral message arise because it will document the message, the sender and the timing of the message. Written follow-up is an important part of documentation, which will be further discussed in Chapter 4.

ACTIVE LISTENING

The process of communication is based upon passing information from one person to another. A critical part of communication is receiving the information. Listening is of fundamental importance in the communication process.

At least one third of one's communication process should be on the listening end. Communication typically requires give and take. The responsive part of communication requires that the message sent be received and understood. Without that understanding, the message cannot be acted upon or responded to appropriately. With faulty hearing of the message, the response will likely be incorrect, inappropriate, or irrelevant. Or, there may be no response, at all.

Active listening is the conscious process of securing all the information relative to the message through hearing and observation. Active listening is a skill, which means that it can be learned, or improved through training and practice. Active listening should become a habit for the supervisor. It should become an involuntary and unconscious action any time the supervisor detects that a message is being sent. With practice, active listening will improve over time.

Different types of information are conveyed in communication. A fact is a piece of information presented as having objective reality. Facts are conveyed by words and are typically presented with assurance in such a way as to dispel questioning. That is, the speaker uses supporting tones, gestures, facial expressions and general body posture that conveys the confidence that what he or she is saying is correct and true. Skilled observers can look at the face, eyes, and posture and determine with great accuracy whether the speaker feels that what they are expressing is the truth or if they are trying to mislead by stating as a fact something that is not, or might not, be true.

Thoughts, theories, and conjectures are conveyed in such a way that they elicit consideration and evaluation. Often the intent is that the listener participates actively to consider the validity of the statement. The intent may also be to engage in a discussion or it may be to place a question in the mind of the listener so that the listener can ponder the statement at his or her leisure.

Emotions are conveyed primarily by nonverbal means. The words must be supported with, or confirmed by, tone, expressions, and general body language. A person can say "I love you" without communicating the message if the message lacks the requisite expressions from the sender. On the other hand, the message "I love you" can be expressed very well without the words. Not only is the existence and validity of the message expressed in an emotional communication, but the intensity of the feeling is also a fundamental part of the communication.

Being able to link words with tones, expressions, and body language is a powerful component of oral communication that can produce strong positive or disastrous negative outcomes. This is one of the key differences between oral and written communication.

Factors Affecting Listening

A number of factors affect the ability to listen effectively.

The *topic* is important. It can be simple or complex. It can be within the listener's sphere of experience, or it can be far outside that experience. The nature of the topic has a significant impact on a listener's ability to listen.

The listener's *context* is very important as well. Does the listener consider this topic to be important or unimportant? Is it interesting or uninteresting? Why is the listener listening to the message? Is it for entertainment, as an assignment, or for self-improvement? Each of these will make it easier or more difficult to listen.

The way in which the message is *delivered* also influences the listener's ability to listen. Is the message sender a skilled communicator? Is the message delivered with the use of gestures and other body language? Does the sender provide examples or illustrations of the message? Is the message presented in a logical fashion, or is it disjointed?

The *physical environment* affects the listener's ability to take in the message. Is it delivered in an open space or in a congested environment? Is there noise or are there other distractions? Is there clean air, or is the atmosphere oxygen depleted, or does it have a high level of contaminants?

Finally, there are *personal factors*. How wide awake is the listener? Is the listener alert? Is the listener having trouble focusing? The listener's receptivity will again have a significant impact on his or her ability to hear, digest, and understand the message.

Enhancing Active Listening

Prior to the communication encounter, the listener needs to prepare to be an active listener. The listener should not come in with assumptions about the message she is about to hear. Also, she should not allow the sender to bring assumptions, such as what she knows or how she feels about the message.

The listener should be prepared to interact with the message's sender but restrain herself from interrupting. Let the sender complete the message. This guideline may need to be modified somewhat should the sender not be a skilled communicator. If the sender displays a rambling style of communication, the listener may need to lead the sender toward a conclusion of the message.

An active listener works to eliminate prejudice and preconceptions. The message should be separated from the sender. The listener should recognize that valuable understanding can be achieved from any type of person, and it is the responsibility of the listener to get past stereotypes to focus on reception of the message.

The listener should be prepared to seek the understanding in the message in spite of the possibility that the means of conveying the message, and even the message itself, might be flawed. Supervisors often work with people who do not have strong communication skills. It is the supervisor's responsibility to pull the message out of a poorly skilled communicator.

An active listener is patient and prepared to not react quickly to strong messages. Emotion and prejudices in the sender and in the sender's message are filtered out to enable focus on deciphering the content of the message. If no clear positive message is found, the active listener must be prepared to disregard the flawed message and draw out the sender's intended message.

Finally, in preparation for oral communication, the previously discussed physical barriers to communication must be eliminated. The listener should seek a good listening environment and needs to ensure that she is receptive.

During the delivery of the message, full attention should be given to the speaker. The listener should maintain eye contact and observe gestures, expressions, and other body language. The listener should maintain an alert posture. If possible, the listener should enter into a dialogue with the message's sender, and maintain an active mind, evaluating the message and preparing to respond with questions or another appropriate response. However, preparation of a response should not distract the listener from hearing the message.

After the exchange of understanding, the active listener needs to follow up by summarizing the important points, in writing, if possible. The listener should seek clarification if needed. Complex topics should be synthesized into a few key points that can be retained. The listener should thank the other participant in the communication and arrange for the next communication, if appropriate.

CUSTOMER RELATIONS

Customer relations is included within this chapter because construction supervisors have a major impact on customer relations, and most of the contact between the customer and the supervisor is oral. As the representative of the contractor at the job site, both legally and in practice, the supervisor will have ongoing interaction with customers throughout the project. It is very important that the supervisor understand who the customer is and how to enhance relations with the customer.

Who Is the Customer?

The customer can be defined as:

Customer: A person with whom a merchant or business-person must deal.

This might be paraphrased as *one with whom I have (my company has) a business relationship*. A customer might also be considered as one with whom value is exchanged on behalf of the business. This can be a useful way in which to consider a customer because it enables consideration of who adds value to the job and, hence, who is a customer.

There are many customers associated with a construction job, the most obvious being the project owner. The owner brings the job in the first place. The owner also funds the job and makes decisions throughout the job that are important to it.

The design team also adds value to the job. The design team provides the design that is the basis for the job. They provide ongoing clarifications and answers to questions. They are responsible for approvals necessary for getting paid and authorizations for changes to the contract. Thus, the design team members are customers and need to be respected as strong contributors to the success of the project.

For subcontractors, the contractor with whom they have a contract is a very important customer. The contractor is the communication link with all other project participants. The contractor also participates in the approval process for payment applications and for changes to the contract. The contractor takes the lead in planning and scheduling the project and has the responsibility for coordinating the work of all participants on the project.

There is another important group of customers with whom the supervisor deals, and that is those who work for the supervisor, either directly or indirectly. Craft laborers add significant value to the job through the work they do. Suppliers of materials and equipment also add value that can significantly affect the project's outcomes. Those in this group are not generally considered customers, but treating them as customers can help maintain good relations with people and entities that are key contributors to the success of the project.

Within the company, there are other customers away from the job site. The project manager facilitates much of the work off the job site and provides significant support to the ongoing operations on the site. Other people in the company office, including the cost estimator, the purchasing agent, and the payroll and accounting people support field operations in many ways and hence should be respected as customers.

In short, most people with whom the supervisor interacts can be considered customers and should be treated with the respect of a customer. An attitude of customer service can significantly improve the various relationships throughout the job and have a very positive impact on the job's outcomes.

Respecting the Customer

The proper way to treat a customer is with strong professional respect. Supervisors should seek opportunities to interact with their customers as often as possible. Interactions should demonstrate an appreciation for the value the customer adds to the job. Customers' needs and expectations should be identified. To the extent possible, the needs should be satisfied while the customers' expectations are exceeded.

Exceeding expectations does not mean providing more expensive solutions to customers' needs. It does mean providing what is due the customer at a higher level of service than expected. Construction is a commodity. Many companies can provide the same product, and a high-quality product is a minimal expectation. However, the way in which the product is provided will distinguish one construction company from another. Providing a high-quality product with the highest level of service and responsiveness to the customer is what will distinguish one contractor from another in a highly competitive market.

THE ART OF NEGOTIATION

Negotiation is a high-level skill of great value to a supervisor. Supervisors are in almost constant negotiations, whether it is negotiating with workers to make work assignments, negotiating with suppliers for delivery of materials or equipment, negotiating with other contractors for working space, negotiating with designers for changes, or even negotiating with owners for payments.

There are many approaches to negotiating and entire books have been written on the art of negotiating. This book will introduce a few basic concepts, but skilled supervisors will want to get additional education and training in negotiating skills.

Negotiating can take place in teams, as two individuals, or on a one-to-many basis. Since supervisors will be primarily negotiating one on one, the focus of this section will be on one-on-one negotiations.

Negotiation has been defined as:

Negotiation: A problem-solving process in which two or more people voluntarily discuss their differences and attempt to reach a joint decision on their common concerns.

This definition and the process outlined below come from training materials developed by Collaborative Decision Resources, Boulder, Colorado.

You can learn a lot from this definition. First, the objective of a negotiating process is to resolve problems. Note that it is important to reach a solution, not just to continue endlessly in discussion. Second, it is a voluntary process. Coercion is not a legitimate part of negotiation. Third, it is based upon a discussion of differences. Education and mutual understanding is the basis of resolving the differences. Finally, the solution is reached jointly. It should be a mutual agreement, not one side imposing a dissatisfactory solution upon the other.

The process of negotiation can be broken into a relatively simple four-step progression:

1. Identify the issues. The more clearly the issues are identified and the more narrowly they are defined, the better the chance for a successfully negotiated solution. The chances of a good solution will improve if complex issues can be broken into a series of simple issues, each issue being negotiated in turn.

2. Educate each other about needs and interests. Notice the emphasis on education, not persuasion. The focus of this educational phase is to identify what is important to each side. Often, issues are resolved at this stage when each side recognizes that what is important to one side is not important to the other, so each can give up something of little importance to themselves in order to gain their primary objective.

3. Generate possible settlement options. No commitments are made at this stage. In a nonthreatening environment, each side is working to put on the table solutions that they could live with. It is all hypothetical at this point and, hence, nonthreatening.

4. Bargain over the details of a final agreement. When a solution is identified that might be acceptable to each side, the focus becomes nailing down the details in a manner acceptable to both sides. When that is done, the negotiation process is successfully completed and what is left is to implement the negotiated agreement.

The outcome of a successful negotiation will have certain characteristics. It will resolve conflicts, so that the two sides have agreement on the issue. It should leave both parties something of value. It will build a basis for future negations. Once the sides have experienced success in one negotiation, this builds a model for future negotiations on what might be more complex issues. Finally, it should bring closure to a cycle of negotiations. When completed, the agreement needs to stand and not be reopened at a later date.

For a successful negotiation, certain prior conditions need to be met.

Interdependence: Each participant must have something that the other values. Interdependence is one of the most important preconditions. If one party can determine what is of value to the other party, they then have something to exchange. Asking the other party directly what they would value most in the negotiated solution might be the easiest and most effective way to determine what each party values. Often, what one party wants is easy for the other party to give, and the negotiation is off to a good start!

Readiness to negotiate: If the other person will not negotiate, the first step is to get them into the mood to negotiate. Otherwise, negotiations will go nowhere.

Means of influence or leverage: There must be an incentive for each party to participate. The cause for a negotiation is generally that one party has something the other party needs or wants. The party without some leverage is in a weak position to achieve a positive outcome. One party's leverage might be what they have that is of value to the other party. On the other hand, one party may actually be in a position of power but would rather negotiate than dictate a solution, which provides some negotiating leverage to the other side.

Initial agreement: Some common basis is needed to start the process. There must be some common ground for agreement, and this needs to be established before moving forward into a successful negotiation.

Will to settle: Some people prefer the contest rather than the solution. The more people love conflict, the less likely they are to push for a resolution.

Potential for success: There must be a reasonable chance of achieving a positive outcome. In a low-stakes game, the potential for success needs to be high enough to make the effort of negotiating worthwhile. As the stakes get higher, more investment of effort is justified to try to negotiate in a situation with a lower likelihood of success. However, in any case, there must be some likelihood of success or parties will not start the process.

A sense of urgency and deadline: Without a deadline, the negotiations become simply an intellectual discussion that can go on and on without the need to come to a conclusion. In the construction environment, there is always a deadline and it is generally nearer, rather than farther off.

Willingness to compromise: Reaching a solution requires give and take on both sides. If both parties are not willing to give and take, a positive outcome is highly unlikely.

External factors favorable to settlement: The solution negotiated by each party must be authoritative. If it can be overridden by higher authority on one side or the other, then a true resolution by the parties in the negotiation has not been achieved. A construction project involves many parties, both on the job site and external to the site. Any bilateral agreement must not cause problems elsewhere, or further negotiation is needed.

SUMMARY

In this chapter, the following key points have been presented.

- Communication is the process of passing information and understanding from one person to another.
- Communication is critical, time-consuming, and pervasive.
- Improving communication skills is important to the supervisor.
- Improving oral communication begins with identifying and removing or mitigating barriers.
- Various techniques can be used to improve the effectiveness of oral communication.
- Active listening is an essential part of oral communication.
- The supervisor is in as particularly advantageous position to maintain good customer relations on behalf of the company.
- Negotiation is a critical skill for any supervisor.

Learning Activities

1. Practicing active listening

Active listening is a critical element in oral communication.

Develop a simple checklist of means you can use to improve your active listening skills. This checklist might be subdivided into:

- Tasks to prepare for a communication event
- Tasks to improve the listening experience
- Tasks to follow up on an oral communication

In a classroom setting, divide into pairs and have one participant relate an event or story while the other practices listening. The listener should follow the checklist and then complete the practice session by relating to the speaker what was heard. Participants should then reverse roles. Finally, the pair should work together to develop a consensus checklist based on their experience in the exercise.

For an individual, take advantage of the next opportunity for oral communication to use your checklist. After the encounter, revise the checklist to improve it, and continue the process whenever the opportunity arises for oral communication.

2. Improving customer relations

Identify a customer with whom you would like an improved relationship. Develop a plan of action for the improvement. This should start by either identifying the next anticipated encounter with this customer or by determining how an encounter can be created. Next, determine in what way(s) you would like to improve the relationship.

Examples might include determining a new or enhanced service you can provide the customer, or establishing a more regular and open channel of communication. Then write out specific steps that you can take to achieve the improvement(s).

After the encounter, review your plan and how well it was achieved. Modify the plan and repeat the process until you are satisfied with the relationship with this customer. Chose another customer with whom you would like to improve relations and initiate the process with this new customer.

CHAPTER 4

WRITTEN COMMUNICATION AND DOCUMENTATION

INTRODUCTION

In Chapter 3, the importance of communication in the construction process was discussed. Much of the communication engaged by the supervisor is oral. However, the supervisor also deals with an abundance of written communication and documentation.

This chapter will consider three broad categories of written communication and documentation that are important to the construction supervisor, including:

- Written communication developed by the supervisor on the job
- Written communication relevant to the job developed at the company level
- Written communication developed outside the company that affects job site operations

The chapter begins by discussing why communications are put in writing and then gives an overview of the functions of written communication and documentation on the site. Discussion then focuses on each of the three broad categories of written communication noted above. Finally, brief comments are made on organizing construction documents and on the process of obtaining information and documentation.

WHY WRITE THE MESSAGE?

There are a number of reasons why communications are written instead of, or in addition to, being transmitted orally. First, written communication is more supportive of mutual understanding than is oral communication. As described in the previous chapter, oral communication is subject to many flaws that can distort or block the understanding the communicator is trying to convey. The voice may not be heard or only partially heard. Unfamiliar words, concepts, or terms can distort the meaning of the message. Accents may be unintelligible. As a result, the message the communicator thinks is being sent is often not received as it was intended. Generally, there is not a confirmation that the message received has the same meaning for the listener as was intended by the sender. However, if the sender writes the message and gives it to the receiver, there is a much higher likelihood that the sender and the receiver understand the message in the same way. If the message is going to multiple people, the likelihood for misunderstanding or misinterpretation with an oral message increases rapidly, as does the need for a written format.

The second reason why communications are written instead of, or in addition to, being transmitted orally is that the written word provides an unchanging record of the message. The message crafted today and written down will be the same tomorrow, next week, and next year. Questions down the road about what she thought she heard or what he thought he said can quickly be dispelled by referring to the written message.

Finally, a written message gives an indication of approval. If a written message is sent from one party to another and the receiving party does not respond on a timely basis that the message is flawed or incorrect, it is assumed that the receiving party was in agreement that the message is correct. It becomes very difficult long after a written message was sent for the receiver to respond and say that the message was incorrect.

Meeting minutes illustrate this point very well. Minutes are sent to all parties in a meeting to document what went on in the meeting. If a party receiving the minutes does not read them and respond on a timely basis that the documentation is flawed, the assumption is made that the minutes are a clear and accurate representation of what went on in the meeting. Disputing the accuracy of the minutes at a later date is very questionable because recollections of what actually happened in the meeting fade quickly. Likewise, if an oral discussion between two parties is documented in writing subsequent to the discussion, it becomes very difficult for either party to later dispute what was said in the discussion unless a timely response was made to the written documentation.

THE FUNCTIONS OF JOB SITE DOCUMENTATION

Job site documentation serves three primary functions. The first is to gather data. Examples of this include collecting data on how many hours a craft worker has invested in a specific activity, or collecting data on quantities of a certain type of work that has been installed in a given period of time. Such documentation is used to determine the status of the job and often goes into reports on the status in terms of cost, time, and physical progress. Collection of this information will be discussed later in this chapter, as well as in Chapter 18. Use of the resulting reports will be discussed in Chapters 12 and 13.

The second function of job site documentation is to transmit information to an entity off the job site or to another entity on the site. Many people want a variety of information about what is happening on the site. The general contractor wants progress reports from the subcontractors. The owner wants to know how the project is progressing. The contractor's office needs to know about progress on the contractor's work on the site and investment of the contractor's time, money, and materials. Designers want to know about functionality of the systems they designed. Many people want to know about safety infractions, incidents, and accidents.

The third function of job site documentation is to build a chronology of the construction process. The job log, updated schedules, record drawings, and minutes from coordination meetings all provide documentation on how the job was built. It is essential to build this documentation as the job progresses. Otherwise, it will be lost, or if there is an attempt to build this documentation at a later date, the record will be incomplete or erroneous.

IMPROVING WRITTEN COMMUNICATION AND DOCUMENTATION

Since writing is not a craft skill that workers learned during apprenticeship, supervisors often are not skilled writers. As a skill, the ability to write can be learned and can be improved. Following are some guidelines for improving writing skills for supervisors.

A specific time and place should be set aside to write. Concentration and time are required in order to develop a high-quality written document. Some supervisors like to write early in the morning before the day's work begins. Others like to write at the end of the day, filling out the jog log or time cards after craft workers have gone home, yet in a timely manner so that the information is fresh in their mind. Some would prefer not to extend the day prior to the start of work or after work is finished and like to use the lunch hour to isolate themselves and get the writing done. Establishing a regular writing routine supports getting the written documentation done in a timely manner.

All available digital tools should be used to improve writing. Most word processors have spelling checkers and grammar checkers that require little time and can add significant value by correcting spelling and improving grammar. Other tools may also be available. It should be remembered that the utility of digital tools is limited. For example, a spelling checker will not distinguish between appropriate and inappropriate uses of context-sensitive words such as "to," "too," and "two," or "there," and "their." It will also accept words that are correctly spelled but that are not the right word because an improper letter was used, changing the meaning. For example, if the writer is talking about a chain saw, but typed in an "m" instead of the "s", the document would talk about a chain maw, which is obviously incorrect, but the spelling checker would not recognize maw as an improper word and would not suggest a correction.

Business writing should be simple and succinct, using just enough words to ensure that the message will get across. A good statement to remember in business writing is that *brevity is beautiful*. Simplicity is also important. Sentences should be short. Words should be common. Business writing is not creating literature. It has a purpose, which is to communicate clearly and to document accurately. Business writing is not meant primarily to entertain, although good business writing should be interesting.

With electronic media, proofreading should be done continuously. What is written should be reviewed as the document moves forward to ensure that grammar and details are correct and to find better ways to communicate succinctly. When completed, the entire document should be proofread to view the document from a more strategic level that can ensure that the overall document flows well.

Another person should proofread important documents. It is very difficult to read one's own writing and pick up all the errors. Another set of eyes is very useful to detect errors that the original author will tend to overlook.

JOB SITE COMMUNICATION AND DOCUMENTATION

The supervisor develops many documents at the job site. In fact, one of the management skills a supervisor must master is writing. In the field, the supervisor's writing skill focuses primarily upon getting the right information recorded in an unambiguous format that can be understood by all. For field documentation, completeness, accuracy, legibility, and clarity of message are the important characteristics. Appropriate language is also highly important. Spelling and grammar, though never unimportant, are of less importance in field-developed documents.

Supervisors are often reluctant to write information down because they lack skills that would enable them to write as effectively as a highly educated professional. However, it is essential to get the supervisor's observations and facts about the job in writing. Often field documentation plays a key part in legal cases, so the supervisor must document activities in such a way that the message is complete and the integrity of the document is not called into question. Such writing will protect the company's reputation, as well as its legal position.

Of the many documents developed in the field by the supervisor, four of the most important documents will be reviewed here:

- The job log
- Time cards
- Record drawings
- Field authorizations

The Job Log

The job log is the primary record of the construction process. Many supervisors keep their own job diary, which documents key information about the part of the construction process in which they are engaged. However, each contractor on the site (prime and sub) should have a single document, developed throughout the duration of the job by the contractor's highest supervisor on the site, that documents the construction process as it relates to that contractor. This is the job log for the contractor.

The job log has a number of uses. First, it *documents major events* on the site. A major event could be an event that consumes a significant amount of time and/or resources. Or, it could be a small but key event that has a significant or long-term impact on the job. An important part of keeping a good job log is identifying and evaluating what is important enough to go into the log and what does not rise to the level of being recorded in the log.

Next, the job log *identifies responsibility and establishes accountability* for what happens on the site. This might relate to decisions that were made, or it could document how work was assigned or directions that were given. It is important to know not only what happened but also, where possible, who was responsible for what happened.

The job log should *trace the history of events* on the job. Any professional familiar with the type of work the contractor does should be able to read the job log like a novel, seeing, on a day-by-day basis, how the work has progressed.

The job log should *document working conditions* on the site and any conditions that will have an effect on the work. This could be as varied as describing the weather, and particularly changes in the weather, or describing conditions where work is being done, such as the condition of the ground or congestion in the area. The log should also document abundance or, more importantly, shortages of resources required to get the work done. This might include worker shortages; appropriate, well-functioning, and safe tools and equipment; and the right materials in the right amount. It should also document availability, or lack of timely, accurate, and complete information and how this is affecting the work. It is particularly important to record when unanswered questions or the lack of information is slowing the work.

Finally, the job log should explain *why events occurred*. Much of the factual information about events, such as what happened and when it happened, is documented elsewhere, but explaining why things happened is often missed, although it is very important. Such accountability is difficult to determine at a later date unless it is documented at the time an event occurs. It is important to be accurate in this analysis. The object is not to place blame or to build a case for something anticipated to come up in a dispute. The intent is to provide an objective, unbiased account of why events occurred. If it is determined later that the documentation is biased or leaves out important facts or includes questionable information, the validity of the entire document is jeopardized.

The contents of a job log are listed in Figure 4.1.

Certain guidelines need to be followed in order to protect the integrity of the job log as a strong legal document. If it is in a physical format, the log needs to be in a

---

Job Log Contents

- Date
- Weather
- Job Accidents
- Safety issues
- Number of Workers
- Major Equipment Used
- Major Events
- Visitors to the Job

- Subcontractor Progress
- Non-conforming Work
- Decisions
- Narrative of the Daily Activities
- Verbal Instructions
- Questions for office
- Signature and Date

Figure 4.1   Job Log Contents

format that restricts late entries or alterations to the original document. This can be achieved by having it in a bound book format. This can also be achieved by having the information for each entry (each day) entered onto a multipart form that is torn apart and distributed at the end of the day after the day's entries are complete. Pages should be numbered consecutively so that pages cannot be inserted or removed later. All entries need to be in ink. Any corrections should be made by crossing out the old and adding in the new, with an indication of who made the change and when. Since the job log is such an important document, it will be addressed again in Chapter 18, where more detailed guidelines for keeping the job log are provided.

As computers become more widely available in the job site office, it is becoming more common to create the log in a digital format. Log security and integrity continue to be critical even in a digital format. Because of the ease of modifying or altering digital files, it is even more important to develop a standard procedure to ensure that the file is not opened and altered after the date on which log entries are made. This could entail such measures as e-mailing a final copy to others off the site at the end of the day and printing out a copy to put in a secure notebook.

There are many standard job log forms that can be purchased for use on a job. On the other hand, many companies feel it is important to develop their own form to better fit the specific company and job culture. If a company wants to develop a standard form, this is easily accomplished by bringing together a team of supervisors to design the company's standard job log form. This process would start with the supervisors developing a list of desirable characteristics of a good job log, then developing the form that demonstrates these characteristics. Such a list of desirable characteristics might include:

- Minimize the work required to fill out the log
- Ensure important information is not forgotten

DAILY JOB LOG				
1.	Job		2.	Date
3.	Weather		(Temperature)	
4.	Manpower	Others	Subcontractors:	
	Foreman		Sheet Metal	
	Mechanics		Insulation	
	Apprentices		Control	
	Laborers		Sprinklers	
	Helpers			
5.	Rental Equipment on Job			
6.	Deliveries Received			

Supplier	Carrier	Item(s)	Damages/Shortages

7.	Delays

8.	Verbal Instructions Received	
		From
9.	Safety	
	Accident Information	
	Is an Accident Report Attached	
	Safety Meeting Information	

10.	Visitors	Company/Organization	Purpose	Time In/Out
	Name			
	Comments			

11.	Photographs: Number of photographs attached
	Is the subject described on the back of each photo?
12.	Questions/Comments for Office

	Date	
13.	Signature	

Figure 4.2 Example of a Daily Job Log Form

- Minimize the time required to fill out the log
- Minimize narrative required

Much of this can be accomplished by using a fill-in-the-blank format and check-off's rather than having to write from scratch. An example of a job log form is shown in Figure 4.2.

This and several other forms in this chapter are taken from the **Alliance Project Management Manual,** © 1998 Mechanical/Electrical/Sheet Metal Alliance, All Rights Reserved, Printed in the USA, First Edition—March 1998.

Time Cards

Time cards or time sheets are the primary record of the investment of field labor. Most foremen consider time cards as a way to document for the office how much time individual crew members spend working so that the payroll office can prepare paychecks for the week's work. It is true that paychecks for craft labor are generally produced using the information on time cards. It is also true that accurately prepared paychecks are important. However, this is the least important use for time cards.

Time card data are entered into the company cost control system so that periodic reports (generally weekly) for each job can be developed documenting how much and where labor was invested in the job for the given period. This information, together with a record of quantities of work completed with that labor, provides the basis for the job cost control system. This will be discussed in detail in Chapter 13.

Time card and units installed data also form the basis for developing company unit costs for use in estimating. These unit costs provide company estimators with invaluable information needed to accurately estimate the cost of future jobs. This information is very company specific and represents one of the most closely held, proprietary pieces of information any construction company has. Accurate cost estimates are of fundamental importance to the success of a construction company. There are many databases of standard unit costs available for all types of construction work. However, only company-developed unit costs tell accurately how much it costs for a specific company to do its work. The company cost database also contains information on the time it takes workers for this company to accomplish tasks. Hence, the information is also quite important for schedulers and project managers who develop the project schedule, as well as for the estimators.

The cost control system is the fundamental source in telling the supervisor and the project manager how well the work is going in terms of budget and time. If time is not recorded accurately at the crew level, the cost control system will be inaccurate, resulting in overlooking production problems or in seeing production problems where they do not exist on the job. Such misinformation can be very damaging to a project and disastrous for the company.

The most damaging result of inaccurate time recording is the impact on the estimating process because of inaccuracies in the company's pricing database. If recorded times are inaccurate, unit costs will be inaccurate, resulting in estimators pricing jobs too high, causing the company to lose out on the job, or pricing jobs too low, leading to the company taking a job without adequate money in it. In either case, over time, this will severely damage company operations.

Accurate time keeping is one of the most important responsibilities of construction supervisors, generally at the foreman level. Accurate time keeping is also the weakest link in the entire cost control and estimating processes. Supervisors often record time inaccurately, either intentionally or unintentionally.

Unintentionally recording time inaccurately is normally the result of one of two situations. The first occurs when supervisors do not realize the importance of accurately recording labor time and they become careless. The second happens when time is not recorded on a timely basis, resulting in the supervisor's being unable to remember accurately which activities each crew member was working on and for how long he or she worked on it.

Intentionally recording time inaccurately is generally the result of the supervisor recognizing that work is not progressing as efficiently as it should be and so recording a few hours of worker's time to another work item.

The problem of inaccurate time recording needs to be addressed in several ways. First, supervisors need to know how important accurate time recording is and why it is important. Next, there needs to be a recognition that job cost reports showing work at less than acceptable production levels are regarded as opportunities for improvement rather than excuses to blame the supervisor. Finally, there should be a company policy stating that accurate time recording is considered critical to the company, and inaccuracies will elicit a strong reprimand from the company. Because of its critical importance, accurate time keeping is again addressed in Chapter 18.

An example of a weekly time sheet is shown in Figure 4.3.

An important part of accurate time keeping is the use of a cost code system. Cost coding assigns a unique code number to each task in which the company workers are engaged on a job. This cost coding system flows throughout the entire job. It is used to estimate the job initially. It is then used as the basis of allocating hours spent in installing the work, as well as the cost of equipment and materials required in the installation process. Finally, it is the basis for the entire procurement process whereby equipment and materials are purchased, their shipment is tracked, and, as stated above, installation hours are recorded. Cost coding will be further explained in Chapter 12.

To avoid time recording errors, in addition to the mitigations identified above, several procedures can be put in effect. First, a daily time card can be used instead of a weekly time sheet. This will cause the foreman to record the time at the end of the day rather than waiting until the end of the week.

Second, certain accounting controls can be put in place. As an example, the payroll office that enters the time card information into the computer can enable only work activities that are currently being executed in the field for data entry. Upcoming activities are only activated in the system when they are about to begin, and activities that are completed are deactivated. This prevents recording work in future work items when a current item is showing too many field hours. It also prevents recording hours in completed work items that were completed under the budgeted labor hours.

Third, it is very important for supervisors up the line to recognize production problems that are identified in the cost control report by accurate time recording as opportunities for improvement, not opportunities to place blame. Foremen are pain averse, and if they know that they will be reprimanded every time a production problem is identified, they will find ways to avoid the pain and hide problems. If they are rewarded for accurate reporting, especially when the reporting has pinpointed a problem that can be resolved, they will tend to more accurately record the data.

WEEKLY TIME TICKET

EMPLOYEE NAME: _____   EMPLOYEE NO: _____   WEEK ENDING: _____   DATE: _____

DAY	JOB NAME	JOB NUMBER	CRAFT	COST CODE	G/L ACCOUNT NUMBER	TYPE*	HOURS				FOR OFFICIAL USE ONLY		
							REG.	OT.	DBL. TIME	TEMP	OTHER HOURS	SHOP	OTHER RATES
TUE													
TUE													
TUE													
WED													
WED													
WED													
THU													
THU													
THU													
FRI													
FRI													
FRI													
SAT													
SUN													
MON													
MON													
MON													

TOTAL HOURS BY CLASSIFICATION

TOTAL HOURS

* TYPE FIELD (F), SHOP (S)

APPROVED BY FOREMAN

EMPLOYEE SIGNATURE

Figure 4.3   Example of a Weekly Time Sheet

As already stated, it is very important that the foreman, or whoever is recording the time, knows the critical importance of time keeping, not only to the current project but also to the entire life of the company.

Record or As-Built Drawings

Record drawings, or as-built drawings, are the primary record of the final product. Record drawings show how the job was built. There are always small changes throughout the job that are done in response to job site conditions and situations. There are also changes that are originated by the owner, by the design team, or by the contractor. Any change from the original design should be recorded by the supervisor on a set of drawings for later incorporation by the designer into the final set of record drawings.

Record drawings are of vital importance to the entire construction team. The owner and the facility manager need accurate record drawings so that when later changes to the facility are contemplated, the drawings will tell them exactly what the physical plant is that they own. Record drawings also show specific routings for systems built in the field so that when problems crop up with a system they can be tracked down. The designer needs record drawings that show any modifications to the details of the design. Major changes are formally processed and can be easily incorporated by the designer, but the day-to-day details of how the job was built in the field are often lost if not recorded on the record drawings. Contractors need the record drawings when estimating or building renovations or revisions to the facility. Without accurate record drawings that show the physical layout of the building, especially behind walls or over the ceiling, it is very difficult to price, plan, and execute change work accurately and efficiently.

To get a good set of record drawings, a dedicated, clean set of construction drawings should be kept at the job site for the express and single purpose of recording as-built conditions. Starting on the first day, anything that is built differently than shown on the construction drawings should be noted on the record set. It is vital to develop record drawings as changes are made and as the project progresses. It is impossible to develop an accurate set of record drawings after work is completed. Chapter 19 will address the disposition of as-built drawings as the project is closed out.

Field Authorizations

Field authorizations are the primary record of contract changes. A field authorization is a form that is used to document an oral request for a change that is issued in the field. Many times a contractor is asked to make a field change without going through the formal change order process. The request often comes on the spur of the moment and is required to be completed right away for work that is ongoing and in circumstances when there is not time to go through the formal change order process.

No change or additional work should be executed without an authorized signature. When the supervisor is asked to do something different or additional work that is not in the base contract, the supervisor should always ask for a signed statement defining the work that is being requested. The person making the request can provide a signed field authorization, and if it is signed by the authorized person, work can proceed immediately. More often than not, the requesting authority does not have a written, signed authorization. At that time, the supervisor can use his company's field authorization form, fill in a few blanks, get the requesting person to sign it (if authorized), and then work can proceed.

The form can be quite simple. It will have blanks to fill in for such information as time, date, project, and who is requesting the work. It should also have a place to briefly describe the work requested. There will generally be some language on the form that talks about executing the work described on the form at the request of the authorized person who signs the form. The language will also state that additional cost and time to execute the work will be calculated and a change order processed to adjust time and cost on the contract, if appropriate.

The field authorization form is a valuable tool, enabling the supervisor to take control of a situation in which field changes are requested. It documents what was requested, and by whom, and it establishes the basis for a future adjustment of time and cost, should they be required. If the requesting party is not authorized, or not willing to sign the field authorization form, the field authorization process forces the parties back into the prescribed procedures for the job. This relieves the supervisor of assuming responsibility for decisions or for doing work outside his jurisdiction. It also minimizes confrontation if the company has the policy that the supervisor cannot do any work that is not specifically shown on the contract documents without signed authorization.

Figure 4.4 provides an example field authorization form used by the city of Port Angeles, Washington.

COMPANY OFFICE COMMUNICATION AND DOCUMENTATION

A significant amount of job-specific written documentation is developed at the company level. This documentation is not developed by the supervisor, but the supervisor often provides information for the documentation. Also, field work is often affected by such documentation.

Written communication developed in the company office is typically more formal than that developed in the field. It often goes outside the company, so it has a significant impact on the company's image. It is machine produced (where as much of the field documentation is hand written) and is written on company letterhead. Not only are completeness, accuracy, and timeliness important, but format, style, grammar, and spelling become very important. A significant portion of such communication is moving into digital format and electronic transmission, so electronic communication will briefly be addressed later.

Sample Only

City of Port Angeles, WA
Public Works Department Policy and Procedures
Design Clarification/Variation Request/Field Authorization Form

PROJECT NAME:_____

PROJECT NUMBER:_____

ORIGINATOR:_____

ITEM:_____

REFERENCE DRAWING OR SPECIFICATION:_____

DESCRIPTION OF REQUEST:_____

DATE REPLY REQUIRED:_____

CRITICAL TO SCHEDULE: YES_____ NO_____

PROJECTED COST EFFECTS:_____

PROJECTED TIME EFFECTS:_____

COST INCREASE [APPROX. VALUE $] INCREASE IN CONTRACT TIME [DAYS]

NO CHANGE N PRICE NO INCREASE IN CONTRACT TIME

COST DECREASE [APPROX. VALUE $] DECREASE IN CONTRACT TIME [DAYS]

ORIGINATOR SIGNATURE: DATE:

OWNER/ARCHITECT/ENGINEER

VARIATION REMARKS:

CLARIFICATION:

It is our opinion that this work is not an extra to the contract and direct the contractor to perform the work described per the contract documents. If you consider the above to be an extra to the contract, you must prepare and submit a detailed cost proposal and justification.

It is our opinion that this work will result in a need for a change order to the contract amount and/or time. You should prepare a detailed cost proposal and justification as soon as possible since the work will not be authorized until such data is approved. If such work is critical to the timely execution of the contract, you are authorized to proceed subject to the work not exceeding $ and days to the contract time. Final cost and time changes shall be subject to the review and approval and execution of a change order. Payment for the work herein authorized will not be made prior to the execution of a change order.

REVIEWED BY:	SIGNATURE:	DATE:
ARCHITECT/ENGINEER		
CITY INSPECTOR		
CITY ENGINEER		
PUBLIC WORKS DIRECTOR		
CITY MANAGER		

Figure 4.4 Field Authorization Form

As with field communication and documentation, there are many different documents produced in the company office; however, only four of the most important will be considered in this book:

- Memoranda
- Letters
- Minutes
- Reports

Memoranda

A memorandum is an individual informal written document for distribution within the company. It is used in lieu of an oral message because the message is too extensive or too complex to trust to a simple oral message. Also, the message may be important enough that it merits special attention. It also may be critical that details are not lost or forgotten and that the message can be referred to later. The message is not urgent enough to require immediate delivery, since it takes a bit of time to write and deliver a memorandum. If urgency is a factor, the message can be delivered immediately orally with a follow up memorandum. A memorandum can also be easily addressed to multiple parties, or delivered to a single party with copies to others.

Memoranda have a simple introductory format that provides the following standard information:

- The person(s) to whom the memorandum is addressed
- Who it is from
- Who will be sent copies
- The date of transmission
- The subject of the message

This initial information is followed by the body of the memorandum.

If memoranda are written often, the writer should have a standard form in the computer, with the introductory fields in place, that can be pulled up and completed quickly. Figure 4.5 shows a simple memorandum form.

The memorandum requires no salutation and instead of a signature, it is initialed next to the name of the person who originated it. The body of the memorandum should be written briefly and succinctly.

Memoranda are quick, simple internal written communications that can be very efficient and very effective.

Letters

A letter is an individual, formal written document for distribution outside the company. It is typically sent to a single person but can be copied to others. Rarely, a letter is sent to several people, but if it is sent to multiple parties, each person needs to be recognized with full address information at the top of the letter. If a number

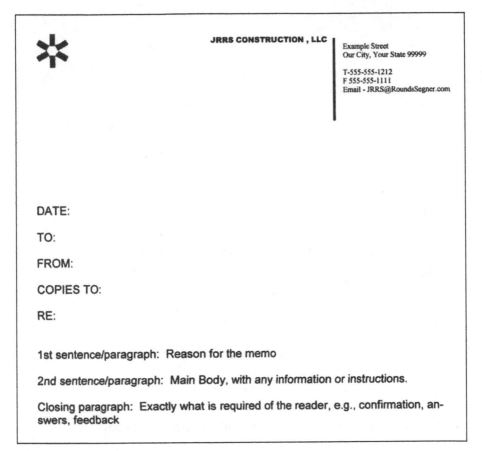

Figure 4.5 Memorandum Form

of people are being sent a letter, rather than recognizing all parties in a single letter, multiple versions of the letter are typically sent, one to each individual. With computerized systems, producing multiple copies of a letter addressed to different individuals is not a difficult task.

Typical uses for a letter are to document or clarify oral conversations, to communicate complex information, or to give emphasis to a message. Letters use company letterhead. When using company letterhead, the writer should take special care to remember that the letter represents the company and should it be written in such a way to represent the most positive image possible for the company.

Letters begin with the full address of the recipient, a date, and salutation. Often they will contain a reference line before the salutation, indicating the main topic of the letter. They end with a formal closing and, where applicable, lists of attachments, enclosures, and parties to whom a copy will be sent.

The body of the letter should be as brief as possible. It should not contain jargon. When abbreviations or acronyms are used, they must be clearly defined on the initial use.

JRRS CONSTRUCTION, LLC

Example Street
Our City, Your State 99999

T 555-555-1212
F 555-555-1111
Email - JRRS@RoundsSegner.com

Month, Day & Year	**DATE**
Ms. Jane Jones, Superintendent	**NAME OF RECIPIENT**
Jane's Street	**ADDRESS OF RECIPIENT**
Jane's City, State and Zip Code	
RE: THE CONTRACT	**SUBJECT**
Dear Ms. Jones:	**SALUTATION**
Our team has determined . . .	**BODY TEXT**
We very much appreciate . . .	**COMPLIMENTARY CLOSING**
Sincerely,	
	SIGNATURE
John C. Smith, Project Superintendent	
cc: Person One	**COPY LIST**
Person Two	

Figure 4.6   Business Letter

The important thing to remember about a letter is that it represents the company. It goes outside the company and could easily become public information. If there is any unease about sending the letter out, someone else in the company should proofread it prior to sending it. Once the letter is in the mail, it cannot be recalled.

Figure 4.6 provides an example of a business letter.

Minutes

Minutes provide an unchanging record of what occurred in a meeting. Minutes should be produced for every meeting, whether a brief, informal meeting, a higher-level meeting planned well in advance, or a periodic meeting that occurs repetitively at regular intervals. If a meeting is important enough to have, it will be important enough to be documented through minutes, even if they are brief and simple.

Meeting minutes record history. People soon forget what occurred in a meeting and recollections will vary widely from one participant to another. If reference is made to a meeting at a later date, the only valid way to recall what occurred in the meeting is through the minutes.

Minutes should be taken by a designated person. Preferably the person taking the minutes is not the person facilitating the meeting because taking accurate and complete minutes requires dedicated concentration, which cannot generally be given by the person facilitating the meeting. Minutes can be taken later if the meeting is recorded, but generally, it is better to make notes throughout the meeting and then to develop the minutes as soon as possible after the meeting. For regular, formal meetings, often a clerical person who is not a participant in the meeting is invited in to take notes and produce the minutes.

Notes should be turned into minutes as soon after the meeting as possible. With quick turnaround, the note taker can remember what a brief note means. On the other hand, if days pass before the notes are turned into minutes, much is lost, the minutes are more difficult to produce, and they will not be as accurate or complete as they would have been if written immediately after the meeting.

If the meeting is recorded for the purpose of taking more accurate or detailed minutes or for any other reason, participants need to know that the meeting is being recorded, and they need to give their permission for the recording. Often, discussion is hindered in a meeting if participants know that it is being recorded.

The minutes should be on letterhead and should start with the following information:

- Date, time, and location of the meeting
- Purpose of the meeting
- A list of participants

If this is a routine meeting or one of a cycle, the participant list might be more extensive, including, in addition to normal participants in attendance, normal participants not in attendance and guests at the meeting who are not normal participants.

At the end of the minutes should be a statement of the date by which any comments or corrections to the initial set of minutes are to be received and the manner of submittal of these corrections so that they can be included in the final version. The cutoff date and comment process discourage late comments and establish the point at which the minutes become the authorized version of what occurred in the meeting. After the cutoff date, the final set of minutes should be completed as soon as possible. They are then distributed to all participants and other interested parties.

If regular meetings are held periodically, the final set of minutes should be distributed prior to the next meeting. Alternatively, if subsequent meetings occur at short intervals, the minutes from the previous meeting can be reviewed at the beginning of the following meeting with a request for changes and corrections at the meeting. In that case, the final minutes are distributed after the subsequent meeting.

For a meeting that is one of a cycle of meetings, the minutes should provide information about the next meeting, such as time, location and, if relevant, the focus of the next meeting.

Figure 4.7 shows a template for a set of minutes for a project team meeting.

Reports

Reports summarize the status of the job or some aspect of the job. They are sometimes written by a supervisor in the field, but more often the participation of the supervisor is in the form of either providing information to go into the report or using information from the report to more effectively supervise the job.

Reports are repetitive in nature and focus on specific topics. Some reports, such as testing reports for installed equipment are written once for a job but are developed for virtually all jobs of a common type. Many reports are produced periodically throughout a job, updating the status of various aspects of the job. Examples of periodic job reports are progress reports and reports on the current status of cost and schedule. Other reports are written multiple times on a single job, but only when triggered by events. An example would be an incident report that summarizes the results of an investigation after a safety infraction.

Companies should have standard formats for reports and standard procedures for producing the reports. Procedures address such things as:

- How information is to be collected
- How the report is to be written
- How the report is to be distributed
- How the report is to be archived
- How long the report needs to be kept

Any standard report should be periodically reviewed by the company to ensure that it is still used and to seek ways to improve the report and the processes for developing the report. In reviewing reports, companies will often find that a report is no longer required or no longer relevant. They will also find that simple modifications can improve the utility of the report or the data collection process.

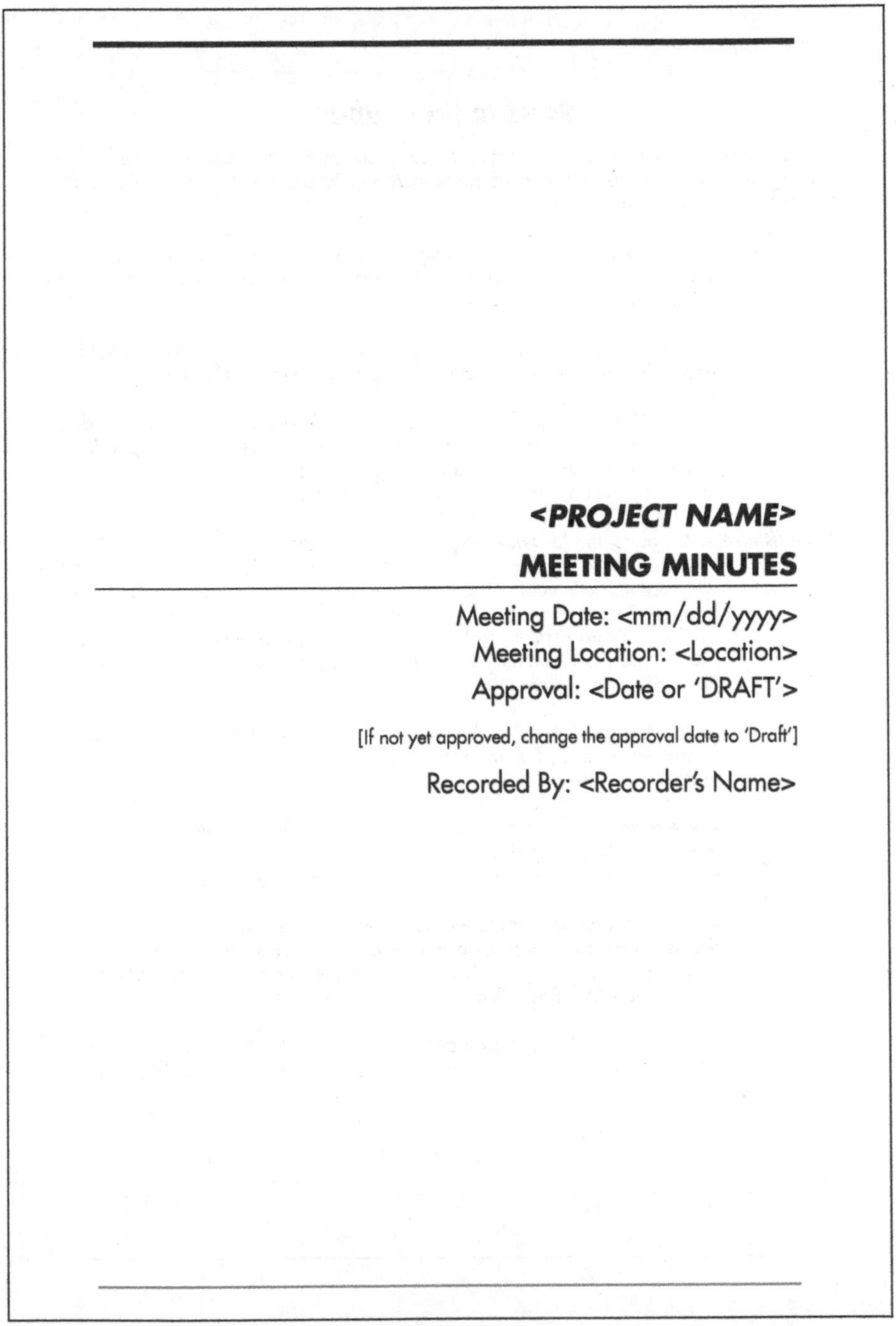

<PROJECT NAME>
MEETING MINUTES

Meeting Date: <mm/dd/yyyy>
Meeting Location: <Location>
Approval: <Date or 'DRAFT'>

[If not yet approved, change the approval date to 'Draft']

Recorded By: <Recorder's Name>

Figure 4.7   Meeting Minutes

Notes to the Author

[This document is a template of a Meeting Minutes document for a project. The template includes instructions to the author, boilerplate text, and fields that should be replaced with the values specific to the project.

- *Blue italicized text enclosed in square brackets ([text]) provides instructions to the document author, or describes the intent, assumptions, and context for content included in this document*

- *Blue italicized text enclosed in angle brackets (<text>) indicates a field that should be replaced with information specific to a particular project.*

- *Text and tables in black are provided as boilerplate examples of wording and formats that may be used or modified as appropriate to a specific project. These are offered only as suggestions to assist in developing project documents; they are not mandatory formats.*

When using this template, the following steps are recommended:

1. *Replace all text enclosed in angle brackets (e.g., <Project Name>) with the correct field document values. These angle brackets appear in both the body of the document and in headers and footers. To customize fields in Microsoft Word (which display a gray background when selected) select File->Properties->Summary and fill in the appropriate fields within the Summary and Custom tabs.*

   After clicking OK to close the dialog box, update all fields throughout the document selecting Edit>Select All (or Ctrl-A) and pressing F9. Or you can update each field individually by clicking on it and pressing F9.

   These actions must be done separately for any fields contained with the document's header and footer.

2. *Modify boilerplate text as appropriate for the specific project.*

3. *To add any new sections to the document, ensure that the appropriate header and body text styles are maintained. Styles used for the section headings are Heading 1, Heading 2 and Heading 3. Style used for boilerplate text is Body Text.*

   Before submission of the first draft of this document, delete this instruction section "Notes to the Author" and all instructions to the author throughout the entire document.]

Figure 4.7   *(Continued)*

<Project Name> Meeting Date <mm/dd/yyyy>

1 ATTENDANCE

Name	Title	Organization	Present
<Name>	<Title>	<OPDIV/Bureau/ETC>	<Y/N/Phone>

*[The usual list of attendees should be detailed here. Any guests can be added. Groups with representatives from multiple organizations (Departments, Operating Divisions, Bureaus, etc.) should detail which organization **each** attendee represents. Attendance should be marked as yes, for those attending in person, no for those absent, and phone for those attending by teleconference or other remote method.]*

2 MEETING LOCATION

Building:
Conference Room:
Conference Line:
Web Address:

3 MEETING START

Meeting Schedule Start: <HH:MM>
Meeting Actual Start: <HH:MM>
Meeting Scribe: <Name>

4 AGENDA

- <Agenda Item 1>
  <Notes on discussion>
- <Agenda Item 2>
  <Notes on discussion>
- <Agenda Item 3>
  <Notes on discussion>
- <And so forth...>

[Exactly method of note taking on discussion can vary from one recorder to another. Most important is capturing the essence of the conversion. Major points raised, and by whom, should be recorded faithfully, although there is no need to capture them word for word. If the agenda is handled out of order, rearrange the agenda items to indicate the order in which they were actually handled.]

5 MEETING END

Meeting Schedule End: <HH:MM>
Meeting Actual End: <HH:MM>

Figure 4.7   *(Continued)*

6 POST MEETING ACTION ITEMS

Action	Assigned To	Deadline
<Action Item>	<Assignee>	<mm/dd/yy>

7 DECISIONS MADE

[Document any decisions made during the meeting

- Decision 1
- Decision 2

Etc.]

8 NEXT MEETING

Next Meeting: <Location><Date><Time>

Figure 4.7 (*Continued*)

Electronic Mail

Business e-mail messages are not notes to friends. They are professional correspondence and should be treated with as much respect as memoranda and letters. The spelling checker that is available with most e-mail programs should always be used. Longer and more important e-mail messages should be written with a word processor and then copied to an e-mail form. Care should be taken not to get caught up in using texting abbreviations that tend to come naturally in a digital environment for those who use extensive texting.

The belief that e-mail messages are restricted in distribution and fleeting in duration is a mistake. E-mail messages often take on a life of their own, with one person copying to a list of contacts who then forward it to others. When deleted from one computer, these messages exist in many other computers, as well as servers in the e-mail system.

Writing joking messages thinking that they will not be sent or will later be deleted is another mistake. Such messages often do get into broad distribution and have, in some cases, had disastrous consequences for individuals and their companies.

The following tips for writing professional e-mail messages were extracted from an electronic article by Dennis G. Jerz, associate professor, English, New Media Journalism, Seton Hill University, Greensburg, Pennsylvania. The article can be found on his Blog at **http://jerz.setonhill.edu/writing/e-text/e-mail.htm# message#message**.

Top 10 Tips for E-Mail

Write a meaningful subject line

Keep the message focused and meaningful

Avoid attachments

Identify yourself clearly

Be kind (professional)—Don't flame

Proofread

Don't assume privacy

Distinguish between formal and informal situations

Respond promptly

Show respect and restraint

COMMUNICATIONS AND DOCUMENTATION DEVELOPED OUTSIDE THE COMPANY

An extensive amount of documentation is developed by others for use on the job. The supervisor might participate in developing some of these documents and will use others to manage and coordinate the work. Three categories of documentation are identified below.

Design Documentation

Design documents are developed by the design team to provide various models from which the project can be built. The drawings are a graphical model of the project. The specifications are a written model of the project. Throughout the project, modifications and clarifications are made to correct, improve, and add detail to these models.

Production Documentation

Production documents are developed by the construction team to support the construction process. Many schedules are developed at a variety of levels. As described in Chapter 14, the overall project schedule is the responsibility of the general contractor or construction manager and should be developed with the participation of various subcontractors and other project stakeholders. Once developed, the project schedule is periodically updated, and these updated schedules provide an invaluable record of how the project progressed from start to finish.

Short interval schedules are developed throughout the project to provide more detail and a narrower focus than the project schedule. Whereas the project schedule provides a strategic view of the overall project, short interval schedules are used as tactical tools and focus on a specific time frame or a limited scope of work.

Minutes from project coordination meetings throughout the duration of the project are developed by the general contractor or construction manager, and they document the short interval planning process, as well as identification and resolution of coordination problems among the various entities working on the job.

Progress reports are developed by the general contractor or construction manager based upon their own work, their observations, and progress reports from the various subcontractors.

Procurement Documentation

Procurement documentation starts with approval drawings and submittals provided by material and equipment suppliers. These documents pass through the various subcontractors to the general contractor and on up to the designer and owner for review and approval. They are then returned through the submittal chain to the suppliers with corrections or a request to resubmit.

Installation drawings are developed by various suppliers and are required for installation of major pieces of equipment, as well as for materials, such as structural steel and reinforcing steel.

Purchase orders and requisitions establish the contractual basis for procurement of materials and equipment. Shipping documentation focuses on getting materials and equipment from the manufacturer to the job site.

Warrantees and guarantees, together with operations manuals complete the procurement documentation. These must be provided for all installed equipment and turned over to the owner with the completed facility.

ORGANIZING CONSTRUCTION DOCUMENTS

A great deal of documentation is developed for a project, much of it used at the job site. The documentation does no good if it is not organized so that it is accessible when needed.

The documentation needs to be accessible to the appropriate people. Accessibility should be easy so that it can be retrieved when it is needed. It should be cross-referenced so that related documentation can easily be retrieved. A document log is quite helpful in managing the documentation. The document log should include such information as date received, sending party, file location and how long the documentation should be kept. Figure 4.8 illustrates an example of the layout of a document log.

The company should have a standard filing system and filing procedures. The procedures will address such issues as what documentation stays at the job site, what goes to the main office, and what should be in both sets of files. They should also define who is responsible to do the filing, when filing should be done, and how long specific types of documentation should be maintained. Appendix II provides an outline of a simple filing system that can be easily adapted for a field office.

OBTAINING INFORMATION AND DOCUMENTATION

A fundamental problem on construction sites is not being able to get information when it is needed. Decisions the supervisor must make are dependent upon a steady flow of information. Supervisors must know how to get information with the following characteristics. It must be:

- Timely
- Authoritative
- Complete
- Correct

Information is time sensitive. As it becomes late, it rapidly loses its value. When information is not available, workarounds must be employed to make up for the lack of information. Wrong decisions or decisions that are not the best are made, resulting in errors and inefficiencies.

Document Log

Job Number_____

Job Title_____

ID	Date Rec.	Date Sent	Type	Form/To	Content	File Code	Discard

Notes:
1. Use company specific Indentification Code Number
2. Fill in either Date Received into the office or Date Sent from the office
3. Type examples: Letter, Report, Memorandum, RFI, etc.
4. Sending party or receiving party
5. Content briefly describes the topic of the correspondence
6. File code number indicating file location
7. Date after which document can be discarded

Figure 4.8   Document Log

Information must come from an authoritative source, one that knows that the information provided is the correct information. Authority is also important so that plans and decisions can be based on information that will not be changed or withdrawn. Partial information is often worse than no information at all because it can be misleading and give a false sense of confidence that the appropriate information has been received when, in fact, some is missing. Incorrect information is also troublesome because decisions are made and actions taken that later turn out to have been incorrect and damaging.

It is helpful to understand where information comes from. Information dealing with meeting the owner's needs should come from the owner or the owner's representative. That representative could be the designer or construction manager. Technical information dealing with aesthetics, performance, and serviceability of systems comes from the designer. Information about the production process comes from a contractor. Information about materials and installed equipment comes from manufacturers, suppliers, and distributors. Seeking information from the wrong source can result in delays in getting the information or in getting incorrect information.

There are a number of things the supervisor can do to improve the flow of information:

- *Timely identification* of the need is essential. If the need for information is identified early so that the person responding has time to research the response, there is a much higher likelihood that the information will be received on time and it will be of better quality.
- *Clear definition* of both what information is needed and when it is needed will support the process.
- *Clear delineation* of who is responsible for the information needed will help ensure not only that appropriate information is obtained but also that the response will not be changed.
- *Proper communication* of the need and possible solutions will assist the responder in providing good information. It will also tend to make the responder more responsive in the future.

Several proactive steps can be taken to improve information flow. First, the supervisor must know procedures that have been established for information flow on the specific project. Developing an information flow chart for the project and defining contacts to whom questions can be addressed will support this.

Expect information to come through standard job procedures. When it does not, quickly ask for the information and indicate that you expected the information to come as a normal and routine part of the process. Communicate that unless told otherwise, you will expect this information to be forthcoming in the future. Expressing your expectations for information flow early in the project will make information flow better as the project moves forward.

SUMMARY

In this chapter, the following key points have been presented.

- There are many reasons to put communications in a written format.
- Job site documentation serves to gather data, transmit information, and chronicle the history of the construction process on the job.
- Written communication is a very important skill for the supervisor, and it can be improved through learning and practice.
- Four key job site documents were reviewed.
- Four types of offsite company documents were reviewed.
- Three categories of documentation developed outside the company that support job site operations were discussed.
- It is important to organize construction documents for easy access.
- Supervisors need to be skilled in getting information when it is needed.

Learning Activities

1. Develop a standard company procedure for one of the following:

   Maintaining the job log

   Filling out time sheets

   Maintaining as-built drawings

   Start by listing objectives for the procedure. Then formulate the objectives into specific statements.

   As an example, objectives for the job log include that it needs to be:

   Current

   Accurate

   Complete

   Procedure statements might take the following form:

   The job log will be updated daily and will be current at the end of each day.

   The job log must accurately represent conditions and activities on the site.

   The job log must address all items listed on the standard job log form.

2. Obtaining timely information

   As the superintendent for the electrical subcontractor on a new middle school, you find that often when an electrical crew is assigned work based on the project schedule and the weekly plan developed at the weekly coordination meeting, the work site is not ready for the crew when they arrive. The problem seems to be that the general contractor is constantly changing priorities but neglecting to inform the subcontractors.

   Develop a simple bulleted list of steps you can take to improve the flow of information as a basis for your crew-planning needs.

CHAPTER 5

TEAM BUILDING AND MAINTENANCE

INTRODUCTION

Team building is an important tool for the construction supervisor. In Chapter 2, it was recognized that, whereas craft workers work with the tools of their trade, supervisors primarily work with and through people. Managing and organizing people to perform at a high level and to effectively carry out the tasks assigned by the supervisor are fundamental supervisory responsibilities.

Coordination of the activities and efforts of various entities on the construction site is also an important responsibility of the supervisor. The foreman must coordinate the activities of the crew with those of other crews in the work area. Higher-level supervisors coordinate the activities of multiple crews within their span of control, and they also work with supervisors from other companies to coordinate the work of all crews toward efficient project execution. The most effective way to coordinate the activities of individuals or crews is by forming them into teams.

Team building is not a classical craft skill taught through apprenticeship. Team building is a management skill that must be learned by most supervisors after they have completed craft training. Some new supervisors have a certain amount of natural ability to bring workers together to perform as a team. The ability to lead fellow workers and to coordinate their activities, forming them into a cohesive work unit, is one of the qualities that brings an individual craft worker to the attention of management, resulting in management moving this person into a supervisory position. However, natural team builders are not found often enough to meet the continual demand for new supervisors. In addition, those with natural team-building skills can improve the skills they already have to enhance the performance of their teams and the teams with which they interact.

Thus, team building is a skill that needs to be taught to most new supervisors and that can be improved even in those that have some natural team-building ability.

This chapter will begin by recognizing the need for effective construction teams. It will next define the meaning of a team in the construction context and then move on to review several team-building theories. It will discuss practical team building and maintenance of an ongoing team. Finally, it will review successful practices in a team environment.

THE CRITICAL NEED FOR EFFECTIVE CONSTRUCTION TEAMS

As the construction industry changes, the demand increases for highly effective teams to successfully execute projects. Among the drivers are:

- Changes in the workforce
- Changes in the breadth of company services
- Changes in project owner expectations
- Changes in technology incorporated into the product
- Changes in technology available to execute construction
- Changes in project organization

Construction has always been highly dependent upon the cooperation of individuals and entities to achieve project goals. Only in very rare cases does an individual procure land, design the facility to go on the land, and then build the project from beginning to end. When this happens, the project is usually a small residence in a rural area.

Projects of any complexity require many people, with specialized skills, each to contribute to the project. Commercial construction projects are executed by a large number of focused specialty companies, employing workers with narrow craft skills in a wide variety of areas. The trend is toward more specialization and the employment of larger numbers of lower-skilled workers.

Consider changes in the workforce. Craft skills are becoming more specialized with varying levels of skill. Take, for example, the carpenter. The broadly skilled carpenter of several generations ago that could work throughout a project from layout to finish work has given way to the framer or rough carpenter, the finish carpenter, the cabinet maker, and other such classifications of carpenter. There are fewer highly skilled carpenters who excel at such complex skills as work layout and finish carpentry, and there are more less-skilled carpenters who can competently perform elements of the work in a narrower area if supported by a few highly skilled craft workers. This trend will continue as prefabrication becomes more pervasive throughout the industry. A few skilled craft workers are required for layout and quality control, but most craft workers can be less-skilled component assemblers, assembling components manufactured offsite. This trend will continue because it drives the cost of construction down, while providing access to a broader pool of potential workers who are less skilled.

For many companies, areas of specialization are becoming more narrowly focused. Also, new companies are starting up in niche areas. For example, the electrical industry has expanded to include not only companies providing traditional electrical work in power and lighting but now companies that specialize in low-voltage work in such areas as electronics, controls, security, data, and communications.

In related parts of the industry, companies are working to broaden the services that they can provide. An example is in material supply and distribution, where traditional suppliers are expanding their services to provide logistics support to help contractors better manage their materials on the site. Thus those who traditionally only entered the site occasionally to deliver materials are becoming part of the job site team, involved in job site operations in new ways.

The bar for the traditional measures of a successful construction project based upon time, budget, quality, and safety is continually being raised. Owners are demanding the execution of projects faster, at lower cost, with zero defects, and zero tolerance for accidents. The understanding of objectives defining the successful project for contractors is also being refined. For example, more stress is being placed on profitability, not just being "under budget." New objectives are being recognized, such as maintaining strong customer relations with an expanding understanding of who the customer is. Chapter 3 described the importance being ascribed to customer relations.

Additionally, projects are becoming more complex, embracing new technologies, in terms of both materials and equipment used in the project and construction tools and equipment employed to execute the project. Newly defined outcomes for completed facilities include such things as energy efficiency and minimized environmental impact. New technologies, rapidly gaining acceptance on the design side, are driving changes on the production side as well. Building Information Modeling (BIM) is making great changes in the way construction is executed from concept through design and construction and on into facility operation and maintenance.

New approaches to organizing and managing construction projects are being developed and more broadly embraced. Public sector owners are gaining more latitude in how they can procure construction services and package construction projects. Implementation of new concepts of project organization and management, such as the application of lean principles to construction operations, as described in Chapter 15 are changing relations among project participants.

All of these changes are making an understanding of teams continually more important. The supervisor must understand how to organize teams, how to operate within a team, how to maintain a team's effectiveness, and what teams can do for the project.

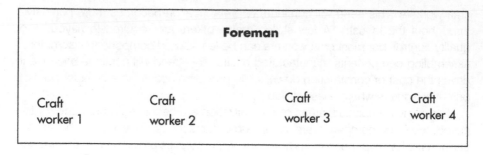

Figure 5.1 The Basic Crew as a Team

THE CONSTRUCTION TEAM

The concept of a team in the context of a construction project is very simple but has extensive ramifications. A construction team can be defined simply as a group of two or more individuals engaged in a common activity.

Certainly, one of the most basic teams in construction is a craft crew. This could be under the leadership of a lead craft person, or a foreman. It can include craft workers of varying skill types and levels. Figure 5.1 illustrates a simple crew based team.

As the project grows in size, companies will employ multiple crews, each operating as a team, in which case the company field operation could be considered a super-team under the leadership of a general foreman or superintendent. Figure 5.2 illustrates the company field team or super-team.

On most projects, there will be various specialty contractors, each with their own team structure, but as a whole, they form the project field team. This team is typically under the leadership of the project superintendent. Figure 5.3 illustrates the project field team.

A subset of the project field team is composed of the supervisors of the various trade contractors that form the supervisory team for the project. This is a horizontal concept of team across the project. Figure 5.4 illustrates the project supervisory team.

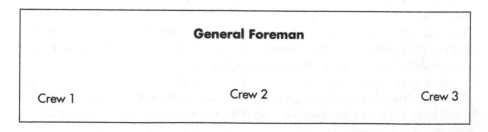

Figure 5.2 The Company Super-Team

Project Superintendent

Specialty Contractor 1   Specialty Contractor 2   General Contractor   Specialty Contractor 3   Specialty Contractor 4

Figure 5.3 Project Field Team

One might also consider a vertical team concept encompassed within each construction company engaged on the site. It starts, again, with the crew under a foreman, and expands to the super-team described above with multiple crews supervised by a general foreman or superintendent. The company team for a project would include participants from the company office, such as the project manager, the purchasing agent, the yard manager, and various other company-level people from such divisions as accounting, payroll, estimating, and human resources. Figure 5.5 illustrates the company project team.

Another concept of the project team moves outside the context of a single company, or the context of field operations and would include managers from the various trade contractors, the general contractor or construction manager, and designers, as well as the project owner. Figure 5.6 illustrates the highest-level project management team.

It becomes clear that a construction project has a multitude of teams, many interrelated. Team members on some teams participate as team leaders on other teams. One conclusion is very important, however. A clear understanding of teams is of critical importance to the supervisor.

TEAM CHARACTERISTICS

Consider the characteristics of successful team members, successful team leaders, and successful teams. It might help to keep in mind high-performance teams you

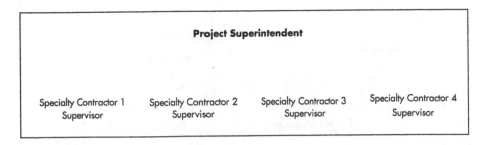

Figure 5.4 Project Supervisory Team

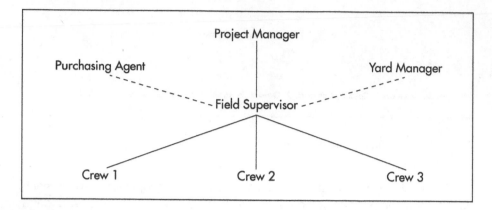

Figure 5.5 Company Project Team

have known. This could be in the context of sports, entertainment, the military, or wherever teams operate.

Characteristics of Successful Team Members

For a team to be successful, it needs to have high-performing members. Some of the characteristics necessary for a highly performing team member include the following:

- Technical skill
- Able to communicate well
- Ability to take direction
- Willingness to give up some personal interests in the interest of team success
- Desire to help teammates operate at a higher level
- A strong work ethic

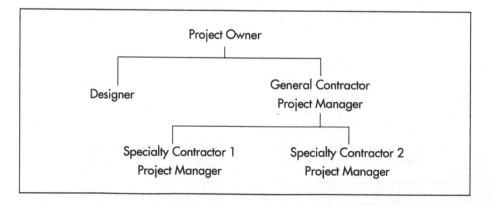

Figure 5.6 Project Management Team

Many sports stars have reported that early in their careers, they sought to gain recognition as a superstar, feeling that this would lead the team to greatness. Only after they discovered that team success depended on team performance, not just on the performance of a single great player, did such stars really become the superstar they desired to be.

Characteristics of Successful Team Leaders

Successful team leaders will display the characteristics of successful team members, but additional characteristics are needed, as well, such as:

- Honesty and integrity
- Respect for others, both on and off the team
- Vision
- Confidence in the team members
- A problem solver that enjoys challenges
- A strong desire to lead

Many successful team members do not want to be the team leader. Many successful team members do not succeed as team leaders because they lack leadership characteristics. Chapter 7 expands significantly on what it takes to be a strong leader.

Characteristics of Successful Teams

Great team performance tends to be a fleeting thing. It is very difficult to maintain a high level of performance over an extended period of time. Certainly, a successful team needs to have successful members and a successful team leader. In addition, the team needs to have such characteristics as:

- The ability to develop well-defined team performance objectives
- Mutual, internal accountability
- High credibility and trust

To perform at a high level, the team members must know exactly what is expected of the team and of each member on the team. Any lack of definition will result in hesitancy, and hence, diminished performance.

Accountability is important in performance at any level. For a team, there is hardly any stronger motivation to perform at a high level than the motivation to support other team members and not to let the team, or any of its members, down.

Credibility and trust are essential to a high-performance team. Each member must trust the other members. Each must trust the team leader and have confidence that following the leader will result in the best possible outcome. When the team makes commitments, the members must believe that the commitments will be kept. Those outside the team must also believe that commitments made by the team will be achieved. Chapter 15 gives more insight into the vital role played by credibility when dealing with the effects of unreliable and reliable planning on a construction project.

TEAM FORMATION

Various theories of team building have been proposed over the years. The predominant one divides team development into four distinct phases: forming, storming, norming, and performing. A variation on this theory adds a final phase of decommissioning, recognizing that teams, especially those involved in projects, are temporary in nature.

Forming

In the forming stage, individuals are brought together to form the team. There are many ways in which teams are formed, depending on the circumstances. Sometimes, team members are brought together by selection, each team member being chosen for a specific set of skills. Many times, teams come together based upon availability, rather than as a result of having the opportunity to select based upon specific characteristics.

During this forming phase, the base level of expectations is defined for the team and for each member. Common ground is identified for the members, such as common goals and similarities among members. An inventory of team member capabilities is made, and these capabilities are analyzed in the context of the team needs. This analysis determines such things as how evenly matched the skills and capabilities of the members are and what unique capabilities are brought to the team by each member.

As the team is forming, team members get to know one another. They begin bonding and developing an initial level of trust. Common goals will be agreed upon and the members will express their dependence upon one another. Leadership emerges either by a formal selection process or through a more unstructured process of the leader beginning to assume some control.

Storming

During the storming phase, team members identify and resolve issues of power and control. As differences are recognized, members try to work out solutions. Leadership begins to exert its influence and team members react to this influence in various ways. Some members appreciate the leadership and fall naturally into a supporting role. Other members might reject the new leadership, either because they envision themselves as the leader or because they do not think the one assuming the leadership role is leading where they want to go or in a manner in which they want to be led. Settling these issues of leadership can often be a stormy process.

During the storming stage, team members tend to act more independently, testing limits and determining what is allowable and what is not, and what the consequences are of engaging in behavior that is not allowed.

Norming

Successful teams move past the storming phase and into the norming phase. Team members begin displaying trust in one another and in the leadership. Communication skills among team members are developed and improved. Team resources are again evaluated and external resources are sought to augment team resources. Based upon team member capabilities, together with a considerable amount of negotiating and consensus building, roles are assumed and assigned. Team processes are developed for various aspects of team operations. The team begins to perform useful tasks.

Performing

Successful teams will achieve the performing phase, where the team works together collaboratively. A unique team identity is developed. Team members begin to feel comfortable with their roles and with other team members. They become interdependent upon one another, trusting their fellow team members to carry their part of the team responsibilities. At this point, the team begins to develop effective and valuable solutions and to produce at a high level.

Not all teams achieve this high level of performance. Some teams never get past the storming phase. Such teams are rarely productive and eventually fall apart or must be disbanded. Many teams operate within the norming phase, not quite achieving the cohesiveness of a great team. They will achieve some level of performance but seem unable to gain that ultimate level of finely tuned team performance.

MAINTAINING TEAM EFFECTIVENESS

Once the team is formed, continued team performance at a high level is not automatic. A number of factors affect team performance.

If the team composition changes, the team as initially developed ceases to exist. The team-building process must be repeated. This can generally be done in an abbreviated manner, but it still needs to be done. A change in team composition could be the result of a team member being reassigned or leaving the project. It could also take the form of a new team member being acquired. In any case, the team must reform.

Maintaining Team Effectiveness at the Crew Level

The fact that team performance is significantly impacted by loss of a team member has several ramifications for work crews. Whenever a worker calls off, there is a significant negative effect on the crew. Even though another worker can be reassigned to the crew for the day, the crew (team) does not have the same composition, and hence productivity is impacted, as well as quality and safety. If a crew member comes in late, the team is again impacted negatively during the time the crew member is not present and until the crew member can be integrated into the day's activities.

Sustaining high-level team performance is a difficult task. For many reasons, physical and psychological, individual team members perform at different levels at different times The team leader needs to be cognizant of the performance of each team member, and if performance is substandard for that team member, the team leader must determine why and what can be done to remedy substandard performance. This is most apparent in ongoing crew performance in construction. The foreman must always be aware of individual worker performance and when that lags, she must determine why and what can be done to bring the crew member back to strong participation on the crew.

Over time, team performance will vary, and if the team is ongoing, team performance will tend to deteriorate. Team performance can be renewed by repeating some of the team building steps enumerated above. Team performance can also be reinvigorated by restating performance goals and objectives, or by establishing new objectives consistent with the team mission and current status.

If team objectives are too distant, team members will lose focus, so if performance is dropping off, one way to renew performance is to develop interim milestones so that shorter bursts of performance can be used. Rewarding team performance with verbal recognition will often encourage continued strong performance. Awarding recognitions of superior performance or strong safety records also helps. Finally, supporting team identity by providing crew or project identifiers, such as hats, tee-shirts, or stickers for hard hats, can also be effective.

Maintaining Team Effectiveness at the Project Level

Each time a new project begins, participating entities assign their work teams based upon capability and availability. However, the project leader rarely gets to select who is on the project team. A partnering process has been developed to jump start the team-building process on new projects.

The project owner is responsible for establishing the new project as a partnering project. Key representatives of all participating entities are brought together for a facilitated team-building exercise. This exercise can last a few hours for smaller, simpler projects of shorter duration, or could last for several days for larger, more complex projects.

Participants are led through a number of bonding exercises that break down barriers and establish lines of communication. The group discusses goals and objectives for the project overall, and for the entities engaged in the project. They recognize significant consistency in goals, such as a desire for a safe project and striving to complete the project quickly, efficiently, and with high quality. A nonbinding contract is drawn up outlining the common goals and objectives and establishing a commitment by all to work to help each entity achieve their goals. The contract is signed by the various entities. Though the document is nonbinding, writing and signing a document reinforces the moral responsibilities and commitments.

The objective of the partnering meeting is to jump start the team-building process. It tends to bring project teams together more quickly and more effectively. However, as the project progresses, the effects of this initial team-building exercise tend to dissipate, so from time to time, a much briefer partnering exercise is carried out to rekindle the partnering commitments.

Without a formal partnering process, individual participants on the project team need to work to establish and maintain strong team relationships throughout the project. No opportunity should be lost at routine meetings to renew and strengthen communication and bonding. Team participants can, from time to time, share a meal or set up specific meetings in the context of which relationships are maintained and strengthened. If a relationship becomes frayed, every effort should be made to repair the connection and reestablish the strong team bond.

Maintaining Team Effectiveness at the Company Level

Wise company owners understand the importance of building strong team relationships across the company's human resource pool, especially with supervisory and management personnel. The company should provide opportunities for supervisors and managers at various levels to get together, both with others at the same level, and with those at other levels.

It has long been a tradition in construction that project managers meet periodically (generally weekly) to discuss their projects and learn company news. At these meetings, they can share successes, thus propagating lessons learned. They can also identify problems on their jobs that might be resolved from the experience of others.

It is much less common, though not less important, for supervisors to have the same opportunity. One company owner relates his practice of having all superintendents come into the office for a short meeting at the end of the week. They share work progress on their jobs and maintain bonds with one another that enable them to call upon one another whenever a problem arises in the field that requires some consultation. This also debriefs supervisors at the end of the week with the expectation that they can leave their work at the office and enjoy a good rest with their families over the weekend, with the added benefit of preparing them for the week ahead.

Such a meeting is generally not possible with lower-level supervisors or supervisors dispersed over a wide geographic area. However, with the availability of modern communications technology, such a meeting could be held in a virtual or digital environment with the same benefits. The supervisory corps of a construction company is a great internal resource, and it is important for companies to develop ways to employ this resource for the benefit of the company.

PRACTICAL TEAM BUILDING

Formation of the team varies depending upon how much latitude the team leader has in selecting team members. When team members can be selected, the team leader identifies the qualities team members should bring to the table based upon the mission of the team. The pool of candidates for the team can then be evaluated to determine which potential members bring the best set of characteristics. This can improve the formation of the team, but even with a relatively large pool of candidates, it is not practical to expect to find the ideal set of team members that display all the right characteristics for the team.

At the crew level, it would be nice if a foreman could choose among craft workers to select those with the requisite skills and compatible personality traits. However, in most practical settings, team members cannot be selected and the team leader must form the team based upon either team members who are available or team members that have been assigned by others. In this situation, the forming step of resource evaluation, that is, identifying what each team member brings to the table, becomes very important.

Whether the team leader has choice or must accept either team members that are available or have been assigned by others, a forming process is followed. Team members are introduced to the team, or introduce themselves to the team. Where possible, some bonding time should be allowed so that team members can begin to get to know each other. They may exchange work-related information, such as what their work background and experience are, who they have worked for, what projects they have worked on and what parts of the country they have worked in. If time permits, this might be a good time to ask about what they feel are their strengths and within the team mission, as they understand it, what part they would like to play. This does not have to take a long period of time, but it can provide valuable information for the team leader, as well as a degree of bonding for the team members.

The team mission needs to be clearly defined and explained to the team members. Goals and objectives for the team are laid out. At the crew level, this may be simply laying out one or a few specific tasks for the crew to work on today. At a higher level, the goals might be weekly tasks. For example, for a systems contractor the mission might be roughing in a specific area of a building by the end of the week. Milestones are defined, which will enable the team to evaluate its progress toward achieving the mission. Metrics are defined that will enable measurement of progress in moving through the milestones.

Ground rules for team performance are laid out. These need to be laid out when the initial team is brought together and repeated whenever the team changes. If a team is ongoing, they might need to be reviewed from time to time. The ground rules define the expected behavior of team members in terms of what is acceptable at a high level, what is acceptable but not desirable, and what is not acceptable. The ground rules will define how performance is to be measured and the consequences if acceptable performance is not achieved.

The ground rules also define the power structure of the team, including who is in charge and what level of authority leaders have. Questions of whether authority can be challenged and, if so, to what extent, are discussed. The question of whether authority can be changed and, if so, how, also needs to be addressed.

These practical steps in team building will tend to be informal in a field setting, but nevertheless need to be carried out if the team is to work effectively.

SUMMARY

In this chapter, the following key points have been presented.

- Construction is a complex activity that is best carried out by teams.
- There are many types and levels of teams operating within a construction project.
- Supervisors participate as members on some teams and as leaders on other teams.
- Characteristics of successful team members, team leaders, and teams were reviewed.
- A four-step team-building process includes forming, storming, norming, and performing.
- Once established, teams need to be maintained and reinvigorated from time to time.
- Practical team building is important, even in an informal field environment.

Learning Activities

1. Building the crew into a team

   A standard process for building a crew of craft workers into a well-performing team is not typically found in construction. The objective of this exercise is to design a standard process for company foremen, guiding them through the team-building process with a crew. This is most easily accomplished by focusing on a specific task on a specific job, then generalizing for application to similar situations.

   Start by defining the type of work the crew will do (for example, the company is a concrete specialty contractor and the crew is a forming crew). Define the crew. How many workers are on the crew and what crafts are involved (for example, three carpenters and two laborers)?

   Define the steps in team building, starting from when the crew initially gathers. Steps might include: introductions, skill assessment, defining crew goals, defining performance expectations, explaining the process to answer questions, etc.

   The process is complete when work is assigned to the new crew.

   (continued)

Once the initial team-building process has been developed, try to generalize it for other tasks. For example, for the concrete subcontractor, how will the reinforcing steel crew be developed into a team? How will the concrete placement and finishing crew be developed into a team?

2. Developing the field supervisory team for a new project

You are the superintendent for a general contractor about to start a new project. Your task is to develop the field supervisory team consisting of your supervisors and the supervisors of the various specialty contractors. What steps will you take to accomplish this?

Begin by defining the specifics of the project and the entities represented on the team. What work will be self-performed and what company supervisors will be needed to accomplish this work? What work will be subcontracted? Anticipate one supervisor from each of the subcontractors.

What preparatory work is needed prior to bringing the team together for the first time? Examples might include getting the contact information for each supervisor. Developing or getting copies of the company field policies and procedures to review and hand out at the meeting. Inviting key participants who will contribute to the team but not be a part of it, such as the owner's representative and various design specialists to participate in the initial team meeting.

Lay out the agenda for the initial team meeting.

The exercise concludes with closure of the initial project field supervisory team.

CHAPTER **6**

MAINTAINING THE RELATIONSHIP BETWEEN THE EMPLOYEE AND THE EMPLOYER

INTRODUCTION

A critical responsibility of any construction supervisor is maintaining a strong and positive relationship between the employees and the company. Chapter 3 dealt with the importance of the supervisor's role in maintaining good relationships with customers—those outside the company. The supervisor also has a responsibility to maintain good relationships between the company and the internal customer, the employee. The supervisor is the face of the company to the workers. As the link between the company and the workforce, as you saw in Chapter 2, the supervisor is a part of the management team, and the first line supervisor, the foreman, is the direct link with craft labor. Thus, the supervisor plays a key role in employee relations.

NASA describes employee relations in the following way:

Employee Relations involves the body of work concerned with maintaining employer-employee relationships that contribute to satisfactory productivity, motivation, and morale. Essentially, Employee Relations is concerned with preventing and resolving problems involving individuals which arise out of or affect work situations.

NASA's Goddard Space Flight Center Office of Human Relations (2001)

This chapter will deal with employee relations in the context of the construction company. It will begin by looking at the supervisor's role in evaluating worker performance. It will then consider what current federal law says about employee relations by considering issues of discrimination and harassment.

EVALUATING PERFORMANCE

Supervisors at all levels have the responsibility to evaluate the performance of their workers. This is accomplished in both formal and informal formats. The primary purpose of performance evaluations is to maintain an open line of communication between supervisor and worker so that each can understand the expectations of the other and be informed about progress in meeting those expectations.

Evaluations at higher supervisory and management levels are typically formal. Formal evaluations are also required for apprentices or novices in training programs. Formal evaluations are carried out on a scheduled basis, but these should not exclude ongoing informal evaluations when needed or when an opportunity presents itself.

Much of the evaluation of craft workers is informal on an ongoing basis unless a specific problem has been identified, at which point a formal write-up may be called for. Some companies do have a formal evaluation process for all employees, including craft workers.

Objectives of Performance Evaluations

A performance evaluation, whether formal or informal, should accomplish several objectives for both the evaluator and the person being evaluated. It should provide an opportunity to discuss performance from the point of view of both the employee and the supervisor. It also offers an opportunity to monitor progress toward the achievement of current performance goals and to recalibrate those performance goals as necessary. Whereas formal evaluations are scheduled on a regular basis and focus broadly across the full spectrum of performance, informal evaluations typically are done on the spot and consider a single performance area. Informal evaluations tend to take place much more often than formal evaluations.

Both the employee and the employer should have clearly defined objectives when entering into the evaluation process. The employee's primary objective in a performance evaluation should be to align her performance objectives with those of the company. To the extent that an employee can align with company objectives, the result will be positive for both employee and employer. Second, the employee should identify opportunities for improvement and map out a strategy to accomplish the improvements.

Employee Creating Opportunities

To illustrate an employee identifying opportunities for improvement and mapping out a strategy, consider a worker working on the site of a major industrial client that stresses safety. The worker understands the importance of a safe work environment to herself and to her family and her fellow workers. She also recognizes the particular interest her company has in worker safety. She has already completed an OSHA 10 course but has learned that a more extensive course is the OSHA 30 course. In addition to helping her perform her job more safely, an OSHA 30 certification will also better qualify her for consideration for a promotion to supervisor the next time an opportunity comes up.

She looks for available courses and among them finds one that is offered on four consecutive Saturdays over the next month. She presents this to the supervisor, who is delighted with the suggestion and offers to have the company pay for the course if the worker takes the course on her own time.

This scenario is good for the worker and for her family. It is good for the company both because safety will be improved on the job and because the worker will be better qualified, and it is good for the client who will benefit from a safer work environment at the facility.

The supervisor's objectives should be to clearly and concisely communicate the company's expectations to the employee. The supervisor is in a particularly good position to help his or her workers recognize the role they play in company operations and how they can have a positive impact on the company. The supervisor should also listen closely to the employee to find out what the employee's expectations are so that the employee and the company can embrace common expectations and seek ways to align those expectations that are not aligned. Finally, the supervisor should work to identify impediments to effective performance and to develop means to remove or mitigate those impediments.

Benefits of Performance Evaluations

Benefits derived from evaluations accrue to both the employee and the supervisor. For the employee, evaluations provide an opportunity to learn of, or to clarify the supervisor's expectations and to get an evaluation of the degree to which the employee is achieving the expectations. This feedback to the employee is recognized as an important element in the motivational process. Evaluations also create an opportunity to define expectations for future performance and to map out a process for realizing those expectations. Finally, the evaluation process provides the employee with opportunities to voice concerns formally and in an appropriate environment.

The first benefit to the supervisor, who represents the company, is to have an opportunity to evaluate individual employee effectiveness, consistent with the mission of the company and the expectations of the supervisor. Performance evaluations also provide an opportunity to identify ways to strengthen employee performance, thus building value in the company's human resources pool. Craft labor is such a major component of the value of any construction company that any opportunity for the company to enhance that value is important.

Well-executed and -documented performance evaluations should lead to improved worker performance. They will also help the company to avoid disputes over disciplinary actions. However, if a disciplinary action must be taken, a properly documented record of consistent performance evaluations can strengthen the company's position in defense against claims or legal actions that might result.

The Process of Evaluating Performance

In evaluating performance, positive reinforcement is very important. Reinforcing positive performance strengthens already good performance and can mitigate the detrimental effects to the employee of corrective elements within the evaluation. Positive reinforcement motivates employees, builds company loyalty, and helps to develop the worker's potential. Acknowledging the contribution of the employee to the project and to the company gives the employee a sense of value and connects them with the success of the project and the success of the company.

When giving a low performance rating, the specific activity in question must be defined clearly. Reasons leading to the rating must be accurately and completely described. The low rating should not be a surprise to the employee, nor should it be in question, if proper and timely notification was given and appropriate documentation has been developed over time. The concerns must not be diluted to avoid confrontation or to cushion the message. On the other hand, accusations must not be falsely created or strengthened to "build a case." A clear, accurate and succinct presentation of the infraction and underlying circumstances helps the employee to recognize the problem and to understand why the low rating was given. An accurate and honest presentation will also support the company later, should a claim arise or legal action be taken.

In evaluating negative performance, certain elements are essential:

- Describe the infraction.
- Cite the rule that was broken.
- Give specific examples of the employee's behavior that triggered the low rating.
- State performance expectations after the performance is corrected.
- Suggest how performance can be improved to an acceptable level.
- Lay out the consequences if performance is not improved.
- Describe the employee's appeal process.
- Agree on a time when performance will be reevaluated to determine if improvements have been made.

These elements are essential whether performing an informal or a formal evaluation.

Informal Evaluation Based on Tardiness

Consider the situation in which a craft worker is often tardy to work. The discussion might take the following path.

"Joe, you are a good craft worker and an asset to the company; however, you are late to work at least twice a week. Company policy states that 'tardiness more than once a week without a legitimate excuse will result in a warning, a reprimand, and possible dismissal.' Your tardiness causes the entire crew to be late and decreases the amount of work the crew accomplishes for the day.

In future, when you must be late for work, let me know beforehand when you will be late, why you will be late, and how late you will be. If it is not possible to give me prior notice, provide a written explanation after the fact citing why you were late and whether this situation is likely to arise in the future.

I will need to write this up for the record and will keep an eye on your work schedule over the next couple of weeks. If you would like to talk about this with a company officer, or if you feel there is something the company can do to help you, please make an appointment to talk with Jim Smith, the general superintendent."

Formal Performance Evaluations

In a formal performance evaluation, it is important to include as much positive reinforcement as possible. Not only does this shine a positive light on the overall performance of the worker, but it will support the worker in seeking to improve a few negative behaviors in order to deliver better overall performance.

Objective measures of performance enhance the performance evaluation process. Objective measures support consistency in evaluations. They are more likely to be applied evenly across the workforce and across time. They also tend to take emotion out of the evaluation, enabling the supervisor to focus on fact. Examples below show forms used to evaluate craft workers, apprentices, and supervisors.

Figure 6.1 shows an evaluation form used by Cannon & Wendt Electric Co., Inc. in Phoenix, Arizona, to evaluate the performance of their craft workers. This form is used by first line supervisors (foremen) to evaluate all craft workers every six months and is the basis upon which the company sets the wage scale. Given that this evaluation serves a very important function, the process is formalized and standardized. If a worker is deemed to be performing at a significantly different level than his wage scale indicates (typically at a higher level), the supervisor can evaluate the worker at any time so that the scale can be appropriately adjusted.

Figure 6.2 shows an evaluation form used for apprentices developed by the Electrical Joint Apprenticeship and Training Committee.

Prior to meeting with a worker to carrying out a formal performance evaluation, the evaluator should prepare for the meeting. The evaluator will first provide the person being evaluated with a packet of information to help him or her prepare. This will include the job description for the position held by the person being evaluated,

CANNON & WENDT ELECTRIC CO., INC.

EMPLOYEE PERFORMANCE EVALUATION

Employee Name/No._____

Class_____

Project

Date: _____

Supervisor: _____

Last Hourly Rate: _____

Type of Work: _____

Performance Measures: The employee's performance should be measured by the criteria listed below, as well as by any additional items you believe to be appropriate, which should be added to the list by you.

NOTE: "AVERAGE RATING"=15.0 POINTS

EVALUATION NO. (1 TO 20)

1　Positive, safe work ethic
2　Neatness of work (does their work meet Company Standards?)
3　Amount of work produced (compare to labor units)
4　Organization of work (lost time or lost motions)
5　Quality of skills (mechanical ability)
6　Accuracy of work (does job correct the first time)
7　Technical capability (electrical knowledge)
8　Assumes responsibility in completing assignments (how much supervision is required?)
9　Self-motivation (waits to be told what to do?)
10　Attendance and punctuality (starts and quits at work station)
11　Attitude (understanding of industry problems-both labor and management)
12　Other_____ J.W. Course = 5 points maximum (not safety courses)
　　_____ Personal appearance (does customer and company approve?)
　　_____ Leadership ability
　　_____ Subtotal Item 12 to Maximum 15 points

TOTAL:_____

Evaluate the employee's performance on his/her job over the last _____ in terms of the performance criteria listed above. Opposite each criteria, insert the evaluation number you feel best describes his/her performance.

Evaluation = _____ points (each point = .14 per hour for JW and .14 per hour for IJ)

	I.J. 45%	81.0 =	11.16
	I.J. 50%	90.0 =	12.40
	I.J. 55%	99.0 =	13.64
	I.J. 60%	108.0 =	14.88
	I.J. 65%	117.0 =	16.12
	I.J. 70%	126.0 =	17.36
	I.J. 75%	135.0 =	18.60
	I.J. 80%	144.0 =	19.64
	I.J. 85%	153.0 =	21.08
	I.J. 90%	162.0 =	22.32
	I.J. 95%	171.0 =	23.56
Average JW Rate 180		180.0 =	24.80
		190.0 =	25.80
Sub-Foreman +5%		200.0 =	26.04
Foreman Rate +12½%		208.0 =	27.90
		210.0 =	28.18
		220.0 =	29.00
		230.0 =	30.00
General Foreman Rate +25½%		236.3 =	31.00
		240.0 =	32.00

Pay: $_____ per hour

Effective: _____

_____ AGW

_____

Employee's Signature

Figure 6.1　Example Evaluation Form for Craft Workers

an agenda for the evaluation meeting, and a request for the person being evaluated to gather certain information about performance during the period under review.

Next, the evaluator will pull together as much factual data about the worker's performance as possible, such as performance quality, quantity, and timeliness. This information should be documented throughout the period between evaluations to expedite the process of preparing for the evaluation meeting. Comments or informal evaluations received during the period will be summarized. The

**Apprentice
Evaluation Form**

**PHOENIX ELECTRICAL JOIN APPRENTICESHIP
AND TRAINING COMMITTEE
FOR THE ELECTRICAL INDUSTRY**

APPRENTICE _____ PERIOD _____

WORKING FOR _____ DATE _____

PURPOSES OF THIS EVALUATION:
To take a personal inventory, to pinpoint weaknesses and strengths, and to outline and agree upon a practical improvement program. Periodically conducted, these evaluations will provide a history of development and progress.

INSTRUCTIONS:
Listed below are a number of traits, abilities, and characteristics that are important for success in business. Place an "X" mark on each rating scale, over the descriptive phrase that which most nearly describes the period being rated. (If this form is being used for self-evaluation, you will be describing yourself.)

Carefully evaluate each of the qualities separately.

Two common mistakes in rating are: (1) A tendency to rate nearly everyone as "average" on every trait instead of being more critical in judgment. The rater should use the ends of the scale as well as the middle, and (2) the "halo effect," i.e., a tendency to rate the same individual "excellent" on every trait or "poor" on every trait based on the overall picture one has of the person being rated. However, each person has strong points and weak points and these should be indicated on the rating scale.

ACCURACY is the correctness of work duties performed.

Makes frequent errors	Careless; makes recurrent errors.	Usually accurate; makes only average number of mistakes.	Requires little supervision; is exact and precise most of the time.	Requires absolute minimum of supervision; is almost always accurate.

ALERTNESS is the ability to grasp instructions, to meet changing conditions, and to solve novel or problem situations.

Slow to "catch on."	Requires more than average instructions and explanations.	Grasps instructions with average ability.	Usually quick to understand and learn.	Exceptionally keen and alert.

CREATIVITY is talent for having new ideas, for finding new and better ways of doing things, and for being imaginative.

Rarely has a new idea; is unimaginative.	Occasionally comes up with a new idea.	Has average imagination; has reasonable number of new ideas.	Frequently suggests new ways of doing things; is very imaginative.	Continually seeks new and better ways of doing things; is extremely imaginative.

Figure 6.2 Example Apprentice Evaluation Form

evaluator should have an idea before the meeting of the employee's strengths and weaknesses and be prepared to review the strengths and weaknesses during the meeting. Specific examples of both strengths and weaknesses are quite helpful. For weaknesses, the evaluator should be prepared to offer suggestions or work with the employee to develop a performance improvement plan addressing the weaknesses. Figure 6.3 shows a checklist for an evaluator's preparation for a formal performance evaluation.

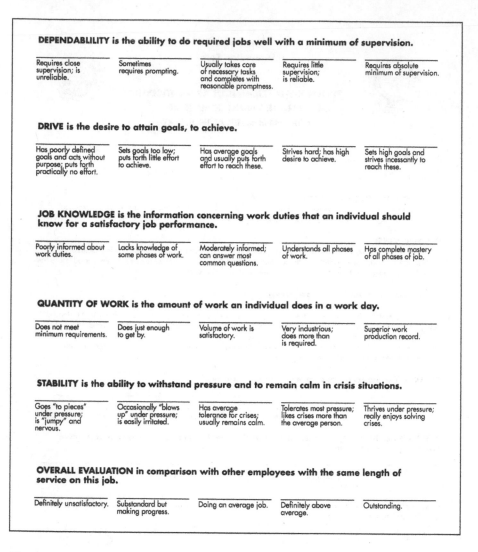

DEPENDABLILITY is the ability to do required jobs well with a minimum of supervision.

| Requires close supervision; is unreliable. | Sometimes requires prompting. | Usually takes care of necessary tasks and completes with reasonable promptness. | Requires little supervision; is reliable. | Requires absolute minimum of supervision. |

DRIVE is the desire to attain goals, to achieve.

| Has poorly defined goals and acts without purpose; puts forth practically no effort. | Sets goals too low; puts forth little effort to achieve. | Has average goals and usually puts forth effort to reach these. | Strives hard; has high desire to achieve. | Sets high goals and strives incessantly to reach these. |

JOB KNOWLEDGE is the information concerning work duties that an individual should know for a satisfactory job performance.

| Poorly informed about work duties. | Lacks knowledge of some phases of work. | Moderately informed; can answer most common questions. | Understands all phases of work. | Has complete mastery of all phases of job. |

QUANTITY OF WORK is the amount of work an individual does in a work day.

| Does not meet minimum requirements. | Does just enough to get by. | Volume of work is satisfactory. | Very industrious; does more than is required. | Superior work production record. |

STABILITY is the ability to withstand pressure and to remain calm in crisis situations.

| Goes "to pieces" under pressure; is "jumpy" and nervous. | Occasionally "blows up" under pressure; is easily irritated. | Has average tolerance for crises; usually remains calm. | Tolerates most pressure; likes crises more than the average person. | Thrives under pressure; really enjoys solving crises. |

OVERALL EVALUATION in comparison with other employees with the same length of service on this job.

| Definitely unsatisfactory. | Substandard but making progress. | Doing an average job. | Definitely above average. | Outstanding. |

Figure 6.2 *(Continued)*

Prior to meeting with the evaluator, the person being evaluated should prepare for the meeting. In a management setting, the evaluator will often ask the person being evaluated to prepare some documentation. Figure 6.4 shows a checklist for a construction supervisor's preparation for a formal evaluation of her performance.

Evaluation meetings should be well organized, starting with some basic information about the ground rules and agenda for the meeting. The evaluator can then ask the person being evaluated for a summary of how he or she felt the period has gone, giving the person an opportunity to present the materials that were requested and giving the person a chance to ask questions. The evaluator can then summarize her evaluation of the person's performance and solicit a response. There should be ample time for questions and the person being evaluated

COURTESY is the polite attention an individual gives other people.

Blunt; discourteous; antagonistic.	Sometimes tactless.	Agreeable and pleasant.	Always very polite and willing to help.	Inspiring to others in being courteous and very pleasant.

PERSONALITY is an individual's behavior characteristics or personal suitability for the job.

Personality unsatisfactory for this job.	Personality questionable for this job.	Personality satisfactory for this job.	Very desirable personality for this job.	Outstanding personality for this job.

PERSONAL APPEARANCE is the personal impression an individual makes on others. (Consider cleanliness, grooming, neatness, and appropriateness of dress on the job.)

Very untidy; poor taste in dress.	Sometimes untidy and careless about personal appearance.	Generally neat and clean; satisfactory personal appearance.	Careful about personal appearance; good taste in dress.	Unusually well groomed; very neat; excellent taste in dress.

PHYSICAL FITNESS is the ability to work consistently and with only moderate fatigue. (Consider alertness and energy.)

Tires easily; is weak and frail.	Frequently tires and is slow.	Meets physical energy job requirements.	Energetic; seldom tires.	Excellent health; no fatigue.

ATTENDANCE is faithfulness in coming to work daily and conforming to work hours.

Often absent without good excuse and/or frequently reports for work late.	Lax in attendance and/or reporting for work on time.	Usually present and on time.	Very prompt; regular in attendance.	Always regular and prompt; volunteers for overtime when needed.

HOUSEKEEPING is the orderliness and cleanliness in which an individual keeps his work area.

Disorderly or untidy.	Some tendency to be careless and untidy.	Ordinarily keeps work area fairly neat.	Quite conscientious about neatness and cleanliness.	Unusually neat, clean, and orderly.

Figure 6.2   *(Continued)*

should be asked if there is anything the supervisor can do to enhance the worker's performance. The meeting should close with a review of what will happen next. Typically, a written summary will be prepared and the person being evaluated will be asked to read and sign the summary, indicating only that he or she has reviewed the summary, not that he or she necessarily agrees.

Figure 6.5 provides an example of an agenda for the evaluation meeting.

Informal Performance Evaluations

All supervisors carry out informal performance evaluations of those working for them. In fact, an ongoing process of any supervisor should be continuous evaluation of

COMMENTS

Major weak points are –

1._____

2._____

3._____

and these can be strengthened by doing the following:

_____

_____

_____

_____

_____

_____

_____

_____

_____

Major strong points are –

1._____

2._____

3._____

and these can be more effectively by doing the following:

_____

_____

_____

_____

_____

_____

_____

_____

_____

Rated by_____   _____
                     (Name)   (Title)
                _____
                       (Date)

Rated by_____   _____
                     (Name)   (Title)
                _____
                       (Date)

Rated by_____   _____
                     (Name)   (Title)
                _____
                       (Date)

Rated by_____   _____
                     (Name)   (Title)
                _____
                       (Date)

THIS REPORT IS YOUR RESPONSIBILITY
Evaluate the Apprentice fairly and honestly.
Discuss each weak point with this Apprentice.

Figure 6.2   (Continued)

their workers' performance. For those receiving formal performance evaluations on a regular basis, there should be no surprise during the formal process since the supervisor should be informing the worker about performance issues (positive or negative) as they come up.

For first-line supervisors who work with construction craft workers, there is no tradition of formal performance evaluations, so informal evaluations take on a much

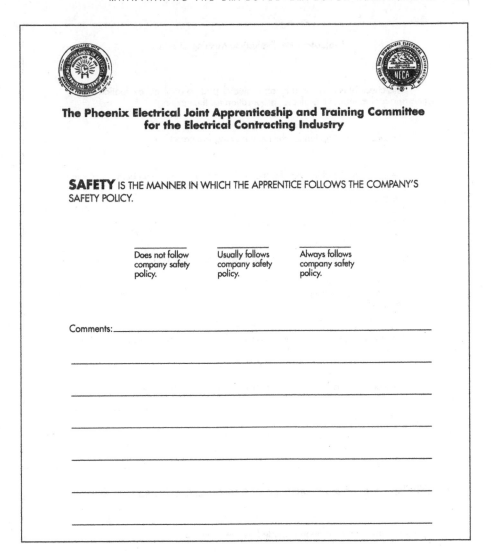

**The Phoenix Electrical Joint Apprenticeship and Training Committee
for the Electrical Contracting Industry**

SAFETY IS THE MANNER IN WHICH THE APPRENTICE FOLLOWS THE COMPANY'S
SAFETY POLICY.

| Does not follow company safety policy. | Usually follows company safety policy. | Always follows company safety policy. |

Comments:_____

Figure 6.2   *(Continued)*

higher level of importance. The supervisor is responsible for continuously evaluating a worker's performance, recognizing excellence and identifying where the worker fails to meet expectations.

To the extent possible, the evaluation should start off on a positive note. This may not be possible if an acute situation arises that calls for immediate corrective action. However, in most instances, the supervisor identifies performance that is subpar and the intent is to bring the performance back up to an acceptable level. In such a case, a few positive words at the beginning will help to couch the interaction in a positive vein. The supervisor should end the discussion agreeing with the worker on what corrective action will be taken.

Evaluator's Pre-Evaluation Meeting Checklist

This checklist is to be used by an evaluator prior to a formal evaluation meeting as a tool to improve evaluator preparation for the meeting.

Procure job description for the person being evaluated.

Set time, date, and location for the evaluation with the person being evaluated.

Time: _____
Date: _____
Location: _____

Develop and give to the person being evaluated a packet of information, including

Meeting time, date, and location
Purpose of the meeting
Job description
Meeting agenda
Copy of the evaluation sheet to be used
List of information to be prepared and brought to the meeting by the person being evaluated

Develop data on the person's performance over the review period.

Figure 6.3 Checklist for Evaluator's Preparation for a Formal Performance Evaluation

---

Checklist for a Supervisor's Preparation for a Formal Performance Evaluation

Review the documentation provided by the evaluator

Prepare information requested by the evaluator

Prepare any additional information that you would like to present to the evaluator

Prepare your own objectives for the meeting

Prepare questions for the evaluator

Figure 6.4 Checklist for a Supervisor's Preparation for a Formal Performance Evaluation

---

Formal Performance Evaluation Meeting Agenda

State the purpose of the meeting

Define ground rules for the meeting

Provide a copy of the job description for the position held by the person

Ask about the person's perception of his or her performance during the last period

Review conclusions from the last performance evaluation meeting

Discuss progress in achieving the goals set in the last meeting

Provide the supervisor's evaluation of the person's performance during the last period

Agree on areas of improvement

Develop a plan for performance improvement

Summarize the remaining steps in the evaluation process

After the meeting, develop a written summary consistent with the oral summary in the meeting

Get a signed copy from the person being evaluated

Note: The signature only indicates that the person has had a chance to review the written summary. It does not indicate that she necessarily agrees with the report.

Figure 6.5   Agenda for Formal Performance Evaluation Meeting

It will help supervisors if they have a checklist in mind so that when they see a problem, they can address it in an organized manner. Figure 6.6 shows an example of a checklist for an informal worker evaluation.

Any informal evaluation should be reinforced with a written follow-up. If a serious infraction has been identified, the written follow-up statement should be signed and dated by the worker to indicate that he or she has been made aware of the infraction. Again, a signature does not indicate agreement with the statement; it just signifies that the employee has been able to review what has been written up.

DIVERSITY AND DISCRIMINATION

The workforce at a construction site is very diverse, and it is important for any supervisor to understand and embrace diversity in the workforce. Increasing diversity together with a better understanding of human rights and concepts of equality has led to an increasing awareness of discrimination. As a result, laws have evolved to protect against discrimination.

Unfortunately, in today's construction industry, problems related to discrimination are very common. This is the result of several factors. Much of the workforce

<u>**Example Checklist for an Informal Worker Evaluation**</u>

When sub-par performance is identified, the supervisor should:

- Call the worker aside

- Tell the worker why he has been called aside

- Describe the performance that is sub-par

- State performance expectations after the performance is corrected

- Explore options for improving performance, which might include:

    o   Training

    o   Different equipment or tools

    o   A different approach

- Agree with the worker on how performance can be improved to an acceptable level

- Lay out the consequences if performance is not improved

- Agree on a time when performance will be reevaluated to determine if improvements have been made

Figure 6.6   Example of a Checklist for an Informal Worker Evaluation

does not understand current standards of what constitutes discrimination. Also, supervisors are not as well prepared to recognize and handle problems inherent in managing human resources as they are prepared to recognize and deal with technical problems. As a result, problems associated with discriminatory practices often are not identified early enough, and when they are identified, supervisors often are not able to deal with them effectively.

Workforce Diversity

The traditional skilled workforce in construction has been predominantly white, male, and Protestant. However, the white, male, Protestant population has not grown as rapidly as the population as a whole, and the construction industry has expanded considerably. This has led to two problems. First, there has been decreased ability to find enough workers to replace those leaving the industry and to meet the labor needs of an expanding industrial sector. Second, there has been an increase in the number of workers coming from nontraditional backgrounds.

Not only are demographics changing in terms of gender and race but diversity in terms of culture, religion, and age is also expanding. To be effective, supervisors must be sensitive to differences in people. They need to understand how to manage a very diverse workforce and to harness the diversity to enhance overall performance. This involves understanding what diversity in the workforce means and how to effectively manage workers with a variety of backgrounds.

Diversity is a good thing and our society loves diversity. We like to go to different restaurants, and we look for a diverse menu when we get there. We enjoy having a broad selection of automobiles to choose from and when we go to the store, we like to have a broad selection of products to choose from. We even like a broad selection of stores within which to find the broad selection of products. Diversity enriches all aspects of our lives.

Diversity in the workplace is also valuable. It brings different points of view and new ideas. Diverse workers bring diverse skills to the workface. If the supervisor understands how to harness the power of diversity, diversity will enable more efficient execution of work.

Unfortunately, a diverse workforce also brings to light stereotypes and prejudices. It is the responsibility of supervisors to help workers to recognize, understand and overcome the biases that they have. It is also the responsibility of supervisors to help their workers embrace diversity in their work environment and harness it to improve production rather than letting the diversity negatively impact their work.

Diverse cultures and background manifest themselves in many ways in the workplace. Examples include the comfort level of people based upon such things as personal space, accents, and gender roles. Other examples are found in basic worker expectations in terms of such things as time management, quality, and responsiveness to direction. Generational differences are observed in what motivates workers and what expectations workers have of the employer and of other workers. Religious differences are seen in practices and in tolerance or lack thereof for certain behaviors.

This diversity can be very beneficial to the job and to the company if workers embrace diversity rather than resisting the changes resulting from the diverse environment. The supervisor plays a key role in helping workers to accept and work within the diverse environment, thus reaping the benefits of diversity and avoiding the pitfalls resulting in conflict and litigation when diversity is not accepted.

Equal Employment Opportunity Laws

Laws have been enacted to help protect employees within the diverse work environment. The laws generally fall under the category of Equal Employment Opportunity (EEO) legislation, and they protect individuals from discriminatory practices in the workplace. It is very important for supervisors to have at least a basic understanding of EEO laws so that they can participate in protecting their workers, as well as their companies and themselves.

Discrimination in the workplace is characterized by treatment or consideration of, or making a distinction in favor of or against, a person or thing based on the group, class, or category to which that person or thing belongs rather than on individual merit.

EEO laws are written to protect people in various categories that tend to draw discriminatory acts. These categories include:

- Race
- Gender
- Age
- Religion
- National origin
- Citizenship
- Disability
- Military status

Laws are written at various levels, including federal, state, and local. The most stringent law is typically the one that governs in any given area. Companies and projects can also have discrimination policies and procedures that need to be observed.

One of the most far-reaching laws prohibiting discrimination is the Civil Rights Act of 1964. Title VII under that law prohibits discrimination based upon race, color, religion, sex, or national origin. Pregnancy was specifically included under Title VII in 1978. Title VII also deals with reverse discrimination, protecting those not in protected categories from discrimination based upon a protected category.

Other laws have been enacted to protect people in additional categories, such as older people (over 40) covered by the Age Discrimination in Employment Act (ADEA) and people with disabilities covered by the Americans with Disabilities Act (ADA). All of these laws are restricted to companies employing at least 15 employees for at least 20 weeks at some time within the last 2 years.

In recent years, immigration has become a very controversial topic. Immigration particularly affects the construction industry because a significant portion of the workforce in construction is made up of immigrants. The Immigration Reform and Control Act prohibits discrimination on the basis of citizenship and visa status. Another act that prohibits discrimination is the Uniformed Services Employment and Reemployment Rights Act, which prohibits discrimination against applicants or employees based upon military status or service obligations. It also provides protection for those returning from service.

Retaliation against those seeking protection under EEO laws is prohibited. An employer cannot discipline or discharge a worker for using the protections afforded by the law, or refuse to hire a person known to have brought charges based upon EEO laws, regardless of the outcome of the legal action.

Reasonable Accommodation

The laws relating to religion and disability have provisions such that, if requested, the employer must make adjustments (reasonable accommodation) to enable an employee to perform the essential functions of their job. The determination of a disability is based upon a record of disability, the perception of a disability, or actual impairment in functioning that affects one or more life activities.

Reasonable is a flexible term. Such accommodation must not place an undue hardship upon the employer or upon other workers. It must not create an unsafe situation. It must not have a detrimental effect upon quality or quantity standards of the job. The outcome must be such that the worker can perform the basic functions of the job if the reasonable accommodation has been made. The accommodation does not need to be the best accommodation or the one requested specifically by the worker. It simply needs to enable the worker to perform the basic functions of the job. An accommodation that would violate a labor agreement clause is not required. In the case of a disability, the employer can request documentation from a doctor.

Family and Medical Leave Act

Another law that is important to the supervisor is the Family and Medical Leave Act (FMLA). This law provides for up to 12 weeks of unpaid leave per year for an employee's own health, to care for ill family members, or for a pregnancy, child birth, or adoption. This leave can be continuous or intermittent, that is, taken in a single block or at various times throughout the year.

Certain conditions are required for FMLA to take effect. The employer must have at least 50 employees employed within 75 miles, and this number must have been reached at least 20 weeks during the previous 2 years. The employee must have been employed for 12 months prior to the time leave is to begin and must have accrued at least 1200 hours during that 12-month period.

If those conditions are met and an employee requests leave, the leave must be granted. Furthermore, benefits, such as medical insurance, must be continued, and the employer must ensure that there is a job for the employee when he or she returns from leave.

Avoiding Discrimination Claims

A number of steps should be taken to avoid infractions of the EEO laws. First, the company must have well-defined, written policies and procedures regarding the various laws. All supervisors must be well trained in understanding the laws and their enforcement, as well as the company's policies and procedures. A supervisor must give due consideration to any request or claim that arises with regard to any of the EEO laws. Finally, the supervisor must enforce all company policies consistently across time and evenly across all employees.

HARASSMENT

Harassment is a behavior often found on the construction worksite that is detrimental in many ways and can be illegal. Harassment encompasses all words, conduct, or actions, especially persistent that are directed at a specific person and alarms or causes substantial emotional distress in that person while serving no legitimate purpose. All harassment is counterproductive, but not all harassment is illegal. Bullying is harassment, but it is not illegal unless it meets the definition of discrimination. On the other hand, harassment associated with a protected class is considered discrimination and is illegal. Sexual harassment is considered a special case of discrimination and is also illegal.

Harassment has numerous effects that are counterproductive, including:

- Negative effects on production
- Negative effects on morale
- Increased turnover
- Detrimental to the company's reputation
- Conducive to legal action

Although counterproductive, harassment only rises to the level of being illegal if it creates a hostile environment. One or a few isolated events that fit into the category of harassment generally are not considered to rise to the level of creating a hostile environment. However, they are early warning signs of a brewing problem and must be dealt with accordingly.

A hostile environment is the result of repeated, unwanted behavior or conduct in terms of a protected category that interferes with work performance or creates an offensive work environment. If a supervisor allows a hostile environment in his work area, not only are productivity, quality, and safety negatively impacted but legal action based upon harassment is likely to ensue.

The supervisor has first-line responsibility for identifying harassment and a hostile environment and dealing with the situation immediately. This may involve direct action, but it may well involve passing the problem up to a higher-level supervisor or manager and eventually to the person in the company who is responsible for human resources issues. Supervisors must know how to respond when confronted with harassment, what the limits are of their capability to deal with the situation, and when to hand it off to higher authority.

As a legal representative of the company, the supervisor puts the company at risk for being liable for a charge of harassment unless the company has protected itself from such a situation. To protect itself, the company needs to have a written harassment policy and to demonstrate that the supervisor did not use the policy, or it must demonstrate that as soon as the company became aware of the problem, it took immediate remedial action in accordance with its harassment policy. Such legal actions by the company to protect itself may put the supervisor at personal risk if the supervisor does not respond appropriately when a harassment situation arises.

The hostile environment could be caused by a supervisor or by another company employee. It could even be created by a non-company-employee, but the company becomes legally liable for the actions of this non-company-employee at the point that it knew or should have known about the alleged harassment and failed to take action to protect the harassed individual. Thus, actions by an individual working for another subcontractor or contractor on the site or by a material supplier, a representative of the designer, or even the owner could put the supervisor and her company in jeopardy.

Many defenses are ineffective against a harassment charge. "I did not know" will not hold up if it can be demonstrated that the company had the opportunity to know or should have known. "It was just a joke" or "I did not think it would be offensive" are not legitimate defenses against a harassment charge. Racial or ethnic remarks that are tolerated within certain circles are often not tolerated by an outsider, so a racial comment by an individual cannot be defended by claiming that the comment was used by others in that community.

Offensive language can be associated with any protected class. It can be racial-, ethnic-, or gender-centered. It can relate to sexual orientation, a disability, or religion. Any language that could be considered offensive has no place in the workplace, and the supervisor must deal with any instance of language impropriety immediately and decisively.

Sexual harassment is one of the most obvious and pervasive types of harassment. Sexual harassment is considered gender discrimination. It can originate from either sex against the other sex, or it can be same-sex harassment. One of the most pervasive types of sexual harassment is offering something in exchange for sex. This could be couched in a positive form, such as a raise or special consideration in return for a sexual favor, or it could be couched in a negative form, such as "you will lose your job if you do not go out with me." An example of same-sex harassment might be asking a new male employee to go along with the guys after work to a "gentleman's club" and berating him when he refuses to go, perhaps based on strong religious convictions. In this case, the hostile environment could be off the job site, but it could still be a situation in which the company is liable because it affects performance on the job and, hence, becomes discernable in the workplace. The refusal to go along could be based upon moral or religious grounds, but it does not matter. If a person does not want to go along, he has the right not to go without any type of harassment. A case of religious discrimination, though not harassment, might involve the same person asked by the company to do work at a bar or a show club, where he does not feel comfortable. A reasonable accommodation might be reassigning the worker to another job.

Prevention of Discrimination and Harassment

Prevention of discrimination and harassment is not difficult. Supervisors must recognize and help their workers to recognize that the workplace is a workplace, not a place of entertainment. All workers must be trained to keep actions professional. Respect must be shown for all others as professional colleagues. Language and conduct must be maintained on a professional level. All workers must recognize and embrace the fact that others have different standards of humor and propriety. All must work to understand, recognize, and respect the standards of others. A company policy outlining standards for nondiscrimination within the company and on company jobs, together with training for each employee, is essential to ensure that every employee understands what is expected.

There will be times when an offense is committed. To best handle these situations, the offending party must apologize, the offended party must accept the apology, and all parties must move on, learning from the situation.

The supervisor must be vigilant to detect a potential discriminatory situation. If someone wants to talk with her off the record about such a situation, that person needs to be reminded that as a company representative, the supervisor must act on anything she hears. If the person would like to talk about it, the supervisor should listen, but then must take action if actionable information is shared. Any observations and discussions must be documented thoroughly and consistently. Any possible incidence of discrimination must be moved up to higher authority in the company. Finally, discrimination issues must never be discussed with anyone not directly involved or in the upward line of communication to higher authority.

If workers and supervisors follow common-sense procedures with respect to discrimination and harassment, in addition to the policies and procedures of the company and the project, they should be safe from disciplinary or legal action. Even more importantly, the job should be more productive and safe, and the work experience will be much more enjoyable.

SUMMARY

In this chapter, the following key points have been presented.

- Performance evaluation is a key responsibility of the construction supervisor.
- The performance evaluation process benefits both the company and the employee.
- Performance evaluations can be either formal or informal.
- Diversity in the workplace can have great value.
- Increasing diversity in the workplace and enhanced understanding of discrimination have led to a legal structure to protect worker rights.
- Supervisors are in a key position to identify and deal with discriminatory practices and situations.

Learning Activities

1. Development of a supervisor evaluation form

   When carrying out evaluations, evaluation forms are important to ensure consistency from one employee to another and across time. They are also important to provide documentation of the process.

Choose a supervisory position to evaluate. Describe the position, including such information as level of supervision and industry sector. Develop an appropriate evaluation form for this position.

Start by identifying the characteristics of this position that are important to evaluate. Describe each characteristic clearly. Determine an appropriate rating scale. Then determine the relative value of each characteristic. The form should begin by describing the evaluation process and then listing the characteristics. Each characteristic should be followed by a column for the rating and a second column that has the relative value. A third column should have the product of the two (rating and value). The third column should be summarized at the bottom to give the employee a single numerical rating.

2. Discrimination in the workplace

By researching construction literature, identify an instance of discrimination in the construction workplace. Write a case study based upon this instance. The case study should describe the situation, the way it was handled, and any ramifications from the situation. The case should then reference the law that has been broken, and suggest what could have been done to avoid the situation and how the situation should have been handled in the field when it surfaced.

CHAPTER 7

MANAGING THE HUMAN RESOURCE

INTRODUCTION

Among the numerous resources that the supervisor manages—dollars, tools, equipment, job site space, and time, as we have discussed them—the most important, the most valuable, and the most complex resources are the craft workers on the project. No other aspect of management presents as many complexities and challenges as managing the human resources, the individual human beings who compose the construction workforce.

In addition to managing craft workers, the supervisor also interacts on a daily basis with numerous other people within the company, as well as with a great many others from outside it. It can be said, therefore, that the supervisor spends a great deal of time interacting with all kinds of people.

It is a fact that construction supervision involves the application of human relations skills, and supervisors should recognize that the longer they remain in a supervisory capacity, and the further they advance in supervision and management, the less they will rely upon the skills of their craft in their work, and the more important their human relations skills will become. The supervisor's development of human relations skills is absolutely vital to success in management.

It follows, then, that understanding some of the elements of human behavior, leadership, and motivation will be extremely valuable to the supervisor throughout his career. In this chapter, we will examine some of those concepts.

THE SUPERVISOR AS A LEADER

Supervisors and other managers do not typically think of themselves as leaders. Yet the concepts of leadership and the management of construction share many elements in common. Leaders make plans, for the both the short term and the long term: so do supervisors. They devise methods to implement those plans; so do supervisors. Leaders get their goals accomplished and their plans fulfilled through managing and directing other people; so do supervisors. Leaders make decisions, and they have responsibility for the outcome of those decisions; so do supervisors. The more aspects of leadership one examines, the more parallels one discovers between leadership, and supervision and management.

How Is Leadership Defined?

Leadership can be defined and characterized in many different ways. Numerous books and references have been written on this subject. Perhaps one of the best ways for supervisors to develop the concept of leadership is to think of a person, or perhaps several people they have worked for, whom they intuitively recognized as good leaders as well as good managers. This person would have been very effective at all aspects of leadership, supervision, and management, would have been a great manager, and would have been very successful. Almost without exception, people genuinely enjoyed working for this person. This person may well have been a role model and/or a mentor.

Next, the supervisor should develop a list of the attributes or characteristics that this person demonstrated that caused the supervisor and others to recognize him or her intuitively as an effective leader. These attributes can be thought of as defining characteristics that distinguished that person being thought of as an effective leader.

As was discussed in Chapter 2, a large number of supervisors and people aspiring to become supervisors have been asked to perform this exercise in supervision and project management programs conducted in numerous venues over a long span of time. The list of attributes that were cited most commonly are summarized in Figure 7.1.

The responses are those of several thousand participants in construction project supervision courses in various parts of the United States and Canada, in response to the instruction: "List what you consider to be the distinguishing attributes or characteristics of a person whom you have worked for, whom you intuitively thought of as a great leader and a great supervisor." This is the same list of defining characteristics of outstanding leaders and supervisors that was examined in Chapter 2, and it represents the most commonly cited attributes that were shared by these people as they thought of their role models in leadership and supervision.

While no rank order is ascribed to the characteristics in this listing, and while no statistical analyses were performed with regard to the responses provided, it is noteworthy that these attributes were provided with an amazing degree of consistency. Further, those who provided the responses consistently noted that they

ATTRIBUTES OF THE BEST LEADERS AND MANAGERS AS DETERMINED BY PRACTICING CONSTRUCTORS	
Good People Skills	Teacher
Good Communicator—Verbal and Written	Dedicated
Respectful—Treats People with Respect	Develops People
Charismatic	Consistent
Organized	Problem Solver
Knowledgeable	Goal Oriented, Goal Setter
Leads by Example	Good Planner
Earns the Respect of Others	Disciplined
Open Minded	Optimistic
Confident	Calm
Honest	Fair
Good Decision Maker	Motivator
Delegator	Humble
Sincere	Provides Recognition
Follows Through	Trustworthy
Visionary	Accessible, Approachable
Willingness to Share Information	Team Builder
Crisis Management—Can be Counted on in a Crisis	Sets High Expectations
Good Listener	Personable

Figure 7.1   Attributes of the Best Leaders and Managers

and others absolutely loved working for a leader and manager who demonstrated these characteristics. They noted additionally, that they and others considered these mentors and role models to be extremely effective as leaders in addition to being great managers, and these leaders and managers were, almost without exception, very successful. Many described the leader whom they portrayed with these attributes as being "the best of the best."

It is suggested that effective leadership in the construction industry can be defined in terms of these attributes, as provided by practicing constructors, far better than in a definition drawn from a text or reference on leadership. It is suggested further that the logical conclusion for supervisors to draw is that, if they wish to improve their human relations skills, and that if they wish to become admired and emulated as a successful leader and manager, they would do well to consider the attributes described in this listing as specifications for further growth and development toward fulfilling that goal.

The assertion is that people *can* aspire to the cultivation and practice of personal attributes and traits that they wish to incorporate into their lives and into their relations and interactions with other people. People *can* develop their leadership and management skills in the fashion they aspire to. Cultivating these qualities in his or her daily life could very well contribute to having the supervisor being described by those whom he or she leads, as the best and most effective of supervisors, "the best of the best."

STYLES OF LEADERSHIP

Those who have studied leadership have defined and described a number of different styles of leadership, and have described some advantages and disadvantages that are inherent in each style. Four styles of leadership are described here, along with some of the benefits and drawbacks of each style: autocratic leadership, democratic leadership, participative leadership, and situational leadership.

It is suggested that, as you read about the characteristics of each style of leader, along with the associated benefits and drawbacks of that leadership style, you think of leaders and managers whom you have known who demonstrate that leadership style. Next, it is suggested that you look inward to determine your own personal leadership style, or the leadership style you would most wish to emulate. Leadership styles, like other personality traits and human relations attributes, can be developed and cultivated in the fashion the supervisor thinks best, and in the direction in which he wishes to grow as he develops the style of leadership that he believes will be most effective for him.

Autocratic Leadership

The autocratic leadership style has been, and remains today, very commonplace in the construction industry among leaders and managers. Those who lead with an autocratic leadership style frequently utilize expressions such as, "I am in charge, and subordinates will do as I say." They believe that leadership, authority, and management decisions, and management responsibilities, are centralized in the leader. There is no doubt who is in charge, and everyone knows what the expectations are.

While autocratic leaders may sometimes take advice and input from others, more commonly they act on their own beliefs. Usually they are convinced, and often they are absolutely convinced, that their analyses and decisions are the best. Additionally, autocratic leaders usually do not like for their authority, or their decisions or actions, to be questioned.

Many successful construction industry leaders, company owners, and managers have practiced this style of leadership. Autocratic leadership has been found to be quite successful at times and, in fact, offers a number of advantages that some people find very attractive. For example, there is never any doubt as to who is in charge or where responsibility and authority are vested. Decisions are quickly made and implemented. People know what to expect, and they know where they stand. People know exactly whom to look to for direction and for decisions.

Autocratic leadership has also been shown to be very effective in emergencies and in crisis situations. When someone needs to take charge, and when quick and unhesitating decision making and immediate strong action are required, autocratic leadership is very effective.

However, it has also been found, especially in recent times, that autocratic leadership may not always be the best leadership style. With changes in society and changes in the workforce, many workers today are not willing to accept autocratic leadership. Some people grow weary and intolerant of constantly being given direction, and constantly being told what to do and how to do it. Many people prefer to be able to voice their opinions regarding their assignments, work methods, and other aspects of their performance in the workplace rather than simply following directives that they have been given.

It has also been demonstrated that autocratic leadership may stifle initiative on the part of workers, and it may lead to such attitudes as, "The boss does all the thinking; I do not get paid to think." Such a mindset deprives the workforce, as well the project and its management, of the creative and innovative thinking and the initiative that many workers will demonstrate if provided the opportunity.

It has been shown that those who perform skilled craft work every day frequently develop methods and processes that are quite effective. They develop insights into doing the work that others, including those in management, simply do not possess. Autocratic leaders, by the fact that they always believe their methods and decisions must be the best, frequently do not provide an opportunity to tap into this wellspring of useful information, nor for the workers to be able to bring forth and express what they know.

Moreover, autocratic leaders frequently generate frustration and demotivation in the workers, which may manifest itself in a number of different ways. It has been shown that demotivated workers are not as productive and do not work as safely as motivated workers.

Although autocratic leadership decidedly has some benefits to offer, many believe that it may not be the most effective leadership style for use in all situations and with all people today. As the workforce and the work environment have evolved and changed, so have other management styles. Autocratic leadership is not the only leadership style supervisors have at their disposal today.

Democratic Leadership

As leadership styles evolved, many believed that the solution to the shortcomings of autocratic leadership was in democratic leadership. Democratic leaders consistently take input and advice from those whom they lead. They hold discussions, conduct focus groups, hold committee and task force meetings, and in other ways seek to gather information and input from workers. They usually base their decisions on what the workers think is best. They operate by seeking consensus from those they lead.

This leadership style has the advantage of allowing workers to express their opinions and to be able to bring forth their knowledge. It allows workers to feel involved in decision making. Further, it allows the workers to feel empowered and to feel they have a voice in decisions that affect them. Many workers are happiest in such an environment, because they feel motivated by this philosophy.

While democratic leadership has a number of noteworthy advantages, it has also been found to have some significant shortcomings. For example, decisions often take a long time to be made as discussions take place and as consensus is built. Additionally, the workers who are contributing to the decision making often do not have all of the information necessary to be able to make the best decisions.

In addition, democratic leadership has actually been found to cause some workers frustration, because they believe leaders should make decisions and workers should follow directions. Some workers have expressed this attitude in terms such as, "I get paid to work. Others get paid, and get paid more than me, to provide direction and to make decisions. Additionally, they have the title of supervisor. Why am I doing their work?"

So, despite the advantages inherent in democratic leadership, many have evolved away from this leadership style as they have sought the best method for leading and managing. Many managers have gravitated to the participative leadership style, which is the topic of the next section.

Participative Leadership

A leadership style that has been shown to be quite effective, and that a great many leaders and managers have adopted today, is participative leadership. This leadership style appears to provide a blend of characteristics and advantages that results in leaders who practice this style being extremely effective. Participative leadership and management are very widely recommended today in leadership development programs of all kinds.

Participative leaders regularly make it a practice to seek input from those whom they lead, with regard to decisions that affect them. They seek, and genuinely consider, the thoughts of the workers with regard to work layout, methods, and procedures to be followed. They ask for the workers' input, counsel, and guidance on matters of all kinds in the workplace.

Such a leader carefully cultivates communication channels with the workers so that the workers feel comfortable in sharing their thoughts and in bringing forth their ideas. The leader is at ease, and the workers are comfortable providing their input as to work methods and other decisions that affect the workplace environment. This allows the craft workers to feel empowered and provides the leader with the benefit of their experience and judgment.

When decisions are to be made in the workplace, participative leaders do thoughtfully consider the input received from the workers. Moreover, they let it be known that the workers' input is valuable, and that this input has been taken into account.

However, participative leaders also let it be known that, when decisions are to be made, they will make them, and that the responsibility for those decisions rests with them. Sometimes they may agree with recommendations that have come from the workers and may implement those recommendations. At other times, they may act counter to what has been recommended.

Everyone knows where the decision will ultimately be made and where responsibility for the decision resides. When things turn out well, it is the workers who receive the credit; when the outcome is unfavorable, it is the leader who accepts responsibility, for both the decision and for the outcome.

When decisions turn out well, the participative leader is quick to point out that the workers' skill and talent are the cause of the good result. The participative leader is careful, however, never to say that he or she acted on the recommendation of the workers when the result was unfavorable and that, therefore, the responsibility for the bad results is the workers'. Rather, the participative leader would be sure to make it understood that the responsibility for making decisions, as well as the responsibility for the outcome of those decisions, is solely his or her own.

The participative leadership style has the advantage of centralized authority and decision making and responsibility. It also offers the important advantage of allowing workers to feel as though their knowledge and their opinions are valued, and to know that they have a voice in matters that affect them. This allows the workers to feel empowered. They are motivated by the fact that management values their opinions, and that they have the freedom to speak their mind, and to input their suggestions, and to apply their knowledge. Often, they will commit to decisions that have been made based on their input and will dedicate themselves to ensuring that the decisions turn out well.

Leaders who can maintain the balance between allowing workers to voice their input into decision making and the mindset that decisions are made by the leader are the most effective at participative leadership. Participative leadership combines some of the features of the autocratic and democratic leadership styles to form a style of leadership where all can be involved but where the leader takes on the mantle of leadership and makes decisions and accepts responsibility.

Participative leadership is very popular today, is very widely practiced, and has been shown to be very effective. Many consider participative leadership to be the best leadership style.

Situational Leadership

A leadership style born of modern management thinking, and one that is increasingly being applied today with great effectiveness, is situational leadership. The essence of situational leadership is the leader's applying the style and method of leadership that is most appropriate for the *situation* at hand.

This leadership style, which was first defined by Dr. Ken Blanchard and Paul Hersey, is described and summarized in the book *Leadership and the One Minute Manager* by Ken Blanchard, Patricia Zigarmi, and Drea Zigarmi (Harper Collins Publishers, 2004), and in the book *Management of Organizational Behavior* by Paul Hersey, Ken Blanchard, and Dewey Johnson (Pearson Prentice Hall, 2008). Both of these books are very highly recommended for further reading by the supervisor.

The description of situational leadership provided here is based on and is paraphrased from these books. Dr. Blanchard describes the essence of situational leadership in these words:

> Effective managers have a range of leadership styles they can use comfortably. They have developed some flexibility in using these styles in different situations. Effective managers also have a knack for being able to diagnose what their people need from them in order to build their skills and confidence in doing the tasks they are assigned. Finally, effective leaders communicate with their people—they are able to reach agreements with them not only about their tasks but also about the amount of support they will need to accomplish these tasks.

Implied in the descriptive definition of situational leadership is the thought that the application of this method relies upon the supervisor knowing his people and their capabilities and their preferences in the workplace. Some would describe this logic as "different strokes for different folks." However, this thought can be broadened to include the concept of "different strokes for the same folks, as the situation indicates." For example, some people are more developed in some areas of their job than others; they can function independently and without a great deal of direct supervision on some tasks, but on other tasks they may need a great deal of direction and support.

Additionally, supervisors need to have available for their use a variety of leadership styles that can be applied as the situation warrants. In addition, they need to have established, and to continuously maintain, open communication channels with the workers. The importance of effective communication is emphasized in a number of other sections of this book and is a key component of situational leadership.

The situational leadership model holds that the situation at hand is the determining driver for the style of leadership the manager will employ. Four styles are defined as subsets of situational leadership: directing, coaching, supporting, and delegating.

When the *directing* style of leadership is utilized, the leader provides specific direction for the task assigned and then closely monitors task accomplishment. The leader lays out a step-by-step plan for how the task is to be accomplished. The leader makes the decisions, and the person receiving the direction carries out the assigned task.

Such a style would be appropriate for use by the supervisor with a person who is new to the job site or who is new to the task to be performed. As this worker gains proficiency and arrives at a comfort level with this set of tasks, the supervisor might transition to a coaching style. For those workers not motivated to move forward with advancing their skills, the supervisor might continue to employ a directing style.

The directive style is also appropriate when a decision has to be made quickly and when the stakes are high. In these situations, strong directive leadership is the way to achieve success.

Directive behavior involves clearly telling people what to do and how to do it and then closely monitoring their performance. Watchwords for the supervisor with regard to utilization of the directive style are: structure, organize, teach, and supervise.

It should be noted that the directing style, and the other three leadership styles as well, are based in the assumption that the supervisor believes his workers are responsible and self-motivated, and that they have the desire and the potential to become high performers. This rationale continues with the thought that the workers will be given direction by the supervisor so that they can begin developing their full potential.

This philosophy differs radically from the assumptions that many managers frequently made about the workforce in the past, wherein the presumption was that the workers are not inclined to work, do not seek responsibility, are not reliable, and do not seek to advance. These assumptions provided the rationale in the past that, therefore, the workers needed close supervision and that an autocratic directive style was the best way to manage. We will further discuss assumptions that managers make with regard to the workers who compose the workforce in the section on motivation later in this chapter.

When she employs a *coaching* style of leadership, the supervisor continues to direct and closely monitor task accomplishment, but she also explains decisions, solicits suggestions, and supports progress. Thus, the coaching leadership style involves both some direction and some support on the part of the supervisor.

When utilizing the coaching style the supervisor engages in two-way communication by asking for suggestions and getting input from the workers, recognizing that the workers frequently have good ideas. This style involves listening to people, and providing encouragement and support for their efforts, and then facilitating their involvement in problem solving and decision making. However, in the end it is the supervisor who makes the decision and who provides the directive with regard to what is to be done.

For workers who have some amount of experience and who demonstrate reasonable competence at performing a set of tasks, a coaching leadership style might be indicated. Additionally, for workers who have progressed from needing a directive style, and who have indicated that they wish to continue to grow and advance, a coaching leadership style on the part of the supervisor might provide the next step in their evolution and growth.

When utilizing a *supporting* leadership style, the supervisor facilitates and supports people's efforts toward task accomplishment, and shares decision making with them. Keywords characterizing supportive leadership are: listen, ask, explain, facilitate. A supervisor utilizing this supportive style, supports and encourages people's efforts, listens to their suggestions, and facilitates their interactions with others.

When employing a *delegating* leadership style, the supervisor turns over responsibility for decision making and problem solving to people whom he trusts with the responsibility. The supervisor relies upon the workers' experience and know-how, and provides very little directive leadership. Rather, he realizes that the workers have the ability and the motivation to make independent judgments and to solve day-to-day problems in the workplace. The supervisor allows these talented and motivated workers to exercise their own initiative and to apply their experiential learning in the workplace.

While four different leadership styles are defined by Dr. Blanchard and his colleagues, he is emphatic in noting that no one of these styles is inherently better than the others. Rather, the most effective leadership style the supervisor can employ is the one indicated by the situation—the environment and the tasks to be performed and the people available at that time. Situational leadership has many advantages and many appealing features to offer—it is highly recommended for consideration and adoption by the supervisor.

ELEMENTS OF MOTIVATION

Along with understanding their role as a leader, supervisors do well to consider the concept of motivation. The more fully they understand the factors that motivate them to act and think the way they do, and the more fully they understand what motivates or demotivates those around them every day in the workforce, the more effective supervisors can be in getting their objectives accomplished through other people.

There are a number of questions that thoughtful people have posed and wondered about for a long period of time. Why do we go to work? Why do we continue to go to work day after day, and year after year? Why do some people excel in the workplace, while others attain only minimum levels of accomplishment? Why do some people exhibit drive and determination in the workplace, while others are satisfied with minimally meeting the requirements in their work? Why do some people seek to advance themselves, while others are content with the status quo? Why do some people aspire to become leaders and managers, while others shun and avoid leadership roles?

Questions such as those noted above have perplexed those who study people and work situations for a very long time. The findings of some leading psychologists and research scientists will help us to better understand. As has been emphasized a number of times, the better supervisors understand the human elements of the workplace, including their own motivation and behavior, as well as that of the people with whom they interact in the workplace, the more effective they can be.

Some of the most significant findings in this regard are summarized in the book, *Management of Organizational Behavior* mentioned earlier. Some of the information in the following paragraphs is adapted from this work.

It has been found in many studies that most of human behavior is goal-oriented. People do what they do, or they avoid the things they do, because the behavior is directed at achieving some significant result, or goal.

Activities are the basic element of human behavior. People perform activities of all kinds every day, and often they perform more than one activity at a time. In the workplace, people vary considerably in their ability to perform various activities, and also in their willingness or motivation to perform various activities. People's motivation varies with the strength of their motives. Motives are, in turn, defined as needs, wants, drives, or impulses in the individual that are directed toward the accomplishment of goals which that person has. Motives, which are also referred to as needs, are the drivers for people behaving the way they do and engaging in the activities they do, as they strive to fulfill their goals.

Freud and Subconscious Needs

While many of people's motives are based in their response to identifiable needs, it has been found that frequently people are also motivated by subconscious needs, as well. It was Sigmund Freud who first pointed out that people are not always consciously aware of everything they want. Therefore, they may do certain things in response to motives or needs which they are not aware of at the time, at a conscious level of thought. This subconscious motivation, when it leads to action, sometimes evokes the thought, "Why did I do that?" as the conscious mind seeks to determine the reason for the action. We will focus most of our attention on the conscious motives or motivations for people's behavior, while realizing that some activities are driven by the subconscious mind and its determination of a person's goals.

Mayo and the Hawthorne Studies

Among the earliest studies that (although indirectly) concluded that management needed to study and understand relationships among people were the Hawthorne Studies, which began in 1924. Elton Mayo of the Harvard Graduate School of Business conducted studies at the Hawthorne, Illinois, manufacturing plant of the Western Electric Company.

The first study at the Hawthorne plant was aimed at determining the optimum levels of lighting in the workplace, in order to achieve maximum production from the workers who were assembling electric components. This study was typical of a number of others that were being conducted at the time, where efficiency experts were concerned with finding the optimum blend of working conditions so as to ensure maximum productivity from factory workers.

In the first of Mayo's studies at the Hawthorne plant, workers were divided into two groups. One group, which was called the control group, would continue working under the existing conditions of lighting. The other group, the experimental group, would work under varying conditions of illumination. The productivity of both groups would be measured. It was believed that the productivity of the control group would remain constant and that the production of the experimental group would vary with different illumination conditions in the plant.

When lighting intensity was increased in the experimental group, production increased, as expected. However, the researchers were surprised to find that the productivity of the control group increased also. Another study was conducted in the plant, with the same result. In fact, when the lighting level in the workplace of the experimental group was eventually reduced to the levels of moonlight, the productivity still increased, while that of the control group increased also. The researchers concluded that something other than illumination levels was affecting the changes in the workers' productivity.

In an effort to explain their findings, the researchers decided to continue their studies at Hawthorne and to conduct additional research. They thought that in addition to studying the physical changes in the work environment, they should also examine some of the behavioral considerations in the workers at the plant.

The researchers learned that it was the human behavioral considerations, and not the changes in the physical environment, that were primarily responsible for the production changes that they had observed in the plant. They learned that the workers were pleased with the attention that was being paid to them during the conduct of the research. The workers felt this indicated that management believed them to be an important part of the company.

Additionally, because the workers were studied as a group in the two study groups during the research, they no longer felt that they were working alone and in isolation but rather that they were participating members of a congenial and cohesive work group. The relationships they developed with their fellow workers brought forth feelings of affiliation, competence, and achievement. These needs had not previously been fulfilled in their workplace, and so the workers worked harder and more effectively than they had before. Additionally, since the workers were paid on a piecework basis they discovered, to their delight, that their additional production was earning them additional money.

Mayo was, thus, the first to realize that human factors and group interactions were an important contributory factor to worker productivity in the workplace. He concluded that workers wanted more from the work environment than being treated as insensitive machines capable of producing. Mayo came to believe that management needed to move away from the commonplace presumption that workers were unorganized individuals whose only concern was their own self-preservation and self-interest. He came to believe that management style and structure needed to be responsive to social, and esteem, and self-actualization needs of the workers. Further research regarding this hypothesis was later conducted by others, and they reached the same conclusion as Mayo.

Maslow's Hierarchy of Needs

Abraham Maslow was one of the first to conduct credible research that showed that people's behaviors are motivated by their desire to fulfill various kinds of needs that they feel. Additionally, he was one of the first to research the humanistic, as opposed to task-based, elements of management. Maslow summarized his findings in his "Hierarchy of Needs Theory," which was first published in 1943, and was subsequently incorporated into his book *Motivation and Personality* in 1954. His work is often quoted and referred to today.

Maslow's research indicated that people are motivated to act in order to satisfy needs that they perceive. He noted that the behavior of individuals at a particular moment is usually determined by their strongest perceived need. He found that people act first to fulfill the needs that they feel most strongly. Then, as those needs become satisfied, other levels of needs become important, and then these needs motivate and dominate people's behavior.

Maslow arranged the needs that people respond to into sets, and he organized these sets of needs into a hierarchical array. Maslow's hierarchy of needs is most often represented on a triangular framework today, although Maslow himself did not depict it this way—Maslow's publications simply indicated the hierarchy of needs in a stair-step arrangement.

As the supervisor reads and seeks to understand the needs that Maslow identified, he is urged to "internalize" his considerations, and to recognize how he perceives these kinds of needs in his own life. Additionally, it is recommended that the supervisor "externalize" these findings, in order to see how he perceives these needs serving as motivators for the people around him in the workplace.

The needs that Maslow identified are indicated below. These needs are shown in a triangular arrangement in Figure 7.2.

The five levels of needs that Maslow identified, survival needs, safety needs, social needs, ego or self-esteem needs, and self-actualization needs, are listed below and are defined in the terms in which Maslow defined them.

Physiological or survival needs: These are the most basic of all needs, consisting of the need for air, food, shelter, clothing, warmth, and sleep.

Safety or security needs: These are the need to be free of physical danger and from threats and harm. These needs have been extended to include the

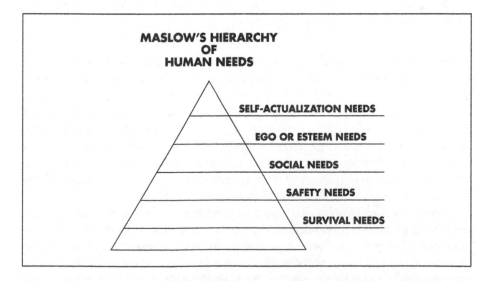

Figure 7.2   Maslow's Hierarchy of Needs

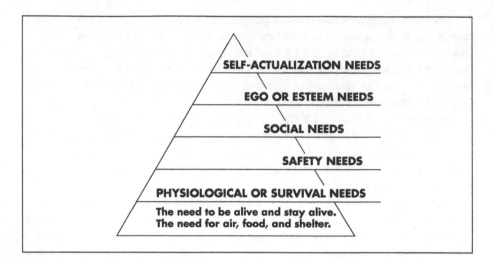

Figure 7.3 Physiological or Survival Needs

need for an orderly and predictable world and also the need to provide for financial security and job security.

Social or affiliation needs: These are the need to belong and to be accepted by others. These include the need to belong to and to be accepted by various groups, including work groups.

Ego, status, and esteem needs: These include the need to be held in esteem, both by oneself and also by others.

Self-actualization needs: These include the need to maximize one's skills and talents and to maximize one's potential.

Physiological or survival needs are indicated at the bottom of the triangle in Figure 7.2 and Figure 7.3, because they are experienced at the highest strength by everyone, or nearly everyone. While some exceptions are sometimes observed, such as when an artist ignores the need for food, water, and shelter while engrossed in capturing a scene on canvas, the physiological needs eventually become dominant and strongly motivate people's behavior. Until these needs are satisfied, they tend be felt most strongly and, therefore, to dominate behavior. As long as these physiological needs remain unsatisfied, people's activities will be focused at this level, and other needs will provide little motivation.

Safety or security needs usually become predominant for a person after the survival needs are fulfilled. They appear in Maslow's hierarchy just above survival needs as de picted in Figure 7.4. After survival needs are met, until an individual feels he has provided for his safety and security, other things seem less important. These needs are characterized as self-preservation needs, and they also include a concern for the future.

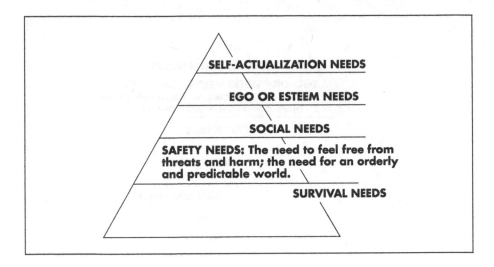

Figure 7.4 Safety Needs

Social or affiliation needs that people feel are based in the fact that humans are social beings and feel the need to affiliate and interact with a group, or more typically, with numerous groups of various kinds. As indicated in Figure 7.5, social needs typically begin to become dominant after survival and security needs have been satisfied. When social needs are dominant, people will seek meaningful relations with others. This is one of the needs that Mayo found the workers responding to in the Hawthorne Studies.

Ego, status, and esteem needs are next in the hierarchy, as indicated in Figure 7.6, and include the need that people feel to be held in high regard, both by

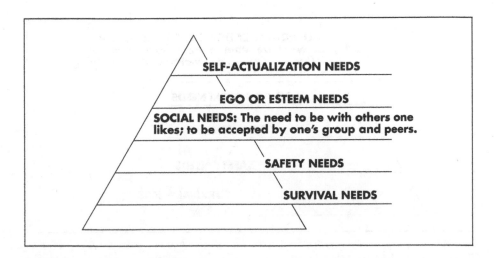

Figure 7.5 Social or Affiliation Needs

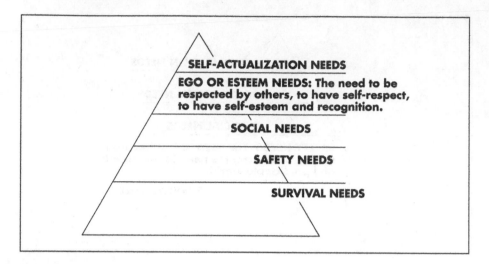

Figure 7.6 Ego, Status, Esteem Needs

themselves and by others. People seek recognition and respect from others, and when they receive it this produces feelings of self-confidence, prestige, and control. People feel as though they are useful and that they have some influence on their environment.

Self-actualization needs are the highest on the hierarchy of human needs, as shown in Figure 7.7. These include the need to maximize one's skills and talents, and to move to self-realization and self-fulfillment. These needs are also felt as the need to reach one's full potential and to be all one is capable of being.

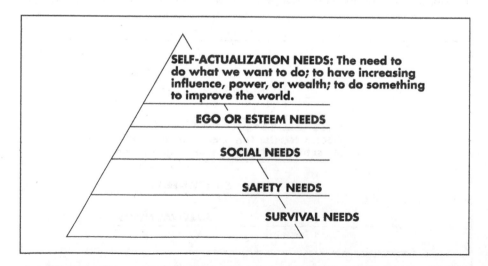

Figure 7.7 Self-Actualization Needs

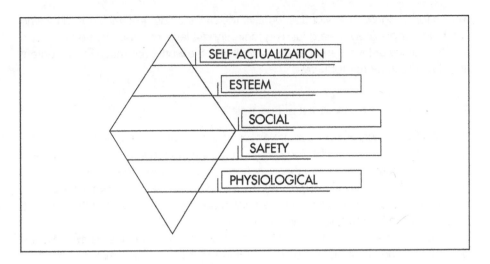

Figure 7.8 Social Need More Dominant Than Others at the Present Time

The way this need is expressed may change over the course of a person's life. A person beginning his career may seek to self-actualize by being the best craft worker he can be. In later years, he may be self-fulfilled in becoming the best supervisor and manager he can be. Later in his life, this person's self-actualization may shift to being a respected and experienced member of a construction company and/or a wonderful grandparent and/or a respected elder in society.

It should also be noted that while Maslow did indicate that people's needs are in a hierarchy, he also said that people might be most strongly motivated to fulfill one need or another that they felt most strongly at some point in time, regardless of its level on the hierarchy. For example, when the social need is most strongly perceived, it well may be the dominant motivator for a time, until it is satisfied or until another need comes to be felt more strongly. This concept is illustrated in Figure 7.8, where the social need is (for the present) occupying the broadest region of the triangle(s) of the person's perceived need.

Later in his career, Elton Mayo endeavored to determine whether his findings at the Hawthorne Plant were in consonance with Maslow's theory, and he concluded that they were. He determined that initially only the physiological and safety needs of the workers were being fulfilled at the plant and that, because there was no avenue for them to fulfill their higher-level needs, they were frustrated and unhappy. Later, as the study proceeded, more of the workers' social and self-esteem needs were being addressed, and they felt much more content in the workplace, and that contentment motivated them to be more productive.

Similarly, the supervisor who can recognize that people in the workplace are motivated in response to various categories of needs that they are feeling has acquired a valuable tool for better understanding the people around him or her, as well as for recognizing why people behave the way they do. And the supervisor who can help people grow toward fulfilling their highest level of needs by helping them satisfy their lower-level needs may very well find the workforce much more content and thus much more productive.

McGregor's Theory X and Theory Y

Douglas McGregor, who was a professor of industrial management at the Massachusetts Institute of Technology, was another researcher who made extensive studies of management and labor practices. He published the results of his research as "Theory X and Theory Y," and included these theories in two of his books, *The Human Side of Enterprise* and *Leadership and Motivation*. Both of these books are recommended reading for the supervisor.

McGregor believed that the traditional structure and management of organizations, with their pyramidal structure, their centralized decision making, and their external control of the work environment, was often employed because it was based on certain assumptions that management makes about the workforce, human nature, and human motivation. His research indicated that many managers believe that most workers prefer to be directed, are not interested in assuming responsibility, and want safety (as defined in Maslow's work) above all.

He concluded that the extension of this philosophy was the belief that money, fringe benefits, and the threat of punishment are the elements that motivate people. Theory X assumptions about the workforce also seem to imply that autocratic leadership is the way to manage people in the workplace, with the keys to success being to ensure control and close supervision of the workers, and to provide structure in the workplace and in work assignments. Figure 7.9 summarizes the Theory X assumptions that many managers, both past and present, make about the workers in the workplace.

McGregor came to believe that Theory X assumptions about the workforce, and about human nature more broadly, are often inaccurate and that they do not apply to all people and should not be universally applied. He reasoned the management by direction and control might not always be best and, in fact, might not succeed, because some people's physiological and safety needs are already reasonably satisfied, and their social, esteem, and self-actualization needs are far more important to them.

McGregor developed an alternative theory of human behavior, which he called Theory Y. This theory holds that management can believe the following about human behavior and about human motivation in the workplace: many people are not lazy and unreliable, and people can be self-directed and creative at work if properly motivated. The assumptions that characterize McGregor's Theory Y are summarized in Figure 7.10.

McGregor believed that the objectives of management should be to help people realize these potentials, and that properly motivated people can best achieve their own goals by directing their efforts toward the accomplishment of the goals of the organization. He believed that in organizations where management operates

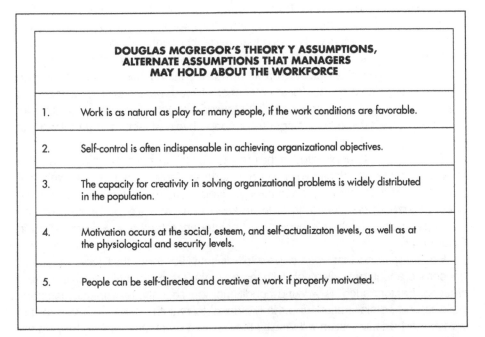

**DOUGLAS MCGREGOR'S THEORY X ASSUMPTIONS,
WHICH MANAGERS MAKE REGARDING PEOPLE IN THE WORKFORCE**

1. Work is inherently distasteful to most people.

2. Most people are not ambitious, have little desire for responsibility, and prefer to be directed.

3. Most people have little capacity for creativity in solving organizational problems.

4. Motivation occurs only at the physiological and security levels of the needs/motivation hierarchy developed by Maslow.

5. Most people must be closely controlled and often must be coerced to achieve organizational objectives.

Figure 7.9   McGregor's Theory X Assumptions

**DOUGLAS MCGREGOR'S THEORY Y ASSUMPTIONS,
ALTERNATE ASSUMPTIONS THAT MANAGERS
MAY HOLD ABOUT THE WORKFORCE**

1. Work is as natural as play for many people, if the work conditions are favorable.

2. Self-control is often indispensable in achieving organizational objectives.

3. The capacity for creativity in solving organizational problems is widely distributed in the population.

4. Motivation occurs at the social, esteem, and self-actualizaton levels, as well as at the physiological and security levels.

5. People can be self-directed and creative at work if properly motivated.

Figure 7.10   McGregor's Theory Y Assumptions

on these theories, there is high productivity and that people will gladly come to work because the work itself is inherently satisfying. McGregor wrote, "Supervision consists of helping employees achieve these objectives: to act as teacher, consultant, colleague, and only rarely as authoritative boss" (*The Human Side of Enterprise*, McGraw-Hill: 1960).

It is important to understand that McGregor also noted in Theory Y that many people have the *potential* to be independent and self-motivated and to work toward the goals of the organization. This does not mean that every worker will be motivated by this management attitude.

Further, he said, Theory X and Theory Y are attitudes or predispositions by management toward people. For example, while a manager may be centered upon Theory Y beliefs, he may find it best to behave in a very directive and controlling manner (typical of Theory X attitudes) at some times and with some people in the workforce, because of these people's attitudes and motivations. Note how this conclusion dovetails exactly with the principles and tenets of situational leadership, as developed earlier in this chapter.

Frederick Herzberg

Frederick Herzberg was a behavioral scientist whose work in the 1950s is still widely accepted and frequently quoted today. Herzberg developed a two-factor theory, called the motivation-hygiene theory.

Herzberg's theory resulted from his research in which more than 200 professional workers in 11 different industries were interviewed. These workers were asked about the kinds of factors in their jobs that made them unhappy or dissatisfied, and the kinds of factors in their workplace that made them happy or satisfied.

The conclusions that Herzberg drew from his research were that people have two different categories of needs. These categories of needs are essentially independent of each other and they affect people's behavior, and their satisfaction, in different ways. He termed these two categories of needs hygiene factors and motivational factors.

The needs that people cited in their responses to Herzberg's surveys that were referred to by Herzberg as hygiene or maintenance factors are those whereby the person is focused primarily on surviving. People are motivated to satisfy these needs, Herzberg concluded, in order to keep alive, and also to avoid pain and unpleasantness. Herzberg's hygiene or maintenance factors are summarized in Figure 7.11.

Herzberg found that when people felt dissatisfied about their jobs, they were concerned about these elements with regard to the environment in which they were working. Factors such as working conditions, job security, reasonable company policies and administration, and the like serve to maintain the status quo for the worker, and dissatisfaction will result if they are not present. However, they do not serve as motivators, nor do they produce satisfaction in the workplace. A person who is dissatisfied because of the absence of one or more of these hygiene factors may look for employment elsewhere where these factors are present. Or, a pay

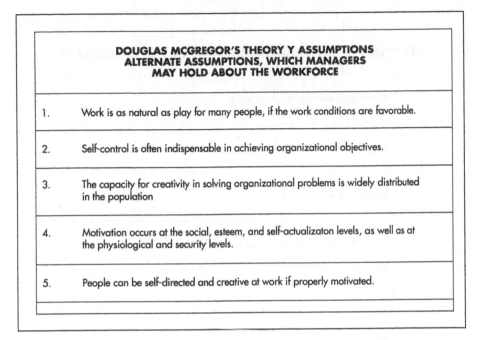

DOUGLAS MCGREGOR'S THEORY Y ASSUMPTIONS
ALTERNATE ASSUMPTIONS, WHICH MANAGERS
MAY HOLD ABOUT THE WORKFORCE

1.	Work is as natural as play for many people, if the work conditions are favorable.
2.	Self-control is often indispensable in achieving organizational objectives.
3.	The capacity for creativity in solving organizational problems is widely distributed in the population
4.	Motivation occurs at the social, esteem, and self-actualizaton levels, as well as at the physiological and security levels.
5.	People can be self-directed and creative at work if properly motivated.

Figure 7.11   Herzberg's Hygiene or Maintenance Factors

increase may prevent employees from quitting, but it will rarely make them work harder.

Herzberg identified another set of factors in the workplace that serve as motivators, spurring people to higher levels of productivity and accomplishment. These were factors that had more to do with the work itself and the opportunities available in the workplace than with the work environment. These motivational factors are summarized in Figure 7.12. Herzberg said in an interview "The things that people said positively about their job experiences were not the opposite of what they said negatively about their experiences. The factors that make people happy all are related to what people did: the job content, and what made people unhappy was related to job environment, job content, and the way they are treated" ("Managers or Animal Trainers," *Management Review*, 60 [1971]).

The supervisor can benefit from this information by noting those factors in the workplace whose presence will cause people to be dissatisfied and unproductive. The supervisor can then seek to minimize or eliminate as many of these factors as possible within his sphere of control.

Additionally, and perhaps more importantly, the supervisor can carefully assess the factors from which employees derive satisfaction. These factors produce motivation, higher levels of satisfaction, and higher levels of productivity and accomplishment. The supervisor can make it a goal to avail himself of opportunities to provide or enhance these factors, which Herzberg terms "motivational factors," insofar as he possibly can.

HERZBERG'S MOTIVATIONAL FACTORS IN THE WORKPLACE	
1.	Interesting and challenging work
2.	Recognition for achievement
3.	Increased responsibility
4.	Sense of importance to the organization
5.	Access to information
6.	Opportunity to do something meaningful
7.	Potential for growth and development
8.	Involvement in decision making

Figure 7.12   Herzberg's Motivational Factors

MANAGER, LEARN TO BETTER UNDERSTAND YOURSELF AS WELL AS THOSE AROUND YOU

As they think in terms of better understanding the concepts of leadership and motivation, supervisors should also be mindful of the value in developing their human relations skills in another dimension. They will do well to seek to better understand aspects of their own personality, their outlook on life, and their approach to life's situations. Additionally, they will benefit in more fully understanding the role that emotions play in their decision making and in their everyday work. Further, the more fully supervisors understand themselves, the more likely it becomes that they will develop the skills to better understand others in the workplace around them, thereby further enhancing their all-important human relations skills.

Myers-Briggs Type Indicator

The Myers-Briggs Type Indicator (MBTI) is a personality indicator survey that has been utilized by a great many people, in order to gain a better understanding of their own personalities and thought processes. The indicator is based on the fact that normal people consistently prefer to use their perception and judgment in certain ways, and these trends can be identified through analysis of a survey questionnaire that the person completes.

The indicator was developed by psychologist Katherine C. Briggs and her daughter, Isabel Briggs Myers, following their 20 years of study of the work of another psychologist whose name was C. G. Jung. Myers and Briggs were interested in and were studying the variety of ways in which people achieve excellence in their lives and were seeking to develop a typology to describe the behaviors of highly successful people. They found Jung's work to provide a means of establishing a characterization of people's innate personality types and the way in which people relate to the world around them.

First published by Educational Testing Services, the MBTI form has been available since 1956. The indicator has been used by millions of people and has consistently been shown to be valid, that is, to measure what is says it will measure. Additionally, the indicator has also been shown to be reliable, that is, to produce the same results when given more than once. Validity and reliability are two very important bases for measuring the usefulness of a research instrument, and the MBTI demonstrates high values for both of these factors.

The Meyers-Briggs Type Indicator consists of a questionnaire that the respondent completes, based in how he or she perceives and prefers to interact with the world. The determinations made from answers to the questionnaire provide useful information for the respondent with regard to his or her own personality type, which may be characterized as any of 16 different variations. The indicator profile emphasizes that there are no "right" or wrong," or "normal or not normal," answers to the questionnaire. Further, the indicator emphasizes that when the respondent's type has been indicated as the result of the survey, no one type is better than any other. These are simply indicators of and names for the person's personality type.

It is recommended that the supervisor complete the MBTI. The objective should be for the supervisor to utilize the MBTI to gain some insight into his own perception, and his own view of the world. This can help him better understand his own thought processes, as well as adding insight as to how he relates to others, both in the workplace as well as in his personal life.

The Myers-Briggs Type Indicator is available from a number of different sources. It is recommended that the supervisor review information that is available on the

Internet at the website of the Myers-Briggs Foundation (http://www.myersbriggs. org). The website describes the indicator in more detail, and also lists a number of ways in which one can take the MBTI instrument.

Temperament Theory

A logical accompaniment to the Myers-Briggs Type indicator and its 16 personality types is a school of thought that is referred to as "temperament theory." This theory has been published in many forms for centuries. However, in 1978 psychologist David Kiersey developed a temperament theory that integrated the ideas of the past with the Myers-Briggs Type Indicator. Kiersey published a book titled *Please Understand Me II: Temperament, Character, Intelligence* (Prometheus Nemesis Book Company; 1st edition, May 1998). It is recommended for reading by the supervisor.

4-Dimensional Management

An accompaniment to the situational leadership concept, which was presented earlier in this chapter, is the concept of "4-dimensional management." This management style is referred to as "DISC Strategies®." The concept is described in a book titled *The 4-Dimensional Manager*, by Julie Straw (Inscape Publishing/Berrett Koehler Publishers, Inc., 2002). The descriptions of DISC that are included here are adapted from this book.

DISC is based on a theory of human behavior that was first developed by American psychologist Moulton Marston. Marston was interested in knowing more about how normal people felt and behaved as they interacted with the world around them.

Inscape Publishing, a Minneapolis-based research and publishing firm, published a book called *Personal Profile System* after conducting further research based on Marston's work. This book was intended to provide people with access to an understanding of their own feelings and behavior in almost any situation.

The research by Inscape Publishing continues today and has included further refinement and development of the *Personal Profile System*. The research has included thousands of subjects from all kinds of backgrounds. This research has likewise been shown to be both valid and reliable.

The book *The 4-Dimensional Manager* is intended to make available to managers of all kinds and at all levels of management, the findings of the research. DISC helps the manager identify how he personally responds to situations around him, as well as how others use various dimensions of their own perception in their everyday environment. DISC reveals how our personalities help us respond to the environment in which we find ourselves. It notes that people demonstrate four different types of behavior: Dominance, Influence, Supportiveness, and Conscientiousness (DISC). It holds that all people have within them elements of all four of these types of styles. However, people tend to utilize one style as their primary response, depending on their assessment of the situation in which they find themselves.

Understanding which style he is most likely to utilize will help the supervisor to better understand elements of his own thinking and behavior. In addition, having characteristics of the other styles available for review will help the supervisor determine the style he might elect to utilize in various situations. Finally, knowing the styles that others may be demonstrating in the workplace, will help the supervisor understand how best to relate to them. *The 4-Dimensional Manager* is recommended for the supervisor's further reading.

Emotional Intelligence

The concept of emotional intelligence offers the supervisor a new way of thinking about life success. The concept is based in groundbreaking research by psychologists John Mayer and Peter Salovey, who first published the terms *emotional intelligence* in an article in 1990. Their research has been advanced by Daniel Goleman, who has for many years been conducting his own research directed to modeling competencies that set star performers apart from average performers in the workplace. Goleman has published a book titled *Emotional Intelligence* (Bantam Dell/Random House, New York, 1995).

Goleman's book defines and explains the role our emotional brains play in our response to and our handling of everyday situations and in our decision-making, as well as in our aspirations for success and excellence. Mr. Goleman traces the fundamentals of emotional intelligence—self-awareness, self-management, social awareness, ability to manage relationships, zeal, persistence, empathy—and our development of the competencies to apply them to best advantage, which are huge components of our self-satisfaction and of our success. The book is highly recommended for the supervisor.

SUMMARY

As supervisors continue to advance their competencies and their management skills, so too should they be constantly seeking to learn more about and advancing their human resources and human relations skills. Certainly the management of the work in the construction industry is multifaceted and complex. However, the variability and complexity of the industry pale in comparison to the diversity and the intricacy of the human beings who function within the industry.

This chapter has provided a degree of insight into the concepts of human leadership and motivation, and goal-oriented behavior, and a better understanding of oneself and the other people the supervisor works with. The human dimension of management is a critically important component of the supervisor's learning and development, and it is also a major determinant of his effectiveness and success.

Learning Activities

1. Ask your supervisor or your company management whether it is possible for you to take the Myers-Briggs Type Indicator Test.

   If it is not possible for you to take the test through your company, you may wish to investigate enrolling to take the test from one of the sources referenced in the chapter.

2. Read at least one of the books recommended in this chapter:

   Leadership and the One Minute Manager, by Ken Blanchard, Ken and Patricia Zigarmi, Patricia and Drea Zigarmi

   Management of Organizational Behavior, by Paul Hersey, Ken Blanchard, and Dewey Johnson

   The Human Side of Enterprise, by Douglas McGregor

   Leadership and Motivation, by Douglas McGregor

   Please Understand Me II: Temperament, Character, Intelligence, by David Kiersey

   The 4-Dimensional Manager, by Julie Straw

   Emotional Intelligence, by Daniel Goleman

3. Observe those in the workplace around you, and others with whom you come into contact on a regular basis, to see the elements of motivation taking place, as described in this chapter.

   Think about your own behavior, and about the things that motivate you to do what you do, in terms of the learning from this chapter.

4. Think about what you can do, within the sphere of your management control in your company, to eliminate or reduce factors that serve as demotivators for the workers whom you manage, and additionally, to enhance motivating factors (as defined by Herzberg) for the workers.

5. Think about what you can do to apply Theory Y motivation factors as defined by McGregor in your workplace.

CHAPTER 8

RISK MANAGEMENT AND PROBLEM SOLVING

INTRODUCTION

Risk management and problem solving are critical skills required for construction supervisors at all levels. Supervisors are in a unique position to protect the company from risk. They must, therefore, keep a lookout for risks that could be damaging to the company and its operations. Once identified, the risks must be eliminated, mitigated, or managed.

One role that the supervisor takes on is being the shield that stands between workers executing the work and any problems or disturbances that might impede workers in any way from doing their job. The effective supervisor will identify and remove or mitigate any problems that serve to impede production.

Risks and problems are similar in nature. In fact, risks that come to fruition become problems. Early identification, together with clear and complete definition, is the critical first step in effectively dealing with both risks and problems.

This chapter begins by defining risk and then discussing a process for dealing with it. The chapter then moves on to problem solving, starting with problem identification, and then considering selection and implementation of solutions.

IDENTIFYING AND DEALING WITH RISK

Unidentified risks are the most dangerous kind of risk. If a risk is detected, there are ways to deal with it, either eliminating it or mitigating it. However, an unidentified risk often leads to problems that cause damage before they can be addressed. The problems, and resultant damage, may quickly grow beyond the ability to control them.

This can be illustrated by considering a house fire. We have learned over the years that if a house fire can be detected early, in many instances it can be extinguished relatively easily. If the fire is growing too fast to enable it to be extinguished, at least people in the house can escape, saving injury and death. As a result of this knowledge, codes have been changed to require extensive use of fire alarms and, in many cases, fire suppression systems in homes.

Timely identification of business risks is just as important to the health and vitality of a business as timely detection of fire in a home. So managing risk starts with risk identification.

Definition of Risk

Risk: The possibility of loss or injury.

Merriam-Webster Online Dictionary

Construction, by nature, is risky. It involves working with dangerous tools, materials, and equipment, so it is physically risky. It is highly dependent upon people, which means that it involves a great deal of variability, leading to an entire class of performance risk.

Construction work is accomplished through tasks that may or may not be repetitive. The work environment on any job differs from that of any other job and on the same job it differs from day to day and even from hour to hour. Hence, the work can involve the risks associated with repetitive activities leading to boredom, but at the same time, it is afflicted with risk associated with unique, nonrepetitive activities replete with many unknowns.

A construction project brings together many different entities that must be focused on producing distinct elements that are integrated into a common product. This merging of a variety of elements produced by diverse entities leads to substantial risk. Often the supervisor and the supervisor's company are in a collaborative, not a controlling relationship with the other entities, which leads to another category of risk. Finally, construction is highly competitive, so it has many business risks.

The following is a review of some common construction risks, together with a discussion of the supervisor's participation in managing them. The risks will be classified in the following six categories:

- Financial risk
- Schedule risk
- Incident risk
- Design risk

- Quality risk
- Business risk

Financial Risk

The most common financial risk is that the cost of executing the job will exceed the budget. Mitigations for this risk include making sure that the cost estimate is accurate and complete and then making sure that the cost control system is effective in detecting at an early stage when actual costs are exceeding those budgeted. Although the primary responsibility for estimating typically rests with others in the company, the supervisor can influence the accuracy of the estimate by providing accurate (or inaccurate) costs for use in the company's cost database. However, once an overrun is identified, often through the cost control system, it falls to the supervisor to bring the cost back within the budget.

Another common financial risk is the risk of nonpayment. This can be the result of many factors outside the supervisor's span of control, such as a burdensome contract or the financial condition of the entity with which the contractor is contracting. However, delayed payments are often the result of dissatisfaction on the part of a client, which may be influenced significantly by the supervisor. Nonpayment may also be the result of an erroneous billing because of inaccurate information provided by the supervisor on completion status of the work. Nonpayment may also be associated with extra work, which was executed without proper authorization, or authorized extra work for which appropriate and complete costs were not documented or were not provided by the supervisor. Thus, much of the risk of nonpayment falls on the shoulders of the supervisor.

Finally, since construction finance is so closely linked to cash flow and hence timeliness of payment, another financial risk is inability to get to the money earned on a timely basis. Though this may well not be a field problem, it could be a field problem associated with lack of or inappropriate documentation from the field, inaccurate estimates and projections of completed work, or customer dissatisfaction because of field operations.

Schedule Risk

Much of the risk associated with schedule is related to field operations. Though some of it is outside the control of the supervisor, much of it falls within the span of control of the supervisor. Schedule risk that is outside the control of the supervisor still has an impact on the supervisor's work.

One of the most common schedule-related risks is that the project, or elements of the project, cannot be started when scheduled. Since completion dates are much less flexible than commencement dates, this schedule risk often results in the requirement to accelerate the project, or certain activities within the project, which becomes the supervisor's responsibility.

Another schedule risk, one that is often not understood, is disturbance to the work flow. If the schedule is modified, changing the sequencing of activities, or if some activities are executed out of sequence, many other activities are affected. This results in disturbances to resource planning for materials, equipment, and craft workers, which the supervisor must deal with.

Finally, scheduling risk may simply involve inability to complete work as scheduled. If this is the case, the supervisor must determine what the impediments are to completion on time, remove them, and ensure that the work will be completed on time.

Incident Risk

Although the goal of any project is to be free of accidents and incidents, they do happen. The risk here may be in the form of damage to, or loss of, materials or installed equipment, or it may take the form of human injury or even death. No matter what the form of the risk, if it is realized, it will have impacts on many aspects of the job, not the least of which are cost and schedule. Chapter 9 expands upon the safety risk on a construction job.

Design Risk

Design risk does not fall to the contractor, unless the project is executed under a design/build project delivery system. However, the impact of design risk is often strongly felt by the supervisor. Design errors and omissions occur on all projects. These can be dealt with in appropriate ways on most projects, however, when an error or omission is identified, the process to obtain a clarification or a redesign can delay the job. Also, changes might be called for that impact work already installed or that affect later work.

One of the biggest design risks to the contractor is lack of constructability; that is, the designer provides construction details that appear in the office to work well, but when considered within the project as a whole or in the context of field operations, the details are inefficient, unsafe, or just do not work. At this point, the supervisor has two options. The first is to request information or a design modification from the designer. The second is to figure out how to make the design work in the field without disturbing the integrity of the design. Both options engage the supervisor and both incur risk to field operations.

A third design risk to the contractor is lack of responsiveness. The designer might take a long time to respond to requests for information or clarifications. The designer might delay the approval process for materials and equipment or the approval process for changes or for payments. All of these can be very disruptive to work flow and require intervention by the supervisor.

Quality Risk

Risks in terms of quality fall into two categories: workmanship and materials. The first risk is that work was not installed in an acceptable manner. It might not have been installed according to the design. It might not have been installed according to code. It might not have been installed in such a way that subsequent work could be completed. Or, it might not have been installed to the satisfaction of a higher-level contractor, the designer, or the owner. Ensuring high-quality installation is a supervisory responsibility.

The other quality risk relates to flaws in materials and installed equipment. The wrong material or item might have been supplied, or an inferior quality item might have been substituted. A damaged item might have been installed, or a dysfunctional piece of equipment might have been installed. It is a field responsibility to identify and deal with obvious defects in material or equipment and ensure that such equipment is not installed. In the case of latent defects, which by nature cannot be detected until later, field operations will be affected when the defect is identified and must be corrected.

In the case of either workmanship or material, quality is a risk that directly impacts field operations and falls under the responsibility of the supervisor.

Business Risk

Business risks focus not on the project but on the company. Although all of the other categories of risk do impact the company, this category relates specifically to risks to the company. Business risk is primarily concerned with the company's reputation. This might be its reputation with a specific customer on the project. (See Chapter 3 for a discussion of customer relations.) Or, it might be the reputation of the company within the industry or for the general public.

Customer relations suffer when a customer is not treated well or not respected, or when the service or product provided is perceived to be inferior. Supervisors must ensure that their behavior, the behavior or performance of the workers for whom they are responsible, and the behavior or performance of a supplier or subcontractor is not perceived as negative by any customer.

Construction projects are often highly visible. The supervisor must ensure that the appearance of the project and the behavior of the workers on the project demonstrate superior quality to anyone observing the project, whether another contractor, another entity associated with the construction industry, or simply a casual observer.

MANAGING RISK

Proper risk management is proactive, not reactive. Some risks can be controlled directly because they fall within the scope of authority of the supervisor. Other risks lie outside the scope of authority of the supervisor, and these risks need to be recognized, tracked, and influenced to the extent possible. The supervisor must be continually vigilant to recognize new risks as they arise or as they become discernable.

The supervisor's risk management plan should have the following elements. *Identify* a risk as soon as possible. *Analyze* the likelihood of occurrence and the potential impact if the risk is realized. *Eliminate* the risk if possible. If the risk cannot be eliminated, *allocate* risk to the entity most able to control that risk. To the extent a risk cannot be eliminated or allocated, *acquire protection* from financial ramifications. Determine and implement *mitigations* for consequences and side effects associated with risks that cannot be eliminated or allocated. *Monitor* ongoing risks to determine whether the risk management plan is effective.

Risk management begins with the estimator during the pre-bid process. It continues during project planning by the project manager. The supervisor continues risk management throughout project execution. Risk management for the supervisor ends only with the end of field operations on the project, although risk management by other company employees may be required well beyond the completion of field operations on the project.

PROBLEM SOLVING

Problem solving is a key competency required of any supervisor. Eliminating and resolving problems is of fundamental importance to supporting workers, maintaining a safe and positive work environment, improving production, and enhancing customer relations.

A systematic problem-solving process is needed to deal with problems, especially the complex problems found on construction jobs. The systematic problem-solving process will incorporate the following steps:

- Detect the problem early
- Define the problem properly
- Analyze the problem
- Develop and analyze creative solutions
- Select the best solution
- Identify side effects and mitigations
- Implement the selected solution and mitigations
- Monitor the effectiveness of the solution
- Redesign or improve the solution
- Learn from the process

Having a problem-solving process is extremely important for many reasons. Problems often arise suddenly. Many problems are acute and a quick response is important. Construction problems are often addressed with a quick or standard solution that does not resolve the problem, but incurs the investment of resources that are wasted and often adds to the problem or causes detrimental side effects.

Throwing Labor at a Schedule Problem

Consider the situation where an electrical contractor has fallen behind schedule in installation of light fixtures. Since the permanent lighting system is to be used in this large, dark building to support the finish work of many of the other specialty trades, installing light fixtures is on the critical path. If not completed when scheduled, it will delay a number of other activities and other specialty contractors, resulting ultimately in delayed completion of the project. The general contractor has become aware of the delay in installing the light fixtures. He has ordered the electrical contractor to use the "standard" solution for a schedule problem by putting more electricians on the job.

This situation creates a number of new problems. First, adding electricians to the light fixture installation activity by starting a second crew will create congestion, as well as requiring additional man lifts and other construction equipment. It will either draw electricians off of other activities, or require bringing more electricians to the job site, both of which introduce additional inefficiencies. So, even if this solution does increase the rate of production, it will cost more per unit to install the fixtures, and although it might fix the scheduling problem, it will have a negative impact on the electrical contractor's profitability.

Further analysis of the problem indicates that the light fixture distributor is working in a just-in-time delivery environment and is providing fixtures at a rate designed to keep a single crew supplied. Increasing the installation rate will require expedited delivery of fixtures at additional cost because of the need to use air freight for some of the fixtures.

An analysis of the cause of the delayed installation of light fixtures shows that the delay resulted from late commencement of installation of the ceiling grid, and speeding up installation of the fixtures will cause the electrical contractor to run out of prepared grid in which to place light fixtures, so the improvement expected to the overall schedule would be only temporary.

If the electrical contractor were to implement the schedule problem resolution required by the general contractor, it would be costly, it would not resolve the problem, and the root cause of the delay would not be addressed, so the problem would not be resolved.

A functional problem-solving process will tend to lead to better solutions that are more likely to resolve the problem more quickly with less investment of resources and fewer damaging side effects. Each step of the process is discussed below.

Detect the Problem Early

Although sometimes problems are anticipated, often they occur undetected and unanticipated. The first indication that a production problem has occurred will often be through the reporting system. An experienced supervisor will have a feeling generated by observations as she walks the site when a production problem begins to manifest itself, but the reporting system, specifically cost and schedule reports, verify that the feeling is real.

Early detection of problems is based upon establishing means to detect the problems as soon as they occur. The supervisor has at hand two powerful tools to detect problems: observation and documentation.

Most supervisors are experienced craft workers. By observing work in process, given their experience, they can identify the existence of a problem or a risk that could become a problem if not eliminated. When they identify the existence of a problem, they might see the problem itself, but it is much more likely that they will see symptoms that indicate that a problem exists. If they see a problem, they can move quickly into the problem-solving process. If what they see is a symptom, they need to go through steps to identify and define the problem before moving into the solution phase.

Another way of detecting problems through observation occurs when supervisors interact with other project participants. These interactions provide an opportunity to identify potential problems relating to the sequencing or coordination of activities. Purposes of the regular planning and coordination meetings include giving all participants an update on current status and the opportunity to review and/or develop the project plan over the next few weeks. Potential or incipient problems can be identified in this environment and either addressed directly in the meeting or noted for future action where they can be dealt with in a more appropriate environment.

In addition to directly observing the work, supervisors also can observe documentation that will reveal an incipient problem. A review of the construction documents, including the drawings and specifications, can reveal potential production or safety problems. These may take the form of errors, omissions, or ambiguities that require clarification. The construction documents may also reveal constructability issues that will impact either production or safety.

Other examples of documentation that can be used to identify potential or incipient problems are progress reports from site operations and submittal and shipping documentation for materials and equipment. In fact, the supervisor should be looking at any documentation available as a source of information enabling detection of potential problems.

Two types of reports are of particular value in detecting problems. Cost reports show when unit production costs are exceeding budgeted production costs, or when projected item costs vary significantly from budgeted costs. Refer to Chapter 13 for more information on cost reports and how to use them. Schedule reports are valuable in determining which activities are on schedule and which are lagging behind. Refer to Chapter 14 for a discussion of schedule reports and how to use them. A third type of report, based upon earned value analysis, enables an evaluation of the current status of the project in terms of both cost and time. Earned value analysis can be valuable to supervisors to help them understand where tradeoffs can be made between cost and schedule to protect the project objectives of being within the budget and on time.

A critical factor in problem detection is timing. Problems, incipient problems, and potential problems need to be detected as early as possible in order to have time to develop appropriate solutions and mitigations. As a rule of thumb, problems affecting ongoing activities need to be detected when the activity is between 5% and 50% complete. If an activity is less than 5% complete, mobilization inefficiencies associated with that activity cause data to be inaccurate. As an activity approaches 50% complete, delays in reporting and the time required to go through the problem-solving process, including implementing the chosen solution, render any solution too late to make a difference. At that point, the problem becomes a history lesson, and it will be a waste of resources to try to implement a solution.

Define the Problem Properly

Since problems often are not detected directly, but rather through their symptoms, the supervisor must determine the problem or problems that are causing an observed set of symptoms. This is typically the case when detecting problems based upon a cost or schedule report. These reports do not show problems directly; rather they show the effects of a problem. That is, cost and schedule reports show symptoms not problems. That the project is behind schedule is a symptom of problems not indicated in the report. When a cost report indicates that production units are not being met, this lack of production is caused by one or more problems, but the report does not reveal what the problem is.

A response that addresses the symptoms without understanding the underlying problem is rarely effective and, generally, costly.

Solution That Addresses Symptoms

The cost report may indicate that a contractor is not meeting the budgeted unit installation costs for an item. The project manager detects this incipient problem and visits the site to inform the superintendent that this work item is over budget, and he must take some of the craft workers off this task and push the rest to work harder. An analysis of the work activity would have shown that the crew is the right size, but they do not have the most effective construction equipment to complete this task. They are using ladders to complete a task that would be completed much more effectively with a man lift; however, the company man lift assigned to this job is being used in another part of the job. Without proper equipment, the diminished crew is not working more efficiently and is also causing the task to lag behind the scheduled time frame. Had the supervisor been able to properly analyze the problem, he could have rented an additional man lift for a couple of weeks, maintained the initially assigned crew, and completed the task at lower cost and more quickly than scheduled.

It is important to recognize that symptomatic solutions generally are not effective in resolving problems. This is clearly demonstrated with a medical example. If a patient has a headache and the doctor suggests taking aspirin to relieve the headache when the problem is a tumor, the headache may be relieved for a time, but the long-term outlook for the patient is not good.

Symptoms are not benign. In many cases, they must be dealt with, as well. Continuing the medical example, if a patient has a high temperature, this is an indication of an infection in the body. The doctor needs to deal with the infection, but a high temperature can also cause damage to the human body, and it can even be fatal, so the high temperature needs to be dealt with while working on the infection or the patient may suffer permanent damage from the sustained high temperature.

In construction, cost overruns, although generally considered symptoms, are detrimental to the job and ultimately to the company. Schedule delays are also symptoms that can be detrimental to the project and to other contractors on the job. Sometimes a temporary fix (a band aid) may be needed while the problem analysis process is ongoing to relieve some of the symptoms and the immediate damage they can inflict while the permanent solution is being sought.

To determine the problem (or problems since the symptoms might be the result of multiple problems) a root cause analysis is often effective. A root cause analysis starts by looking at the symptom or symptoms and asking, "What has caused this?" When that cause has been identified, the question must be asked of that original cause: "What caused this?" The questioning process is continued until the most basic or root cause has been identified, at which point the root cause can be dealt with. If the root cause is dealt with, the problem should be resolved and not return. If a subcause is dealt with, but the root cause remains, it is highly likely that the problem will persist.

This process is a fundamental tool in lean manufacturing that grew out of the Toyota Manufacturing Process. Toyota executive Taiichi Ohno called the process the 5 Whys method because he found that repeating "why?" five times was required to understand the nature of the problem as well as its solution. The application of lean production principles to construction is discussed in Chapter 15.

Improper problem identification is one of the most important causes of ineffective problem solving. A solution thrown at a symptom without proper analysis will rarely be effective but will generally require additional investment of resources and may cause additional problems and further complications. However, if the problem analysis is correct and an effective solution is implemented, the problem should go away and not return. If either the analysis or the solution is not correct, the problem will continue.

Analyze the Problem

Once identified, the problem must be understood. It is important to recognize the nature of the problem.

The first consideration is whether it is an isolated problem that occurred once, as in Figure 8.1, or a trend that occurs repeatedly, as in Figure 8.2.

An isolated problem becomes a history lesson. The negative impact of the problem has already been felt, and it will not have future detrimental effects. Two questions might be considered for this type of problem. First, can the negative impacts that have already occurred be overcome and, if so, is it worth the investment to gain the recovery? Second, what can the supervisor learn regarding why it

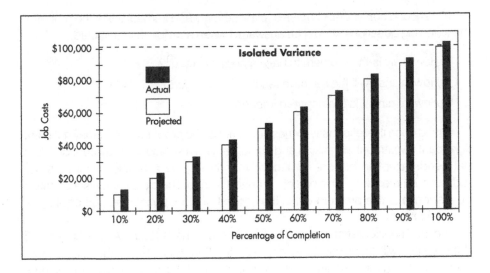

Figure 8.1   Isolated Problem

happened in order to make sure that it does not happen in the future? Other than that, the isolated problem that has already occurred does not merit a significant investment of time.

A trend problem is ongoing. The negative impacts will continue and the effects will accumulate over time. Therefore, this problem merits study and investment in removing or mitigating it. If a problem is determined to be a trend, the supervisor should continue through the problem-solving process.

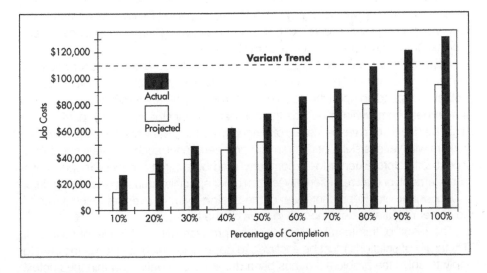

Figure 8.2   Trend Problem

The second consideration relates to the severity and immediacy of the problem. The following questions should be answered when a problem is detected:

- How likely is the problem to have a negative result?
- How intense will the negative result be, if it occurs?
- How imminent is the negative impact?

Based upon the responses to these questions, the problem may or may not merit significant investment in seeking and implementing a solution, it may be a problem that needs attention but not immediately, or it may be a problem that must be addressed immediately. This part of the analysis should help the supervisor prioritize the response to this problem within the context of other duties and responsibilities she might have.

The third consideration focuses on the source of the problem. An understanding of the source will often lead to an effective solution. As stated in the introduction, when a risk turns into reality, a problem is created, so one source of problems is risk. Problems also arise from mistakes or oversights. They arise from lack of performance either within the supervisor's own sphere of responsibility, or by others outside the sphere of responsibility, but affecting the supervisor's area. Another source of problems is lack of communication or lack of understanding.

Understanding the source of the problem will shed light on who can, or must, deal with the problem and what resources might be needed to develop and implement the solution.

Develop and Analyze Creative Solutions

Once the problem has been identified and determined to merit attention, a variety of solutions can be applied to deal with it. Some time should be spent in creative thinking to come up with as many solutions as possible. This can be done by the supervisor, but it often works best as a team activity. Chapter 5 provides more information on developing effective team solutions.

A team with diverse members can come up with a much wider variety of solutions than an individual can. The composition of the team should reflect the nature of the problem. Depending upon the nature of the problem, the team might be composed of members from within or outside the company. Within the company, participants might be drawn from various levels of management and supervision, company craft workers with various skills, or different support personnel, such as safety specialists or cost estimators. External to the company, members might come from supervisory, management, or craft representatives from other specialty areas, material suppliers, or subcontractors. Representatives of the design team or even representatives of the owner might also be important to have on the team.

The result of this team creative thinking process will be a family of solutions. This family of solutions must be analyzed to determine which solution is the best to apply to solve the problem that has been detected and defined within the context of the specific job conditions.

Select the Best Solution

Before the best solution can be selected, thought must be given to what constitutes the best solution. Some examples of criteria used to evaluate the quality of the solution are:

- Easiest to implement
- Quickest to implement
- Least expensive
- Most effective
- Quickest to show results
- Least side effects

After determining the appropriate criteria for quality of solution, a simple evaluation form can be developed so that each solution can be evaluated for each criterion to determine which one comes out on top. To the extent possible, a numerical basis should be used. Each criterion should be evaluated on a numerical scale, and a second number, a weighting factor can be applied if the criteria are not considered to be of equal importance in the evaluation. To improve the evaluation process, guidelines should be provided to help evaluators assign the most appropriate numerical value to the various criteria under consideration. Figure 8.3 provides an example of an evaluation sheet that might be used.

Identify Side Effects and Mitigations

Solutions generally cause side effects. Part of the consideration of a solution is to determine what side effects will be produced when it is implemented. In evaluating solutions, considerations with regard to side effects include whether these side effects are acceptable and how they can be managed, mitigated, or eliminated.

After a solution has been selected, a more extensive analysis of these side effects is required. Side effects will often result in disturbances to budget and schedule, but as with problem identification, the primary focus in terms of analyzing side effects must be on root causes. Some common side effects to the contractor when solving problems include inappropriate or inadequate materials onsite, the wrong tools or construction equipment onsite, insufficient labor availability, or the craft workers that are available are not sufficiently skilled in the appropriated areas. The solution may well produce side effects that impact the performance of other contractors on the site, as well, so other contractors might need to be involved in the analysis. For example, using float time on an activity to resolve a labor resource problem for my company can have an effect on other contractors later in the job by delaying start of their activities or denying them access to the float time that I use earlier on the job. The use of float time to manage labor resources will be dealt with in Chapter 14.

Problem Solution Evaluation Sheet

The purpose of this sheet is to evaluate potential solutions to a problem to determine the most appropriate solution for a problem.

Problem_____

Criteria	Weight	Value	Weighted Value
Easiest to implement			
Quickest to implement			
Least expensive			
Most effective			
Quickest to show results			
Least side effects			
Total Weighted Value			

Figure 8.3   Problem Solution Evaluation Sheet

To mitigate these side effects, common solutions might include expedited material delivery, provision of different or additional tools and equipment, increased labor, adjustments to the schedule, and increased supervisory time necessary to coordinate with other contractors on site.

Implement the Selected Solution and Mitigations

Once the solution has been selected and side effects analyzed, unless the solution is simple and uncomplicated, an implementation process must be designed. The implementation process will consider a number of factors. Are adequate and appropriate resources (tools, equipment, materials, labor) available? If not, they must be procured prior to implementation in the field. Who is involved in the implementation process? Are simple instructions adequate or is some training required? How will other entities be affected and how will they be brought into the process? How will the implementation process be monitored and the level of success be measured? How will improvements to the process be identified and incorporated?

When the implementation process design is completed, implementation can begin. Since time is critical in dealing with an ongoing problem, implementation of the solution should begin as soon as the solution has been selected and the implementation process has been designed. The implementation process should be monitored to ensure that it is working effectively to resolve the problem that was determined to be the root cause. Side effects should be reviewed, as well, to ensure that the mitigations have been effective. If the solution is ineffective or if it is not as effective as anticipated, the problem solution process should be repeated to develop refinements or to select another of the proposed solutions.

Learn from the Process

The final step in the problem solution process is to document what has been done so that others can learn from the success of this problem-solving cycle. Such documentation will help avoid similar problems in the future. It will improve the problem detection process by alerting others that this type of problem can occur and explaining how to detect it early. It will provide a model for resolving such problems in the future.

Problem-solving exercises that have been particularly successful should be formatted briefly as a case study for the company archives. The case can be distributed to other supervisors on the current job and on other jobs. It can be entered into the company's "lessons learned" file to become part of the company's supervisory training program. In this way, resolving a problem on one project can become a valuable component of the corporate knowledge that enables the company to work at a higher level.

SUMMARY

In this chapter, the following key points have been presented.

- Risk management and problem solving are critical supervisory skills.
- Construction risk can be classified into six categories.
- The supervisor should have a risk management plan to identify and analyze risks as they arise.
- Risks can be removed, assigned to others, mitigated, managed, or protected against.
- Problems on the construction site arise from many sources.
- Early detection and proper analysis are critical to resolving problems.
- Applying an appropriate solution to the root cause of a problem should eliminate it.
- The supervisor should be adept at identifying, analyzing, and resolving problems affecting field operations.

Learning Activities

1. Risk analysis

   For this exercise, either use the company and job on which you are currently employed or a fictitious company and job. Specify the company's name and work sector, and the general characteristics of the company and job that might be critical to risk analysis.

   Pick one of the six categories of risk enumerated in the chapter and identify a specific risk within that category that would be likely to occur on your job. Determine how best to detect the risk. Develop an analysis sheet to evaluate that risk should it develop on your job. Determine the best way to deal with the risk, whether that is elimination, reassignment, mitigation, management, or procuring protection. Write a brief report to your supervisor about the risk and how you propose dealing with it.

2. Developing a problem resolution process

   For this exercise, either use the company and job on which you are currently employed, or a fictitious company and job. Specify the company's name and work sector, and the general characteristics of the company and job that might be critical to problem resolution.

   Using the steps presented in the chapter, develop a procedure to identify, analyze, and resolve a problem that might come up on your project. It might be easier to start by identifying a specific type of problem, such as a cost overrun, develop the procedure for that specific type of problem, and then consider how the procedure can be adapted to address other problems, such as schedule problems or safety problems.

   Finalize the exercise by laying out your step-by-step process for dealing with problems on your job that you can keep and refer to in the future. If you do use this in the future, you will probably be able to continuously improve the process as you use it so that what you develop now will be the starting point for an ongoing problem solving process for your jobs into the future.

SECTION III

TECHNICAL SKILLS

CHAPTER 9

SAFETY

INTRODUCTION

Safety is a topic of fundamental importance to the construction supervisor. There are numerous reasons for this. Among the most important is that most supervisors feel a moral obligation to keep their workers safe. There are also reasons tied to costs, schedule, production, quality, and company reputation, all of which can be enhanced by a strong safety record, or significantly damaged by a poor safety record.

Safety is a broad topic around which many courses are designed, certifications provided, and entire degree programs structured. Safety in the workplace is a major concern of government agencies at all levels: federal, state, and local. Safety is also a major concern of sophisticated owners, whether in the public or private sector, as well as the various entities that participate in the construction process.

All individuals that contribute to construction, from the project owner to the laborer working in the field, have a stake in safety. All participants in the construction process have a part to play in establishing and maintaining a safe work environment. Contributors to the construction process, such as material suppliers, equipment providers, lawyers, accountants, and sureties also have a keen interest in construction safety.

This chapter provides an overview of construction safety, focusing on a few key areas. Since safety plays such an important part in construction, any supervisor should have significantly more training and education in safety than it is possible to provide in this book; however, any book on construction supervision would be deficient if it did not recognize safety as an important component of construction supervision.

The chapter begins by addressing why it is important to have an effective construction safety program. This is followed by discussion of the supervisor's responsibilities and activities in support of safety. The chapter will then look at the two major factors that lead to work site incidents: unsafe job site conditions and unsafe behaviors. The chapter considers creating a safe work environment, followed by a discussion of developing a mentality of safety that is a powerful tool in combating unsafe behavior. Finally, the chapter closes with a review of how to deal with a safety incident or accident when it arises.

ACCIDENTS, INCIDENTS, AND SAFETY RISK

An accident is an unwanted, unplanned event that results in injury, illness and/or property damage. An incident is an unwanted, unplanned event that has the potential to cause injuries, illnesses, and/or property damage. Since an incident focuses on potential, many incidents do not cause injury, illness, or property damage and, hence, could be considered a near miss. Incidents are an indication of safety risk.

Chapter 8 provided a broad overview of risk associated with construction. It identified the risk posed by incidents and accidents on the job site as one of the primary areas of risk. Although this was only one of six areas of risk reviewed, it is one of the most critical, and it is directly in the area of interest of the supervisor.

Construction work is very risky. The supervisor has a twofold responsibility with respect to safety. The first part is to keep her area of the construction site safe for all workers in that area and for anyone else entering the area. The second part is to ensure that her workers work safely. The discharge of these responsibilities should be supported by a well-designed construction safety program.

THE IMPORTANCE OF A CONSTRUCTION SAFETY PROGRAM

A construction safety program provides many benefits for the workers, the job, and the company. Following are some of the benefits to such a program. It will:

- Allow workers to go home safely at the end of the work day
- Increase productivity
- Provide higher profit margin for the company
- Provide lower insurance costs

- Enhance the company reputation
- Offer better personnel policy
- Support compliance with the law

In the rest of this section, we will focus upon one of the key reasons, from a company standpoint, why a safety program is important. That reason is cost.

According to the Occupational Safety and Health Administration (OSHA), "Safety is good business. An effective safety and health program can save $4 to $6 for every $1 invested. It's the right thing to do, and doing it right pays off in lower costs, increased productivity, and higher employee morale."

Many factors contribute to the cost of an accident. These factors can generally be grouped into direct costs, indirect costs, and intangible costs. Direct costs include such items as medical treatments for injured workers, temporary wage replacement, and awards for permanent impairment. Indirect costs, those generally not covered by insurance, include such items as lost time by injured workers or other workers affected by the accident, training new or temporary substitute workers, and efficiency costs associated with breaking up and reforming crews. Intangible costs include such things as worker morale and company reputation.

Direct costs are those that are normally associated with accidents and are relatively easy to evaluate. Indirect costs are more difficult to calculate but are considered to be significantly higher than the direct costs. By nature, intangible costs are very difficult to measure.

Accidents are very costly to the project and to the company. There is a line item in the project budget to account for the cost of safety prevention, but there is no line item in the budget to account for the cost of accidents. If a cost item is not accounted for somewhere in the budget, then the only place to account for that cost is from the profit. Figure 9.1 gives an analysis of the estimated total cost of an accident and the sales required to pay for accident costs.

Sales Required to Pay for Accidents					
Accident	Hidden	Total	If your company margin is:		
Costs	Costs*	Costs	1%	3%	5%
$ 1,000	$ 6,000	$ 7,000	$ 700,000	$ 233,333	$ 140,000
$ 5,000	$ 20,000	$ 25,000	$ 2,500,000	$ 833,333	$ 500,000
$ 10,000	$ 30,000	$ 40,000	$ 4,000,000	$ 1,333,333	$ 800,000
$ 25,000	$ 50,000	$ 75,000	$ 7,500,000	$ 2,500,000	$ 1,500,000
$ 100,000	$ 150,000	$ 250,000	$ 25,000,000	$ 8,333,333	$ 5,000,000
* Hidden costs vary from about 1.5 to 6.0 times accident costs					

Figure 9.1   Sales Required to Pay for Accident Costs

The Cost of an Accident to the Company

In Figure 9.1, consider an accident whose direct costs amount to $10,000. The hidden costs are estimated to be about three times the direct costs, so the total cost of the accident, excluding intangible costs, amounts to $40,000. If the company profit is on the order of 3%, which is about the average for a U.S. construction company in good times, the amount of work that must be completed to offset that $10,000 accident is $1.3 million. This would be significant for any construction company, and a disaster for most small construction companies.

One key cost factor for construction companies is the cost of insurance. Not only does the cost of an accident, as illustrated above, have a significant impact on company profitability, but company insurance rates are also linked to the company's safety record. Insurance costs increase significantly when an accident occurs. A measure of a company's safety record, developed years ago by the insurance industry, is the Experience Modification Ratio (EMR). This is calculated by a complex formula, and is expressed by a single number with 1 as the average across the industry. Since the EMR is a multiplier applied to the base insurance rate, an EMR above 1 increases the insurance rate above the industry average and if the EMR is below 1, the result is an insurance rate that is below the industry average. It is not uncommon for safe companies to have EMRs in the area of 0.7, while companies that have had recent accidents score well above 1.

Two important factors associated with the EMR should be noted. First, the EMR is based upon a rolling average, considering the company's safety record for the three years prior to the current year. Thus, an accident will affect the EMR not only this year but for three years into the future. Second, since the EMR has become a respected measure of a company's safety record, many owners of construction are using the EMR as a prequalification factor for those who want to compete for their construction work. So not only does a high EMR directly impact the cost of operation of a construction company, but it also has a profound impact on which jobs the company is allowed to bid.

The conclusion is that just from a purely financial point of view, safety is very important to a construction company. This imposes on the supervisor a very strong responsibility to make sure that his jobs are safe by implementing a strong safety program within his sector of responsibility.

THE SUPERVISOR'S SAFETY RESPONSIBILITIES AND ACTIVITIES

The supervisor has a number of responsibilities with respect to safety. First, the supervisor must be trained and current in safety, especially as it relates to the specific activities in their work area. They need to have their own safety program and be aware of other safety programs and requirements relevant to their work area, including governmental safety requirements, those of the client, those of their own company, and those of the specific job, if they represent a specialty contractor.

Within the context of their work sector, supervisors are responsible for developing and implementing their own safety program, compatible with all of those in the hierarchy above them, including the ones identified in the previous paragraph. The supervisor is responsible to train their workers continuously, both formally and informally. Formal training might take the form of sending a worker to a class or requiring participation in a periodic safety discussion. Informal training happens when the supervisor identifies an unsafe behavior or condition and discusses it with the worker involved. Finally, the supervisor is responsible for enforcement of the safety program.

Derived from the responsibilities, the supervisor will engage in a number of safety-oriented activities. They will ensure that each worker is properly oriented to safety for their work area when they first arrive in that sector. They will continually develop safety awareness through such activities as regular "toolbox" safety meetings, recognition of safety achievements and innovative safety practices, and taking note of and dealing definitively with any safety infractions.

The supervisor will always be on the lookout for unsafe conditions or behavior and enlist all workers in this vigilance. Not only does the supervisor look for safety issues whenever he or she is walking the site, but from time to time, the supervisor should make a safety inspection, looking specifically for safety concerns.

The supervisor should deal immediately and decisively with safety incidents or accidents. These need to be reported to the appropriate authorities immediately and an investigation undertaken. Falling in the category of risk, near misses should also be identified and investigated. The objective of any safety investigation should be to identify and eliminate the root causes so that problems do not reoccur. This takes precedence over finding someone to blame. Responsibility for a safety incident needs to be assigned, but this is not as important as eliminating the causes so that the situation does not occur again.

Finally, the supervisor should be on the lookout for improvements for his safety program or that of any other entity on the site. No safety program is perfect. Continuous improvement in safety leads to fewer accidents and incidents and ultimately to a lower EMR.

CREATING A SAFE WORK ENVIRONMENT BY REMOVING JOB HAZARDS

Creating a safe work environment is approached in two directions. The first is to enlist the creativity and commitment of all stakeholders to prevent hazards from reaching the job site. The second is to identify hazards that have occurred at the job site and remove or mitigate them.

The Stakeholder's Responsibility in Preventing Hazards

Creating a safe work environment is the responsibility of all stakeholders in construction. There are three phases involved in creating a safe work environment: design, training, and enforcement.

Designing for safety takes many forms. The supervisor must design a good safety program for his area of work. Other supervisors must do the same for their areas. Each trade contractor must have an effective safety program for the workers in their company. The general contractor or construction manager must have a sound safety program for the project as a whole within which the trade contractor safety programs operate. More sophisticated owners also have safety programs that they impose upon their construction projects. This is particularly true on jobs that are located within an operating industrial complex or an ongoing business operation.

Designing for safety is also the responsibility of the project design team. Project elements can be designed in a fashion that makes them safer or more dangerous. For example, production that can be done in a controlled environment is typically safer than production that must be carried out on an open job site. Thus, an electrical engineer can design in such a way that electrical installation can be modularized with significant prefabrication in a shop environment rather than building the entire system in the field. Work that can be accomplished at grade level is safer than work at elevation. Therefore, if the structural engineer can design steel structural elements that can be easily assembled in the field with a minimum of field welding, risk to welders that do need to work high up and risk to those working below them can be diminished.

Design for safety also involves construction equipment manufacturers that are continually seeking ways to make their equipment safer to use in the field. Equipment safety also depends upon workers respecting the integrity of safety designed into the construction equipment by not altering safety features to try to gain perceived improvements in efficiency of operation.

Training is an ongoing process that also takes many forms. Craft training at the apprentice level should emphasize safe practices from the first day the apprentice enrolls in the craft program. Journeyman training is required any time new materials or equipment are to be used and any time new or particularly dangerous procedures are to be carried out. Retraining in the use of dangerous equipment or handling dangerous materials should be carried out on a routine basis and particularly if the equipment or materials have not been used recently.

Safety awareness is one of the most effective ways to improve the safety record. Training is very important in maintaining awareness of safety practices and procedures. Routine can be very dangerous if it allows workers to become complacent and to forget to be on safety alert. Periodic safety talks (known by many other names, such as tool box talks or tailgate talks) are a required part of any safety program, one of the reasons being to counter the negative effects of routine. Periodic safety talks are a safety tool that the supervisor needs to take full advantage of.

Enforcement is a primary responsibility of the supervisor. The best practice is to engage all workers in identification of safety infractions on the site, but primary responsibility for enforcement lies with the supervisors. Supervisors need to be continually cognizant, seeking out any unsafe condition or behavior and then taking corrective action. If one of the workers tells the supervisor of a safety concern about a condition or practice, the supervisor needs to respond quickly and positively to investigate the situation. A positive response will keep the worker involved in bringing safety issues to the attention of the supervisor. Any negative response could well set up a barrier to further involvement of the worker.

Craft workers can be engaged in enforcement by commenting to fellow workers when they see unsafe behavior. If a worker sees a coworker engaging in an unsafe practice, she can give a friendly reminder that "there may be a better way to do that." Often it is the case that the coworker may not be aware that he is doing something that is unsafe and will appreciate the concern and friendly reminder.

Removing Job Site Hazards

Creating a safe work environment requires the elimination of job site hazards. This can be done in three steps: recognition, evaluation, and control.

Recognition involves identifying hazards in the work area. There are many ways to identify hazards, among which are the following.

Observation to detect hazards in the work area is the responsibility of anyone in the area. The supervisor and all workers should always be on the lookout for hazards. Such hazards might be associated with tools, with equipment, or with the physical environment. Examples of hazards that might be observed include improperly functioning equipment, a messy or congested work area, or temporary structures, such as scaffolds, that are unsafe or not built to code.

Monitoring is an important part of job site hazard detection. There are many devices that can be used to detect the presence of a danger. Air-sensing devices can detect many types of hazardous materials in the air. Decibel meters can detect when sound levels are dangerously high. There are many types of electrical devices that can detect dangerous electrical current or charges.

Discussion is also important in hazard identification. Workers involved in job site activities on a day-to-day basis often see hazards that they do not report or that they do not recognize as potentially dangerous. If the supervisor talks with workers about hazards that they might have seen or hazardous situations that might occur on the job, the supervisor will often gain information about the existence of these hazards that were not previously identified.

Once identified, the hazard must be evaluated. Three questions need to be considered for any hazard. How likely is a problem to develop out of this hazard? If a problem does develop, how destructive might it be? Finally, how imminent is the problem? The response to the hazard will depend upon the answers to these three questions. If the problem that arises is not likely to be destructive, then the hazard becomes low priority. If the problem is likely to cause significant damage but is not likely to occur, response to the hazard rises in priority, but not in urgency.

If the problem will be destructive, it is likely to occur and it is likely to occur soon, the priority in responding to the hazard rises to the top and the response to the hazard becomes immediate.

When the decision is made to address the hazard the process is similar to the problem-solving process discussed in Chapter 8. A family of potential solutions is developed. Each is evaluated to determine how effective it would be if applied. The cost of the solution is evaluated. Side effects are identified and evaluated and mitigations explored. Finally, the hazard removal process is implemented.

There are many ways to deal with hazards. Following are listed some common means for mitigating hazards and an example of each.

Engineering controls can be used to constrain or remove the hazard. For example, if the hazard is a hole through which a worker could fall, a guard rail system could be designed and installed. Engineering controls can also be imposed by the design team, as discussed above, when the designer considers the relative safety of various installation processes in the design of the element.

Administrative controls put in place processes and procedures that will protect people from the hazard. An example would be to put in place a rule on the site that any flammable material must be stored in a specific location remote from people and from a potential ignition system, and with firefighting equipment immediately available.

Isolation separates the hazard from people or equipment that might either trigger a problem or be subject to damage or injury should the hazard cause a problem. For example, electricians practice "lockout-tag out" procedures on a switch that has been disconnected to turn off power to an otherwise live electrical line while an electrician works on the line. Lockout physically locks the switch in the off position. A tag out places a notice on the switch that it is not to be turned on until the tag is removed by an authorized person.

Substitution uses a different item that is safer in lieu of one that is a hazard. An example might be using a different, more appropriate piece of equipment to execute a task. A larger crane might be brought in to set a piece of heavy equipment at a considerable distance where the combination of weight and reach would be at the limit of a smaller crane.

Process change devises a different way to execute what would otherwise be a dangerous process. For example, it might be found that in excavating a deep trench by slopping the sides to prevent cave-ins, one of the sloped sides might approach an adjacent heavy structure that puts a surcharge on the soil that could overcome the protection afforded by the sloped trench side. Changing the process in the area of danger to use a temporary retaining structure or a trench box to protect against the surcharge would mitigate the problem.

Work practices ensure that the safest practice is used in all instances. For example, in working on a ladder, a worker might have to stand on the top rung to reach the highest work area. Standing on the top rung of a ladder is a bad work practice. A better work practice would be to ensure that a long enough ladder is available to be able to work from the top allowable rung (generally two below the top).

Training should be provided for any procedure that is not routine and commonly performed by the specific worker tasked with an activity. For example, if a worker will be using a type of power tool for the task that is new to her, she must be trained in the proper use of the that tool.

A great variety of *personal protective equipment* (PPE) is available to protect workers from a large variety or work hazards. The supervisor needs to ensure that the appropriate PPE is not only available, but it is used for a task that requires it. For example, protective goggles must be worn whenever there is any possibility that some contamination might enter the eye.

DEVELOPING A MENTALITY OF SAFETY

The other major cause of accidents on the job site, in addition to unsafe conditions, is unsafe behavior by construction workers. There are many effective tools that can be employed to improve the safety behavior of workers. A positive approach is best. This includes creating and maintaining awareness of unsafe practices and the dangers associated with them. It also involves rewarding safe behavior. However, sometimes enforcement is required in the form of reprimands and punishments, or even removal from the job. Chapter 6 provides some guidance and cautions when a supervisor needs to reprimand or remove a worker from the job.

Habitual Safety Practices

One of the most effective tools used to improve safety in construction is developing an attitude toward safety that is so strong safety becomes second nature. An example of such a habitual practice experienced by most people is buckling a seatbelt when getting into the driver's seat of an automobile. This practice is taught by parents from a child's early years both by example and by making sure children's seatbelts are buckled every time before driving anywhere. Most adults feel so bound by the practice of putting on a seatbelt that not doing so is uncomfortable.

Buckling a seat belt is an example of a habitual safety practice. It is ingrained. It is done without thinking. It is part of life 24 hours a day, 7 days a week. It is encouraged by the individual for everyone else, especially family and loved ones.

One key measure of whether a safety practice has become habitual is if it is practiced at home, when no one else is watching. For example, standing on the top rung of a ladder is recognized as unsafe. New ladders even have notices that the top rung is not to be used for standing. If I have a ladder out working on a home project and decide to reach a little higher by standing on the top rung just this once, being very careful, the practice is not habitual!

Examples of safety practices on the construction work site that should become habitual include the following.

Safety orientation for new workers is so important that it is required by OSHA, and has become a feature of any construction safety plan. The likelihood of accidents is higher for workers who have just come onto the site than for workers who have been on the site for some time. This is true whether the new worker is an apprentice or a highly experienced journeyman. When first coming to a construction site, a worker should anticipate safety orientation and should look for specific information from the orientation, including:

■ General rules of conduct expected for safety on that job
■ Safety procedures for the site, especially emergency procedures
■ Hazards unique to this site

Periodic safety discussions ("toolbox talks") have also become standard on construction jobs and should be expected by all workers. These talks maintain awareness of the importance of safety on the site and create awareness of specific dangers now existing on the site or associated with upcoming processes. These discussions should be used as an opportunity by both the supervisor and the workers to discuss current safety issues that workers might have.

Activity safety planning should be built into the planning process for all activities and tasks. Special emphasis on safety is given when an upcoming activity is particularly hazardous, such as the replacement of utility poles on an energized high-voltage power line.

Safety training should become automatic and be expected by workers when preparing to execute any nonroutine task or when preparing to use dangerous equipment.

Continual vigilance should be an engrained habit for any construction worker, both for their own work environment and activities and for those of workers around them.

Safety monitoring should be in place any time a monitoring device is available, or any time a condition might be present that could be detected by a monitor. If the monitoring device is not available, it should be requested by the worker.

Safety reporting is an important supervisory task related to keeping the daily log and developing any periodic job progress reports. Safety reporting should become a routine part of any worker's daily regimen whenever an unsafe condition or behavior is detected.

Emergency preparedness is a very important part of habitual safety practices. When an emergency occurs, an expedited response is required. This means one must act very quickly in a controlled manner that has been predetermined. There is no time when an emergency occurs to figure out what to do.

There are many ways in which to develop habitual safety practices. Habitual safety practice should be taught starting on the first day a new worker enters the construction workforce. Apprenticeship training, formal or informal, should emphasize safety. Workers should be taught what to expect from the employer, but they also must be taught that safety begins with them. Individual workers must take primary responsibility for their own safety and then take responsibility for the safety of others around them. Safety discussions, safety signage, safe work practices, safe equipment and tools, and a safe working environment all contribute to development and maintenance of habitual safety practices.

Recognizing and Correcting Unsafe Behavior

People failure is at the heart of construction accidents. Those accidents that are not caused directly by human beings could have been prevented by human intervention at some point in the chain of events that led up to the accident.

Rules, regulations, and procedures that have been developed over the years to improve safety on construction sites should not cause individuals to lose sight of the fact that they need to take personal responsibility for their own safety. Safe behavior is the result of a safe attitude. With such an attitude, workers will say:

- I take responsibility for my safety.
- I want to make the workplace safe for everyone.
- I appreciate safety comments and suggestions from others and I will learn from them.
- We will work together to make our workplace safer.

Unsafe behavior is the result of either deliberate or inadvertent acts. A deliberate act might be based upon an attitude. For example, the worker feels he is not portraying himself as a "rough and ready" construction worker if he puts on PPE or has to get training on a piece of equipment. It also might be the result of not properly assessing the dangers of a situation, thinking, "I can get away with it this time." An example might be that the worker thinks that he does not need to wear his hardhat in this situation because the situation is not dangerous, and the hardhat is uncomfortable and it gets in the way.

Deliberate unsafe acts cannot be tolerated without at least an informal reprimand. The response to the offense becomes stronger for repeated and more dangerous infractions, rising to the level of suspension or firing for the most egregious and dangerous violations.

Policies, either at the job level or at the company level, should lay out clearly what behaviors are not allowed and the response to such behaviors. All workers should be trained so that they clearly understand what behaviors are not allowed and the responses that misbehavior will draw. In the event such a behavior occurs, the supervisor must take on the role of enforcer and make sure the policies are enforced. Such enforcement must be consistent and carried out each time an infraction occurs. This consistency is important both to reinforce habitual behavior and to protect the supervisor and the company from possible legal action should a reprimand or suspension be issued in an inconsistent manner.

An inadvertent, unsafe act is often the result of the worker not hearing or not understanding what is required. When confronted with the inadmissible behavior, the response might be, "I thought that is the way I was told to do it," or, "I did not realize this would happen."

Inadvertent unsafe behavior is a systemic problem. To eliminate inadvertent unsafe acts, policies, and procedures need to be modified. This modification begins with a root cause analysis to determine what allowed the offensive behavior and then making a change to the system so that such causes are eliminated to prevent future problems.

EFFECTIVELY DEALING WITH SAFETY EVENTS

Supervisors must have a well-defined procedure to provide guidance when accidents happen. Any incident or accident should elicit an immediate response to determine if it constitutes an emergency. If it is an emergency, the emergency response procedure is triggered. If it is not determined to be an emergency, the event requires review. Based upon a determination of the urgency and severity of the event, the response will be different.

Immediate Response to an Accident

If the event is not urgent and does not fall into the category of emergency, any medical issues should be taken care of first. This may involve no more than the application of a simple remedy that can be applied on site so that the worker can return to work. For example, a scrape or minor cut may require nothing more than cleaning the injured area and applying a disinfectant and small bandage. A quick note can be made in the log and nothing more is required.

If there is any question at all about the severity or consequences of the injury, first aid may be required and medical assistance should be engaged. For less severe injuries, this may require a trip to the nearest urgent care facility. As the severity grows, the decision will need to be made to summon medical assistance. The safety plan that is in effect, either at the contractor's level or at the project owner's level, should include well-defined procedures for handling injuries and the supervisor should be very familiar with what they are.

In the event that the accident rises to the level of an emergency, the safety plan should have well-defined procedures in place. Emergencies require expedited response, not impulsive response. That is, response to an emergency should be swift and well thought out, not just quick without thought.

As a general rule, the first step in an emergency is to activate the emergency response system. On many jobs, this entails calling 911, but on larger jobs, especially those located on a campus, there might be an onsite emergency response team that is to be contacted in the event of an emergency. If the event involves structural damage or an unsafe work environment, the environment needs to be isolated and stabilized so that responders do not become victims. For example, in the event of a cave-in that has buried a worker, the excavation must be stabilized prior to trying to reach the injured worker so that a further cave-in does not injure rescuers, or further injure the victim. Once the site of the accident is stabilized, trained first aid providers can begin to stabilize and treat the victim.

While the site and injured party are being stabilized, concurrent activities should be carried out by others. Someone should go to an appropriate location to intercept emergency workers and lead them to the accident location. Others onsite should be notified of the accident. Others offsite should also be notified, such as the worker's home office.

Follow-Up Response to an Accident

If the accident results in a recordable injury, as defined by OSHA, a report must be written. In any event, an investigation should be carried out for any incident or accident with the objective of determining root causes and making adjustments to ensure that such an accident will not happen again. The primary objective is to determine root causes, not to place blame. If there is a person responsible for the accident, it will come out during the investigation and appropriate action can be taken. The objective of an accident investigation is fact finding, not fault finding. The purpose is to learn and make changes so that such safety-related events do not occur in the future.

SUMMARY

In this chapter, the following key points have been presented:

- Safety is of fundamental importance to the construction supervisor.
- Safety is important to all construction stakeholders for many reasons.
- Accidents are very costly.
- The supervisor has the dual responsibility to maintain a safe working environment and to eliminate unsafe behavior.
- All stakeholders contribute to creating a safe work environment.
- Each worker should develop habitual safety practices.
- Well-designed and -practiced procedures need to be in place to address a safety-related event.

Learning Activities

1. Effectively using the toolbox talk

   Regular toolbox safety meetings are a requirement on most jobs. Yet it is difficult to maintain interest and a sense of urgency in these meetings. They become routine and lose their impact.

   Develop a list of means to maintain interest and a sense of urgency in toolbox meetings. Examples of means might include:

   Make each talk relevant to the work the crew is currently engaged in.

   Pass around responsibility for each meeting to different workers on the crew.

   (continued)

The product of this exercise should be a list of means to reinvigorate the periodic toolbox talks. It can be used by the supervisor as a guide in how the supervisor can transform these routine meetings into a vital and effective element of the job site safety program that will support safe behavior and a safe work environment. If provided to the company safety officer, it could become a part of the company safety program, disseminated across the various jobs, and could become a living document that can be continuously improved.

2. Developing an attitude of safety

A key responsibility of the construction supervisor is to cultivate an attitude of safety in the workers. Examples of an attitude of safety include:

I take responsibility for my safety.

I want to make the workplace safe for all.

I appreciate safety comments from others and will learn from them.

We will work together to make our workplace safe for all.

The objective of this exercise is to devise ways in which the supervisor can cultivate an attitude of safety in their workers.

First, expand the list of characteristics of an attitude of safety given above.

Next consider how these might be cultivated in workers. If other supervisors are available, this is a good team exercise.

Finally, develop a report to your company safety officer that summarizes the expanded list and describes how you intend to work on developing an attitude of safety among the workers on your site.

CHAPTER 10

THE CONTRACT AS A MANAGEMENT TOOL

INTRODUCTION

The contract is a very powerful tool that can be used by supervisors at all levels to more effectively manage their work. There are many misconceptions about contracts, especially held by workers in the field, such as:

- It has little to do with field operations.
- Legal training is required to understand it.
- It is for the use of a higher entity to control my work.

The objective of this chapter is to dispel many of these myths and to enable the supervisor to begin using the contract as the very powerful tool it is to better manage and control their work. This will be accomplished by first providing a practical understanding of the contract, then by describing what the contract can do for the supervisor and finally, by laying out how to begin using the contract to exercise more control over the work.

An example copy of a commonly used agreement: ConsensusDOCS 750: Standard Form of Agreement between Contractor and Subcontractor™ is included as Appendix I and will be referred to extensively throughout this chapter. This example document has been chosen because it represents one of the newer forms of agreement; one that was developed over a considerable period of time by representatives of designers, owners, contractors, and subcontractors (the DOCS part of ConsensusDOCS™). Being an agreement between the contractor and the sub-contractor, it illustrates points relevant to supervisors in both categories.

A PRACTICAL UNDERSTANDING OF THE CONTRACT

A contract is a legal instrument. However, it is also a very useful tool to organize complex work into a manageable form. Although the legal ramifications of a contract cannot be overlooked, the intent of this chapter is to focus on the use of the contract as a management tool. The objective is to enable supervisors to understand enough about contracts so that they can use the contract to better control their job and so that they can recognize when contractual problems are arising that require the help of others more qualified in the use of contracts.

The contract accomplishes many things throughout the project. It establishes a business relationship between, or among entities engaged in the project. It defines the responsibilities of each of those entities and defines the scope of work to be performed by each. It sets forth quality standards for the work, either in terms of the end product itself or in terms of how the end product is to perform. It deals with many process issues, such as when work is to be done and how changes to the work are to be handled. It defines financial processes, including who gets how much money and when, as well as the processes regarding how that money is to be paid. It establishes the ground rules for execution of the project. It is intended to resolve problems, such as determining what happens when something unanticipated comes up and how to proceed when something goes wrong.

Much of what the contract deals with takes place in the field or is impacted by field operations. For example, ConsensusDOCS™ Section 4.5 defines responsibility for assigning lay-down areas. The location and size of lay-down areas is generally negotiated between supervisors in the field and has significant impact on field operations throughout the duration of the job, so it behooves the supervisor to know what the contract says about lay-down areas.

Sometimes the contract contains a number of different references relevant to a field situation. For example, communication protocols in the field are established by at least three references in the ConsensusDOCS™. Section 3.6 states that the *Subcontractor shall direct all communications related to the Project to the Contractor*. Section 3.11 defines who the subcontractor's representative is and states that *this representative shall be the only person to whom the Contractor shall issue instructions, orders or directions, except in an emergency*. Section 4.1 defines who the contractor's representative is and states that *The Contractor's representative shall be the only person the Subcontractor shall look to for instructions, orders or directions, except in an emergency*. These three, taken together, establish a very clear, strong, and specific line of communication in the field. This is clearly information any supervisor must have to properly manage the work and relate to other entities on the job.

Other important field issues dealt with in the contract include definition of safety requirements for all parties to the contract and establishment of responsibilities for planning and scheduling the work as well as for coordination among the trades, especially when the work site becomes congested.

Contract: A binding agreement between two or more persons or parties; especially one legally enforceable.

Merriam-Webster Dictionary

The key points of this definition are very clear: an agreement has been reached and each party will abide by the agreement.

Generally, contracts are in writing, but the basic definition does not require that they be written. Writing the contract helps to make it enforceable. If it is written, there is an expectation that the parties understand and agree to the same thing. Also, if it is written, it will not change over time unless modified by the mutual agreement of the parties. If a dispute arises, a third party could read a written agreement and help resolve the dispute, whereas if it is oral, it becomes a discussion of what one party or the other said or thought they heard.

Thus, written contracts are preferable to oral agreements. Even though oral agreements can still meet the basic requirement of the definition of a contract, *an agreement between parties*, it becomes very difficult to enforce an oral contract.

Oral contracts are often found in construction, especially in the informal environment of the work site. Since legal enforceability becomes a problem if the agreement is based only on the spoken word, it is advisable that all agreements, other than the most routine, be written out and signed by the parties. If an oral agreement is reached in the field it is good practice to confirm it in writing.

Field Authorization for a Change

Supervisors are often asked by the customer to install work different from what is shown on the drawings. For example, an electrical foreman might be asked by the general contractor's superintendent to relocate an electrical panel in order to be able to accommodate a new opening in the wall where the panel was supposed to go. In this case, it is a simple task and can be accomplished fairly quickly and easily. The foreman, wanting to avoid confrontation, reassigns a crew to make the change.

Later, it is determined that there was some cost involved, primarily in terms of time to accomplish the work and disturbance to the crew that was already engaged in another activity. When the project manager submits a change order proposal that includes this additional work, the proposal is rejected because there is no documentation.

This is such a common field situation that a very common practice in the field has been developed to accommodate it. When a change to the work is requested, a field authorization form can be filled out and signed by both parties. This form was discussed in Chapter 4 and an example was given as Figure 4.4. The form has blanks to fill in that briefly describe the work requested, the requesting party, and the date and time of the request. Then the work can continue in the field. Later a new written agreement (contract modification) can be used to work out details, like payment for the additional work.

REQUIRED ELEMENTS FOR AN ENFORCEABLE CONTRACT

To expand on the enforceability of the contract, certain elements are required for a contract to be enforceable.

Meeting of the Minds

Meeting of the minds, also referred to as mutual consent, is required to show that the parties actually have agreed on the same thing.

Offer and Acceptance

Offer and acceptance means that one of the parties has made an offer that has been accepted in whole by the other party. A counteroffer is not acceptance. It is a new proposal. In negotiating a contract, offers can go back and forth as many times as required to reach an agreement, but the agreement does not become a binding contract until one party accepts the other's offer without exception.

Mutual Consideration

Mutual consideration establishes that there has been an exchange of value from each party to the other. The value exchanged does not have to be equal, but there does need to be some value exchanged and received on each side.

Performance

Performance or delivery establishes that the action contemplated by the contract must be completed. If the action(s) defined by the contract is (are) not executed, the contract is broken or invalidated.

Good Faith

Good faith requires that each party be honest and not attempt to deceive the other party. The purchaser of a Mustang believes he is buying a car, not a horse.

No Violation of Public Policy

No violation of public policy is allowable in a legal contract. Laws, whether federal, state, or local, prohibit some types of contracts, and hence, even though the contract has all of the elements enumerated above, it cannot be legal and enforceable if it runs counter to the law. For example, many cities require that construction projects over a certain contract amount be procured through the bidding process. A contract between the city and a contractor to execute a large project with procurement based upon a negotiated, rather than a bid, process would not be a legal contract.

COMPONENTS OF THE CONSTRUCTION CONTRACT

The construction contract is extensive and complex. It starts with *the agreement*, which is what many call the contract, but which is really only one of several elements of the contract. There are many standard forms of agreement used in the construction industry. Some of the oldest and most recognized are those published by the American Institute of Architects. Others are published by various organizations, like the Associated General Contractors of America, or the Construction Owners Association of America. More recently, forms of agreement for construction have been developed by coalitions of construction-related organizations with the expectation that an agreement developed by a broader cross-section of the industry will be less biased and more widely accepted. One such coalition involving designers, owners, contractors, and subcontractors has produced the ConsensusDOCS™, already referenced in Appendix I.

In addition to the many general forms of agreement used in the industry, it is not uncommon for owners who engage in a significant amount of construction to develop their own forms of agreement. Many higher-level contractors also craft their own company specific subcontract forms of agreement. Furthermore, most contractors have such forms as purchase orders or field authorizations that become contracts when accepted by the other party, as confirmed by a signature.

Although supervisors are subject to project-level contracts, they rarely get involved in negotiating agreements at that level and often do not see them. On the other hand, they might use purchase orders to buy materials or field authorizations to document a requested change in the work. Thus, more often than not, supervisors do get directly involved with contracts of some kind or another.

Agreements at the project level generally start with a fill in the blank section that enables definition of the parties, the project, the date, and any other information relevant to a specific job.

The next section often defines the contract documents to be included within the contract for the specific job. Among those documents, in addition to the agreement will be:

- Conditions of contract
- Drawings
- Specifications
- Addenda
- Modifications

In the case of a subcontract, one additional item is typically included in the subcontract documents that is very important to the subcontractor, and that is the prime contract that the contractor has with the owner.

Other common elements brought into the contract by reference in the agreement are codes that may be specific for a certain construction discipline or a general building code that governs overall construction for projects in a given geographic area. Finally, there is typically a place in the agreement that allows identification of other documents that will become part of this specific contract. All of these documents govern how the project is to be built, and, hence each is important to the supervisor.

It is clear that the set of construction contract documents is very extensive and complex. These contract documents tell the supervisor what to construct and how the project will run. It is, therefore, very important that the supervisor has a complete set of contract documents for her/his portion of the project and knows how to read, understand, and use these documents. In the event the supervisor does not want the full contract or the supervisor's manager does not want to provide the full contract, it becomes very important for the manager to provide the supervisor extensive information from the contract to support the work of the supervisor.

CONTRACT COUPLETS

It can be helpful in understanding an agreement to remember several things that go together. These will be called *couplets*.

Rights and Responsibilities

A major part of what a contract does is to define the rights, opportunities, and protections; and the responsibilities, obligations, and risks assigned to the parties. To the extent possible, the rights and the responsibilities need to be balanced. That is, if a responsibility is assigned a contractor, that contractor should have the rights that go along with that responsibility. If a risk is allocated to a contractor, the appropriate protections should also be assigned to that contractor.

Construction contracts are hierarchical. That is, there is a prime contract, with the owner, and then a series of subcontracts and sub-subcontracts that assign much of the work to lower-tier, specialty contractors. Often, a higher-tier contractor will write a subcontract such that the protections afforded by their contract remain with the contractor while the subcontract allocates as much risk as possible down the chain to a lower-tier contractor. When negotiating a contract, it is important, to the extent possible, to pass rights, opportunities, and protections in the higher-tier contract that relate to the responsibility assigned to the subcontractor through to the subcontractor.

For example, the contract documents, specifically the drawings and specifications, define the work to be performed by the subcontractor. The subcontractor has the responsibility to develop means and methods to accomplish this work. However, the work is not done in a vacuum. The ConsensusDOCS 750™ Article 5.2 recognizes the primary responsibility of the contractor to develop the project schedule, but it also affords the subcontractor the right to participate in development of the project schedule:

In consultation with the Subcontractor, the Contractor shall prepare the schedule for performance of the Work (the Project Schedule) and shall revise and update such schedule, as necessary, as the Work progresses.

Many contractors prefer to develop the schedule and hand it to the subcontractors. Sophisticated contractors have learned that, not only is it important to the subcontractor to be involved in project scheduling, but it is advantageous to the contractor to maintain this right for the subcontractor.

Written and Unwritten Language

Contracts are long and complex, and there can be many problems buried within the contract. Those experienced in reading construction contracts recognize certain terms and phrases that raise a red flag as a problematic element of the contract. For example, one such term might be *indemnification* or the associated phrase *hold harmless*. The supervisor seeing those words might decide to refer this contract element to the project manager and ask for a simple explanation of what it means in the context of the current project.

However, words that are missing may be just as problematic, but much more difficult to detect. A missing word or phrase can significantly change the meaning and impact of a contract clause.

The effect of missing contract language can be illustrated by the following example. ConsensusDOCS 750™ Article 5.2 states that:

The Contractor shall have the right to determine and, if necessary, change the time, order and priority in which the various portions of the Work shall be performed and all other matters relative to the Subcontract Work. To the extent such changes increase Subcontractor's time and costs, the Subcontract Amount and Subcontract Time shall be equitably adjusted.

The last sentence protects the subcontractor, so that in the event the project schedule is revised or day-to-day coordination is modified in such a way as to be detrimental to the subcontractor, the subcontractor can apply for and expect to be awarded monetary compensation and, if needed, additional time to complete the work without suffering loss. On the other hand, if the last sentence is removed, the contractor can micromanage the subcontractor's work, causing a substantial increase of cost to the subcontractor without being liable to reimburse the subcontractor.

Helpful and Harmful Language

Some language in a contract is helpful to the party reviewing the contract and other language is potentially harmful. The negotiating process for a contract enables the parties to tailor the agreement, somewhat, to eliminate harmful language and include helpful language.

A typical clause that is debated in negotiating subcontract language is a *contingent payment* clause that makes payment to the subcontractor contingent upon something other than successful completion of the work. The contingency is often based upon payment received by the contractor. Subcontractors feel that contingent payment clauses are very harmful and they work hard to get them removed from the contract. Contractors feel that the protection of a contingent payment clause is essential to their company and work hard to make sure it is included. The contingent payment clause is a good example of helpful and harmful language, because it represents helpful language to the general contractor but represents harmful language to the subcontractor.

READING THE CONTRACT

A construction contract can be quite daunting to someone who has never read one before. However, contracts are typically written in plain language and, with some study, much can be gleaned from the contract by construction professionals other than lawyers. Once one becomes reasonably familiar with a commonly used agreement other agreements become less daunting because construction contracts tend to deal with the same issues and have common patterns and use common terms. To illustrate commonly found contract sections, references will be made to ConsensusDOCS 750™, which will be found in the appendix.

A checklist can provide valuable assistance in reading contracts. An example of a checklist is provided in Figure 10.1. This checklist can be used to provide initial help

---

Contract Checklist

Who is your contract with?

What type of project delivery system is defined?

What is the pricing basis?

If other contracts are incorporated, which takes precedence?

What lines of communication are established?

What procedures are established for:

  Requests for Information

  Field authorizations

  Notifications

  On site material storage

Who is responsible for scheduling and coordination?

What safety requirements are established for this project?

Who is responsible for cleanup?

Are there liquidated damages?

Figure 10.1   Contract Checklist

in reading a contract, but as the supervisor becomes more familiar with contracts, she will want to modify this standard checklist to create a personalized checklist.

The first section of an agreement will typically have a number of blanks that allow the specific project and the parties to the contract to be identified. It also allows for entering other information like the contract date. This is found in Article 1: *Agreement* in the ConsensusDOCS 750™.

The next section typically defines what comprises the full set of contract documents (as described in a previous section). In the event of a subcontract, this section will often deal with several issues in the prime contractor/subcontractor relationship. For example, it is highly likely that in merging two complex contracts, there will be instances when they are in conflict with one another. The subcontract will often define which governs: the prime contract or the subcontract in the case of such a discrepancy. Article 2.4: *Conflicts* in the ConsensusDOCS 750™ has such a statement.

The ConsensusDOCS 750™ addresses another very important aspect of the prime contractor/subcontractor relationship, this time in Article 2.3: *Subcontract Documents*. It gives the subcontractor the right to obtain a copy of the prime contract, since the prime contract is now incorporated into and becomes a part of the subcontract. This is an essential right to be protected by the subcontract since contractors are very reluctant to give a third party (the subcontractor) a copy of their contract with the owner. A version of this clause is discussed in more detail later in the chapter when dealing with proper use of the contract to better manage the job.

The next sections typically define responsibilities of the parties. For a subcontracting situation, the first section often deals with mutual rights and responsibilities and it will often incorporate a *flow-through or pass-through* clause that has the purpose of flowing through the rights and obligations that exist between the prime contractor and the owner to the relationship between subcontractor and the prime contractor. This is found in Article 3.1: *Obligations* in the ConsensusDOCS 750™. If properly written, this flow-through clause can be of significant benefit to the subcontractor. However, if the subcontract is written by the prime contractor rather than using one of the more general forms, the flow-through clause often flows down to the subcontractor as much risk as possible, retaining as many protections as possible for the prime contractor. This type of flow-through clause is detrimental to the subcontractor.

Again, for a subcontract, there will be a section that defines the responsibilities of the subcontractor with respect to the contractor. This is found in Article 3: *Subcontractor's Responsibilities* in the ConsensusDOCS 750™. This section lays out in detail what the expectations are for the subcontractor. It deals with issues such as safety, coordination, and the responsibility of the subcontractor to supervise and direct the subcontractor's work.

There is also a section that defines the responsibilities of the contractor. This is found in Article 4: *Contractor's Responsibilities* in the ConsensusDOCS 750™. This is of critical importance to the subcontractor's supervisor, because it deals with many common field situations, such as lay-down areas, terms of notification in the event of unacceptable work, and project scheduling and coordination.

USING THE CONTRACT

As just seen, a great variety of field situations for which the supervisor is responsible are dealt with by the agreement. Many of the chapters of this book deal with topics that are critical to the supervisor that are also addressed in the contract. Among them are:

- Safety
- Communication
- Planning and scheduling
- Cost control
- Managing production

It is imperative that the supervisor know and understand what the contract says about these and other field issues. Knowing and using the contract to supervise the job will support success for the project. On the other hand, being unaware of, or unknowledgeable about, the contract will be detrimental to the success of the project and will impair the ability of the supervisor to successfully plan and execute the work.

Proper use of the contract requires action. The contract cannot effectively be executed in a passive way. For example, the contract provides the parties rights, opportunities, and protections; however, it does not impose these upon either party. One must exercise the rights given or they go away. To illustrate, AIA Document 401 - 1997: *Standard Form of Agreement between Contractor and Subcontractor* developed by the American Institute of Architects protects the right of the subcontractor to have a copy of the prime contract since it is incorporated into the subcontract. On the title page, it states that "A copy of the Prime Contract ... has been made available to the subcontractor." In Article 1.4, it states: "The subcontractor shall be furnished copies of the Subcontract Documents upon request." The subcontractor must "avail herself" of the right to have a copy of the prime contract. The subcontractor must request a copy of the subcontract documents (which include the prime contract) from the contractor. Supervisors must understand that they are required to take action to exercise or protect their rights. They do not automatically happen and can go away if not exercised.

Properly Using the Contract

Proper use of the contract involves several steps. First a copy of the contract or relevant parts of the contract must be available to the supervisor. Lower-level supervisors, those at the foreman level, generally do not have, and do not want to have, a complete copy of the contract. However, if they know what contracts generally deal with, they can ask their supervisor to provide specific information on what the contract for this project says about critical situations that affect their work. On the other hand, it is important for higher-level supervisors and managers to make sure that lower-level supervisors have all of the relevant information provided by the contract and that they understand how to use that information.

Second, the supervisor must be familiar with contract requirements for the specific job and understand what the contract means. Having the contract or specific elements of the contract that are relevant to the foreman's responsibility does no good if the language is not understood. The supervisor must not only understand what the contract means but also understand how to use the terms and conditions of the contract to better manage and control the work. Part of supervisory training should focus on practical use of the contract in the field.

Third, the supervisor must be able to assert rights provided by the contract. This assertion must be in a nonconfrontational manner. Supervisors must be able to ask for the rights provided and resist requests or directions that are contrary to the contract. This may be through the use of standard company forms and procedures, such as the use of a company field authorization form to document additional work requested by the customer. It might also involve negotiating or bartering. Bartering is of fundamental importance to supervisors and can be demonstrated by the following example.

Bartering to Protect Value Provided by the Contract

In Section 4.5, dealing with storage areas (lay-down areas), the ConsensusDOCS 750™ says: *Unless otherwise agreed upon, the Contractor shall reimburse the Subcontractor for the additional costs of having to relocate such storage areas at the direction of the Contractor.* When asked to relocate the lay-down area, the supervisor can request a field authorization to provide the documentation for a change order to get the reimbursement provided by the contract. However, contractors often do not like to pay for such things so instead, the subcontractor's supervisor can refer to section 3.25, which says: *The Subcontractor, its agents, employees, subcontractors or suppliers shall use the Contractor's equipment only with the express written permission of the Contractor's designated representative and in accordance with the Contractor's terms and conditions for such use.* If the subcontractor's supervisor responds positively to the request to relocate the lay-down area and then asks to use the contractor's crane the following week for an hour to offload a heavy piece of equipment, the contractor's representative would be hard pressed to deny the request after just requesting and receiving something of value from the subcontractor's superintendent. Such a request will generally be approved, and the supervisor has earned the value provided in the contract for relocation of a lay-down area while avoiding confrontation and also avoiding a great deal of paperwork and further negotiation down the road.

Functions of the Contract

In discussing use of the contract, it is very important to keep several functions of the contract in mind. One function of the contract is to provide guidance in the event an unanticipated or unknown situation arises, or in the event that something goes wrong. For example, contracts specifically assign responsibility to keep the project clean. This is normally the responsibility of the specialty contractor for work performed by that specialty contractor. Sometimes, the prime contractor will find that the workplace is not kept clean. The ConsensusDOCS 750™ Article 3.13.2 addresses this, as follows:

> If the Subcontractor fails to commence compliance with cleanup duties within two (2) business days after written notification from the Contractor of non-compliance, the Contractor may implement appropriate cleanup measures without further notice and the cost thereof shall be deducted from any amounts due or to become due the Subcontractor in the next payment period.

Such a clause protects the contractor from the situation where the subcontractor does not perform as agreed upon in the contract. However, the notification requirement also protects the subcontractor from the situation where the contractor perceives a deficiency and takes action to make good that deficiency without notification to the subcontractor giving the subcontractor an opportunity to correct the deficiency or explain why he believes it is not a deficiency. This subcontractor protection should protect the subcontractor from blind back charges from the contractor.

Notification is a particularly useful example of protections provided by the contract. The supervisor must be aware of what the various terms of notification are. They relate to such areas as cleanup and nonconforming work. When notified, he must take two actions. First, he needs to immediately begin to deal with the cause of the notification. Second, he must inform the main office that a notice has been received, so that the main office is aware of a potentially damaging situation that might be coming up if not addressed in a timely fashion.

Another function of the contract is to assist the parties in interpreting what the contract itself means, should there be ambiguities. An example of this is found with the flow-through clause that was discussed above. The purpose of a flow-through clause is to tie together two very complex contracts found in the hierarchy of contracts underlying a construction project. It is likely that instances will arise when a subcontract and a prime contract, tied together either by incorporating the prime contract into the subcontract by reference or by use of a flow-through clause, are in conflict.

A common statement found with most flow-through clauses defines which governs in the event of a conflict between subcontract and prime contract. When a conflict arises in the field in which the contractor's superintendent refers to the prime contract, and the subcontractor's supervisor refers to the subcontract, it is important to know which governs. The first thing to do is to see what the contracts say. In most cases, they will have a clear statement that resolves the conflict. For example, ConsensusDOCS 750™ Article 2.4 states that:

In the event of a conflict between this Agreement and the other Subcontract Documents, this Agreement shall govern.

Thus, ConsensusDOCS 750™ establishes that the subcontract governs in resolving a discrepancy between it and any other contract on the project.

In the event that there is no clear statement, there is legal precedent that holds that the stronger contract is the one that is closest to the situation, or the one that both parties have signed. Hence, if no clear statement occurs in the contract documents, the subcontract takes precedence over the prime contract in the event of a conflict.

It should be noted, however, that even though there is legal precedence, a well-written contract will recognize when a conflict is likely to come up and make a clear statement. It should also be noted that legal precedence can be overturned by a specific statement in a contract. Thus, if the contract says that the prime contract governs, the legal precedent is overruled in this case and the prime contract governs.

Finally, a fundamental purpose of the contract is to avoid and resolve conflicts, not to create and foster conflicts. Supervisors must know what is in their contract (and other related contracts incorporated by reference) and must be able to negotiate based upon that knowledge. However, it is easy for discussions about contracts to become emotional with tempers flaring. It is very important that this does not happen. If supervisors reach a point in the discussion where they wave the contract in the face of the other party and yell, "I will see you in court!" the situation is lost and no party wins. The contract should be used to avoid confrontation and avoid litigation. If supervisor becomes emotionally involved, they must hand the situation off to another and step out of the discussion.

In terms of resolving conflicts, it should be remembered that contracts deal with the strength of a legal position. In evaluating a situation based upon what the contract says, the consideration is who is in the stronger position and who is in the weaker position. Contracts are not about right or wrong. They are not about good and evil. They are about being supported by the contract or not being supported by the contract.

In working to resolve a situation based upon the contract, the question the supervisor needs to consider is: Am I in a stronger legal position or a weaker legal position in this situation? If I am in a stronger legal position, I have the option to exercise my contractual right, or I may decide that it is not worth the time, the money, or the potential damage to the relationship with the other party to pursue my rights. If I am in a weaker position, my option is to cut my losses and not throw good money (or good will) after bad. Drop the discussion. Execute what is required and get on with the job.

SUMMARY

In this chapter, the following key points have been presented.

- The contract is a very powerful tool for the supervisor.
- The agreement binds together a complex set of construction documents into the construction contract.

- Supervisors at all levels can benefit from a working knowledge of the contract governing their work.
- Oral agreements need to be followed up in writing.
- Using the contract requires action. Rights, opportunities, and protections can be lost if not used.
- Contracts determine a stronger or weaker legal position. If you are in a stronger legal position, you have options. If you are in a weaker legal position, the option is to cut your losses.
- Contracts are meant to remove conflict, not create conflict.

Learning Activities

1. Using the contract to protect a right

   One of the most common abuses of subcontractor rights is for the prime contractor superintendent to provide an unsuitable lay-down area to the subcontractor and then to require the subcontractor to repeatedly move materials to another location. The objective of this exercise is for a specialty contractor supervisor to gain an advantageous lay-down area initially and to recover costs of later moves. The current situation is that the prime contractor superintendent has assigned an inadequate lay-down area initially and then will require you to move your materials to another lay-down area throughout the project.

   From the point of view of the specialty contractor, review relevant contract clauses in ConsensusDOCS 750™ that provide the right to a suitable lay-down area and to reimbursement for movement from that area. List characteristics of what you consider a suitable lay-down area. List options available to you to collect the value promised for moving your lay-down area. Finally, prepare a case in two parts as the basis from which to negotiate with the prime contractor for your initial lay-down area and for your right to reimbursement for subsequent moves. Part 1 will cover the negotiation for the initial lay-down area. Part 2 will prepare you for negotiations when asked to relocate your lay-down area.

2. Participation in planning and scheduling work

   The supervisor can exert considerable control over tactical (within a limited time frame) work scheduling through active participation in weekly coordination meetings. The objective of this exercise is for a specialty contractor to use your contract rights to give you better control over short-interval planning of your work. The current situation is that the prime contractor unilaterally assigns work based upon his schedule and priorities.

   From the standpoint of the specialty contractor, using ConsensusDOCS 750™ contract clauses relating to scheduling and coordinating the work, review the relevant clauses giving you the right to participate in planning and coordination and the responsibility to plan and coordinate your own work. Identify benefits to all participants to jointly plan the work. Identify loss to the job of not jointly coordinating the work. Define what you need in order to be able to effectively plan your own work. Finally, prepare a case from which to negotiate with the prime contractor to secure the right to plan your own work and to participate in planning the coordination of other work activities that have an impact on your work.

CHAPTER **11**

MANAGING PHYSICAL RESOURCES

INTRODUCTION

In his management of a construction project, the supervisor is responsible for the management of the resources that will be utilized to construct the project. While there are certainly a number of others involved who will provide their input and assistance, the supervisor should understand that the day-by-day management of the set of resources that will be applied to the completion of the construction project, is his responsibility. Included in these resources are the following: manpower; materials, tools, and equipment; the construction site itself; dollars; and time. This chapter will begin to develop an understanding of some tools and techniques for managing these resources. Information will follow in successive chapters that will deepen this understanding.

MANAGING MANPOWER

The talented and skilled people who build construction projects are the most valuable, and the most variable, and the most complex resource whom the supervisor manages. These people compose the labor force on a construction project, and the wages they earn constitute the labor costs for the project.

As has been noted previously, labor costs are a significant fraction of the total cost of performing every construction project. On building construction projects, labor costs typically are 50 percent or more of the total cost of the project.

The effectiveness with which a supervisor manages the craft labor workers who perform the skilled work to construct the projects will have an enormous impact upon the success of those projects and, therefore, will in large measure define the effectiveness and the success of the supervisor. Managing people in the workforce entails the application of conceptual or human relations skills. These skills are decidedly different from the skills that most supervisors learned when they worked as craftsmen. Supervisors do well to understand that the longer they remain in a supervisory capacity, and the further they advance in management, the less they will rely upon their technical or craft skills and the more important their human relations and conceptual skills will become.

Chapter 2 listed and described a number of attributes and characteristics that people in the construction workforce have said they most admire in those who manage the work. Supervisors would do well to emulate those qualities, if they wish to be viewed in the same manner by those whom they supervise. Chapter 7 examined some other important aspects of the human and conceptual skills that are major components of success in managing manpower. Also discussed were some of the elements of motivation, and leadership, as well as some other important aspects of understanding human behavior. In this chapter, some additional important elements of manpower management will be presented.

As is the case with all other resources that are needed for the project, all of the skilled construction craft labor that will perform the work on the project will need to be procured. Depending upon the contractor's business structure, and also on the prevailing collective bargaining agreements that may be in effect in the company, the supervisor may or may not be directly involved in hiring decisions relative to the craft workers who will perform the work on his project. However, whether the hiring of craft labor is performed at the job site or at the company home office, all of the craft labor workers must be recruited, hired, oriented to company policies and procedures, provided safety training, and provided orientation to the job site and to the other workers on the site.

Manpower needs, in terms of both number of workers and the skills that the workers need to possess in order to perform the work at hand, will vary during the course of the construction of the project. The supervisor should plan to be a part of determining how many workers are needed on the project at any point in time, in order to assure that the correct numbers of craft workers and also the correct skill sets for performing the work are present on the construction site as the work is planned and performed.

During the course of construction on the project, it is the supervisor who will convey work assignments to all of the individual workers and crews of workers for each day of work throughout the duration of the project. This means, by definition, that the supervisor must be planning the conduct of the work, in both the short term and for the longer term, at all times. Chapter 2 defined and discussed the management function of planning. In Chapter 14, discussion will be provided with regard to the primary planning tool utilized by the supervisor, which is the construction schedule.

Planning the work, and assigning responsibility for the accomplishment of tasks and activities, means that the supervisor should be well acquainted with the craft workers who compose his crew(s), their personalities, and their capabilities. Knowing the skill levels of the various craft workers, and knowing which workers are most proficient and most productive at which tasks, factor into the supervisor's decisions regarding work assignments. In addition, matters of workers' personalities and disposition, and who gets along well with whom, as well as which crew members make up the best teams, in accord the information provided in Chapter 5, are also important elements of the supervisor's consideration.

Manpower management also means that the supervisor is responsible for upholding all company policies among all of the members of the workforce. Matters such as punctuality, and behavioral considerations, as well as discipline and reprimand as may be necessary, are matters among many others that the supervisor is responsible for. The supervisor must be well acquainted with the provisions of his company's policies and procedures, and must be willing to consistently apply those policies among all of the workers in the craft labor force.

The supervisor is also charged with keeping job site records relative to the members of the workforce. Throughout this book the importance of proper documentation has been repeatedly emphasized—it is a fundamental and absolutely vital supervisory responsibility.

Time cards will be completed daily by the supervisor for each craft worker who performs work on the project on that day. In addition to recording the actual number of hours worked by each craft worker, the supervisor will enter the appropriate labor cost codes for each element of each worker's labor on that day. Further consideration of this topic is provided in Chapter 13.

Additionally, the supervisor will maintain a job log, where he will make entries on a consistent daily basis, which relate to all of the workers on the project. The name of everyone who was present on the job site on each day is recorded in the job log. These entries are accompanied by notes regarding matters such as training conducted, injuries that a worker may have suffered, the occurrence of near misses, warnings, disciplinary actions taken, and a myriad of others. Any and all occurrences on the job site that affect any member of the craft labor workforce are recorded by the supervisor on a daily basis in the job log.

While the supervisor's responsibilities relative to the job log are discussed more fully in Chapter 18, as well as in other portions of this book, the summary guidance that is provided here is: *any* matter which the supervisor thinks may be important relative to anyone who worked on the job site on a particular day should be recorded in the job log. Some supervisors have been heard to say, "How do I know whether a certain matter is important enough that I should write it down?" The best guidance is: if the supervisor asks that question with regard to any matter, then likely the matter is of sufficient importance to merit being recorded, accurately and completely, in the job log. The job log is considered to be the primary record of everything that takes place on a construction project.

Safety Considerations

Safety planning, safety training, and the use of personal protective equipment (often referred to as PPE), as well as ensuring safety in the workplace in accordance with company policy and in keeping with the requirements of the law, are also the domain of the supervisor. Supervisors should be thoroughly familiar with all company safety policies. Additionally, they should certainly understand and know how to apply the provisions of the Occupational Safety and Health Act (OSHA), as well as the provisions of "29 CFR 1926, OSHA Construction Industry Regulations," which pertain to the kind of work that supervisors and their crews are doing.

Materials Safety Data Sheets (MSDS), and their acquisition and filing, are additional important elements of consideration for the supervisor. Workers must be informed regarding MSDS sheets, must be aware of their right to see them, and must know where the MSDS sheets are available for their examination and reference. Training of the workers will be necessary in this regard, and documentation of the training and careful record keeping are required. The importance of documentation, record keeping, and filing on the part of the supervisor cannot be overemphasized.

Additionally, hazardous materials management (HAZMAT) must also be considered by the supervisor. Training is required relative to any materials that may pose a hazard in the workplace, both in the work of the supervisor's company and his labor force, and on the part of other workers and subcontractors on the job site. Again, careful documentation and record keeping are important in this consideration. Additional elements of safety and the supervisor's additional responsibilities in safety management and HAZMAT are further discussed in Chapter 9.

Training

Training is a very important management and supervisory responsibility. Yet, unfortunately, it is a matter that is all too often neglected amid the bustle of meeting project completion requirements. Supervisors and other managers as well, are frequently heard to lament, "While I agree that training is necessary and good, we simply do not have the time to spend. We have a project to build, and a budget and a schedule to keep."

As a matter of fact, it has been found that training often yields efficiencies more than sufficient to compensate for the time and other resources that are expended in conducting the training. Training the craft labor workforce should be considered to be an investment in the skilled people who perform construction work. Like other good investments of all kinds, this investment will continue to yield benefits, and in many different ways, over time.

Training craft workers can include considerations such as the proper handling and installation of new materials and systems, the application of new work methods, proper use and care and maintenance of tools and equipment, safety matters of all kinds, and many other matters. There is simply no end to the list of matters on which training can be conducted in order to develop and improve the workforce.

Training yields tangible benefits in a number of ways. First, the training certainly advances the skills of the craft workers, resulting in greater productivity, and enhanced safety. Additionally, many workers are eager to learn and to advance their skills; they feel valued, and, as has been demonstrated in other sections of this book, they become better motivated by the fact that the company is willing to invest in their further learning.

MANAGING MATERIALS

It is a fundamental and unchanging principle for the supervisor: having the proper materials—the right materials, in the right condition, in the right quantity, in the right place, at the right time—in order for the craft workers to be able to conduct their assigned work, is a basic responsibility of the supervisor. While others in management may provide assistance and input to the process, the supervisor should never lose sight of the fact that this is fundamentally his responsibility.

It is the expectation of the craft workers that when they are assigned a task by the supervisor, the correct materials will be on hand for the completion of that activity. If not, time is wasted, and, in addition, the workers become frustrated and demotivated. Their morale declines, and productivity suffers, and the supervisor's stature diminishes.

In addition to all of these unpleasant and costly results, if the proper materials are not available for the performance of an assigned task, it follows by definition that craft workers will need to be reassigned to other work. This occurrence is itself costly, and additional time and energy are wasted and productivity suffers further.

Materials Procurement

Every material and product that is to be installed on a construction project must be procured and purchased. As obvious as this statement appears to be, it is important for the supervisor to understand this basic axiom, as well as the fact that there may be considerable variability in the processes by which different materials are procured. Procedures for procurement will vary by company policy, as well as by the product being purchased. Some companies have a purchasing agent or another person in the company office, who may purchase some of the materials and products to be installed on a project. Some companies utilize a warehousing system, and furnish materials to job sites from the warehouse, usually on a requisition basis. Other materials and products may be purchased by the project manager, while still others may be purchased by the supervisor. It is vitally important for the supervisor to understand company policies, as well as the procedures to be employed with regard to the procurement of all of the materials to be utilized on the project. Additionally, the supervisor must remain in constant communication with others who may be involved with this process, so that everyone understands who is responsible for acquiring each material and who is responsible for each aspect of materials purchasing.

The supervisor must be sure that one person within the company does not think another is handling procurement of a material, while the other believes someone else is handling the purchase—with the result that no one has made the purchase by the time the material is needed. This can lead to catastrophic consequences on the project.

In the same way, the supervisor must assure that two (or perhaps more) people are not independently purchasing the same material, each believing it is his or her responsibility to do so. Careful planning, constant update and follow-through, as well as effective communication, are required in order to avoid such an occurrence.

Additionally, the quantities of each material to be purchased must be verified against the actual project requirements. While the estimator will have determined materials quantities during the formulation of the estimate for the project, the supervisor must be sure that the correct quantities of materials are purchased in order to actually meet the requirements of performing the work.

Submittals

Some of the materials needed for each project will require the preparation of submittals as part of their procurement. Whether submittals take the form of shop drawings, data sheets from the manufacturer, or samples, or mockups, this process requires special care and attention on the part of the supervisor.

All of the materials submittals required for the project are defined in the project drawings and specifications. All of the submittals that are required should be identified in the drawings and specifications, and should be listed in a submittal log, which can be utilized throughout the duration of the project. The purpose of this log is to ensure that all required submittals have been identified and are furnished as necessary, and that their approval is tracked throughout the entirety of the submission and approval process.

Submittals take time for their preparation and approval. This time must be planned for and allocated as part of the procurement process. Additionally, the supervisor should know that submittal approval process is often iterative, with a submittal sometimes sent forward, and then modified or disapproved, and then re-submitted on a number of occasions during the approval process. It is all too easy to lose track of the status of a submittal, and then to discover, at the time when the material or product is needed for installation, that there will be a delay because submittals have not yet been approved. This is a very expensive occurrence!

Additionally, all communications that take place, both verbal and written, with respect to submittals must be fully documented by the supervisor. Requests for preparation of submittals, all requests for information (RFIs) with regard to any submittal, requests for approval, or for expediting of approval, or for verification of approval, are commonplace elements of the submittal tracking and management process, which must be fully documented.

It is highly recommended that the supervisor and the project manager prepare a submittal tracking log for their use and reliance throughout the life of the project. While various formats may be employed for this log, one of the most basic forms is often the most effective. This entails use of a spreadsheet. An example of such a submittal tracking log is included in Figure 11.1.

As indicated in the example in Figure 11.1, all of the submittals that will be required for the project are listed in the log in the left column. This information is extracted from the drawings and specifications. The page number, detail number, or article number where the submittal is called out in the drawings and specifications can be referenced in a column just to the right of the first. Descriptive notes should be added, to ensure complete understanding and communication with regard to each submittal. The columns to the right of the submittal name and description, and the

JRRS CONSTRUCTION

SUBMITTAL TRACKING LOG

PROJECT: SHINING STAR OFFICE BUILDING
JOB NUMBER: 1654

SUBMITTAL REQUIRED	FABRICATOR, VENDOR, SUPPLIER, DISTRIBUTOR, MANUFACTURER	DRAWING AND/OR SPECIFICATIONS REFERENCE	NOTIFIED TO PREPARE SUBMITTAL DATE	SUBMITTAL RECEIVED THIS OFFICE DATE	APPROVED OR DISAPPROVED DATE	FORWARD TO ARCHITECT DATE	NOTES	APPROVED OR DISAPPROVED DATE	BACK TO ORIGINATOR DATE	PROMISED FABRICATION DATE	PROMISED DELIVERY DATE
Column Base Plates, Structural Steel Columns C12 - C16	Mosher Steel, Houston, Texas. Danny Gilman.	A22, A24, S12, S13, S14, S15	8-20	9-14	9-15. Disapproved: Wrong Bolt Diameters. REVISE AND RESUBMIT	X					
				9-22 RESUBMITTAL RECEIVED	9-23 DANNY NOTIFIED APPROVED BY THIS OFFICE, EMAIL	9-23		10-2 ARCHITECT AND STRUCTURAL ENGINEER APPROVAL RECEIVED.	10-2	11-2	11-6
Precast Weldments, Front Façade	Haco Precast, New Braunfels, Texas. Mark Thurmond.	A11, A21, D-16. SPECIFICATIONS P. 42, 61.	8-20	9-18	9-22. MARK NOTIFIED APPROVED BY THIS OFFICE, EMAIL	9-22		9-28 ARCHITECT AND STRUCTURAL ENGINEER APPROVAL RECEIVED.	9-28 ATTN MARK.	12-3	12-7
Atrium Skylight	Building Specialties, Inc. San Antonio, Texas. Sharon Archer.	A19, A33, D3-9. SPECIFICATIONS P. 241; SPECS, P 312.	8-21	8-28	8-31 SHARON NOTIFIED APPROVED BY THIS OFFICE, EMAIL	8-31	9-18. REQUEST EXPEDITED APPROVAL FROM ARCHITECT. EMAIL	9-30 DISAPPROVED. ARCHITECT NOTE: STAINLESS FLASHING SPECIFIED. REVISE AND RESUBMIT	10-1 NOTE TO SHARON: REVISE AND RESUBMIT.		
				9-22 RESUBMITTAL RECEIVED	9-25 SHARON NOTIFIED APPROVED BY THIS OFFICE, EMAIL	9-25		10-2 ARCHITECT APPROVAL RECEIVED	10-5 ATTENTION SHARON	1-16-2010	1-25-2010
Hollow Metal Door Frames	Shumard Supply. Austin, Texas. Stanley Morris.	A4, A19, A23. SPECIFICATIONS P. 111	8-21	8-28	8-28 STANLEY NOTIFIED APPROVED BY THIS OFFICE, EMAIL	8-28		10-2 ARCHITECT APPROVAL RECEIVED	10-2	1-6-2010	1-16

Figure 11.1 Submittal Tracking Log

relevant drawings and specifications reference, will be given names corresponding to the typical steps in the preparation and approval of the submittal. Typical entries for a general contractor's submittals would include: notification to fabricator to begin submittal preparation; submittal received from fabricator; forwarded to architect/engineer; anticipated approval date; received from architect/engineer, and status, approved or disapproved; sent to fabricator; acknowledgment from fabricator; promised fabrication date; promised delivery date. If a submittal is disapproved and requires modification and resubmittal at any point in the process, additional cells or columns should be added in order to reflect and to monitor these additional steps in the process.

The submittal log can be used both as a checklist and also as a reminder of pending materials purchases for which submittals need to be initiated. This will assure that no submittal is overlooked.

In the cell at the intersection of columns and rows on the log form, the supervisor can insert dates, as well as notes as appropriate, in order to keep the log current and up to date at all times. The principal management objective for the supervisor to remember is: the supervisor must know at all times, the status of every submittal on the project for every material for which submittals are required.

Preparing a submittal tracking log and keeping it constantly updated is the best means of ensuring the outcome noted above. Even if the supervisor is not directly responsible for materials purchases or for submittals management, his awareness of the process will be a valuable assist to the project manager, or whoever it is who has this direct responsibility in the company.

Managing Materials Deliveries

When materials have been purchased and are delivered to the job site, properly managing the delivery, unloading, handling, and storage of these materials will pay dividends for the supervisor. Conversely, neglecting or improperly managing this process results in difficulties and added expenses of all kinds. This is an aspect of materials management that most certainly merits time and attention on the part of the supervisor; yet it is often overlooked. As with most other management matters, attention to detail is an important aspect of the supervisor's proper management of this process.

The supervisor's advance planning should include consideration of all aspects of the materials delivery. Before materials are delivered, delivery arrangements should be confirmed with suppliers, so that surprise deliveries are avoided. Considerations that should be incorporated into the supervisor's delivery planning include: where the delivery truck will enter the site, where it will park for unloading, which person or persons in the crew will unload, what equipment will be used for unloading and handling, what ancillary items will be necessary (slings, pallets, cribbing, etc.). Additionally, arrangements may need to be made for scheduling equipment that will be used for materials unloading and handling. This may include arranging for the use of forklifts, skylifts, or mobile cranes, or it may include scheduling arrangements with the general contractor for the use of the job site crane to make the lifts for unloading.

In addition, advance consideration should be given not only to which workers will perform the unloading, but also to which of the workers will count the items that have been delivered in order to verify correct quantities in the delivery as compared to the order, and to verify that what was delivered is what was ordered in terms of description and specification. Additionally, inspection of the material being delivered is necessary, in order to ensure that no visible damage has occurred. Further, those who are taking part in the unloading and handling should be on the alert for potential concealed damage (e.g., boxes that rattle when they should not be expected to). One of the workers should be assigned the responsibility for signing the delivery invoice when all elements of the delivery have been verified, and for filing the delivery invoice in the onsite filing system for documentation. The supervisor's planning should include delegating each of these important tasks to an individual worker.

This matter of delegation—assigning a task specifically to a person and leaving that person with the responsibility for accomplishment of the task—is itself an extremely valuable supervisory skill. It is a skill that, when practiced by the supervisor, will accrue significant savings of the supervisor's time, and that will also pay dividends in a number of other ways. Delegation is discussed further in other sections of this book.

Managing materials delivery also includes advance planning for what will be done with the material after it has been delivered. Will it be handled by hand or by machine? Have the people who will accomplish this been informed? Is the equipment that will do the handling, available? Will the material be placed in the lay-down area? If so, where should it be located, so that it is not likely to be damaged and so that it can be accessed when it is needed? Does the material require cribbing underneath or protection from the elements? Should a sketch map of the lay-down area be prepared, on which the location of materials can be noted for future reference? Does the material need to be placed in lockable storage or otherwise secured? How should it be arranged or stacked? How high can it safely be stacked? These and numerous other considerations of a similar nature should become part of the supervisor's detailed planning with regard to what to do with materials after they have been delivered and unloaded.

Job Site Materials Handling

The supervisor's materials management plan should also include consideration for getting the material from the lay-down area or other onsite storage, to the workface for installation. Again, careful thought and planning should be provided by the supervisor with regard to matters such as which workers will handle the material, what equipment may be needed, and so forth. It is imperative that the move to the workface be accomplished safely, without damage to the material, and in the most economical manner. The advance planning that the supervisor provides to all aspects of materials delivery and handling will pay dividends and will make the planning effort worthwhile.

Billable Materials

An additional important element of materials management that should become a part of the supervisor's job site materials management plan is making note of those materials whose cost can be included in the project periodic payment requests after the materials have been delivered. Many construction contracts contain a provision that states that materials that have been delivered and properly stored at the job site can be included in the project periodic progress payment billings. Because of the importance of cash flow, the supervisor must be absolutely sure that no opportunity is overlooked to include delivered materials in monthly periodic payment requests. This also reinforces yet again the principle that the supervisor's having available, and having read, the contract for the project will provide benefits for the supervisor and for the project.

In an additional note, sometimes the architect or owner will permit payment for materials delivered to a suitable approved warehouse, in addition to allowing for payment for materials delivered to the job site. If the materials management plan for the project includes warehousing materials in advance of delivering them to the job site, it is certainly worthwhile for the supervisor to check the contract provisions in this regard, and/or to make inquiry of the designer as to whether these materials can be included in monthly payment requests, and to document the response.

MANAGING TOOLS AND EQUIPMENT

The supervisor has a fundamental responsibility to ensure that the tools and equipment that are necessary for the performance of the work on the project are on the job site, are in a safe condition, and are in good working order. Whether it is machinery, or power tools, or any of the numerous hand tools that the craft workers employ, the supervisor must ensure that exactly what is needed is available in terms of tools and equipment, at the time when it is needed on the job site. While others, such as the project manager, may provide their input and assistance in this regard, the basic responsibility rests with the supervisor.

The construction company makes a considerable financial investment in providing the tools and equipment that are necessary for the performance of the work. The company will entrust this investment to the supervisor for his proper management on the job site.

Many believe that if supervisors are able make the craft workers aware of the cost of the tools and equipment that they utilize, this awareness will lead to better care of the tools on the part of the workmen. In everyday communications, as well as in "toolbox talk," or "tailgate talk" communications with the workers, this awareness can be conveyed by the supervisor.

In addition, security measures to help prevent loss and theft of tools must be managed by the supervisor. Construction companies have different policies and procedures in effect with regard to providing security for tools. The supervisor must know what these policies are in his company and must consistently apply them.

Additionally, constant awareness of this matter on the part of the supervisor, and conveying that awareness to the workers, and communicating to workers the importance of their helping ensure tool security, will help achieve success in managing tool security.

Many supervisors have found value in maintaining workers' awareness of the importance of tool security through regular discussions in "toolbox talks." In addition, continual watchfulness on the part of the supervisor, and providing reminders or directives to the craft workers as appropriate for occurrences such as leaving a tool unattended or for mishandling a tool, have been found to be very effective.

Additionally, many supervisors utilize a "gang box inventory sheet" to maintain a listing of the power and hand tools, as well as extension cords and other components that are kept in the job site gang box. This inventory list can be checked on a regular basis in order to ensure that nothing is missing. Similarly, a list or spreadsheet can be maintained of other tools, such as stepladders, extension ladders, and the like, which are not kept in the gang box. Checking these lists regularly and verifying that all tools are accounted for is a function that supervisors should perform on a regular basis as part of their responsibilities. Alternately, this is another example of a task that could be delegated to one of the craft workers.

The use of chains and locks and maintaining a systematic method of key control for the keys to the locks are additional basic elements of a tool security management plan. Making it a practice to check at the conclusion of every work day, that all tools and equipment are in the gang box or secured with chains and locks as appropriate and then ensuring that all locks are in place and secure are basic procedures that should be a part of the supervisor's daily plan of work.

Training and Communication

Additionally, the supervisor should be aware that training of the workers may very well be required in order to ensure productive and safe operation of some tools and equipment. Required maintenance must also be brought to the attention of the craft workers, and a reliable communication system established, so that tools and equipment are properly maintained. Additionally, a communication system must be established and maintained, so that the supervisor can be made aware of, and can make remedy for, any tool that is not working properly and that may be in need of maintenance or repair, or that may be in need of accessory items or attachments.

MANAGING THE CONSTRUCTION SITE

As components of planning and analysis relative to the construction project, the supervisor should include consideration of the construction site itself. Numerous matters that are inherent in the nature of the site, and other elements of consideration as planned and determined by the supervisor, will impact productivity and safety in the work on the site throughout the duration of the project.

Before the work on the project begins, the supervisor should carefully study the site plan or plot plan for the project and, if possible, should visit the site. Included in this consideration are thoughts relative to the size of the site and the amount of space available for lay-down area and job site storage, topography, drainage, adjacent streets, existing buildings, both on the site and on neighboring sites, utility locations, and the like. These will translate to decisions regarding matters such as where fences and gates will be located, both for vehicles and for construction workers; access and egress points for delivery vehicles and for construction workers' vehicles; parking areas for construction workers' vehicles; access and parking for visitors to the site; access, and routing of temporary utility services and the associated electrical panels, disconnects, shutoff valves, and so forth.

The project manual or book of specifications should also be studied by the supervisor for provisions that relate to site management. The General Conditions or the Special Conditions, for example, frequently contain provisions relative to construction boundaries on the site, or to protection of special artifacts or landscape elements on the site.

Additionally, the supervisor should meet with those who prepared the estimate for the project, in order to avail himself of site planning that has already been done, and the determinations that have been made, and the assumptions that were utilized with regard to the site when the estimate for the project was prepared. The importance of "handoff meetings" of this kind, and their value to the supervisor, are discussed in other sections of this book.

Site layout considerations will also include the location and orientation of the contractor's onsite office or job shack, as well as for the offices of subcontractors on the project, and perhaps facilities for an architect's representative or owner's representative. Provision of utilities services for these facilities is also a necessary consideration. Planning should include provision of designated areas and facilities for dressing areas for workers, as well as break areas and lunch areas.

Tower crane location, as well as access routes and setup points for the various mobile cranes that will be utilized during the course of the project, must be carefully planned. Parking areas for mobile equipment that will be used during construction, as well as considerations regarding locations for fueling and for routine daily maintenance of all equipment must be considered.

Additionally, material lay-down areas, as well as secure and weathertight storage areas, must be planned and provided, for both the general contractor and all of the subcontractors. Planning must also include temporary roads onsite. Unloading areas, turnaround areas, and perhaps washout areas must be considered for the vehicles that will deliver materials to the site. Sanitary facilities, and access for the vehicles that will service them, must be planned for. Planning must also include provisions for refuse containers and access for the vehicles that will deliver them and transport them.

The supervisor's safety plan and site planning should include consideration of the location and proximity of all emergency services, and the emergency responders' awareness of the location of the construction site and where they will access and enter the site if they are needed. Contingency planning should also include consideration of a site evacuation plan.

This analysis and planning should be conducted before the contractor mobilizes and moves onto the site. The project manager may provide assistance to the supervisor and may be a source of additional valuable information with regard to site planning and site management. However, it is the supervisor who will be working on the site every day throughout the duration of the project; he should have a key role in site management considerations and should perform his planning accordingly.

Additionally, it should be recognized that as the work on the project proceeds, numerous additional elements of site management will present themselves. These will, in turn, warrant the supervisor's proper care and attention. In summary, supervisors can be sure that the time and energy that they invest in analysis of and planning for the many considerations that are inherent in proper site management yield significant dividends, and will continue to pay dividends throughout the duration of the project.

CASH AS A RESOURCE

The supervisor should know that, in financial terms, cash is a resource and one that must be managed with the same degree of care and effectiveness with which other company resources and project resources are managed. One of the contractor's most important business functions is maintaining positive cash flow, both for the company and for the projects that the company performs. Because cash flow is crucially important to those in company management, its importance to the supervisor should be apparent.

While positive cash flow can be defined in a number of ways, a simple yet effective definition will suffice for purposes of this discussion. Positive cash flow means that current receipts or revenue exceed current expenses.

The concept of a contractor's cash flow could hardly be more important. It is a matter of fact that most contractors who fail go out of business not because their total liabilities exceed their total assets, but rather because their liquid (available) assets are insufficient to meet current (immediate) demands.

There are quite a number of ways in which the management of the construction projects that the company performs affects cash flow. Depending upon company policies and procedures, the construction supervisor may have a greater or lesser role in cash flow management. In most companies, most aspects of this important matter are managed at the level of the project manager and company office-level management.

However, the importance of the supervisor's being aware of the significance of cash flow management, and his taking measures to help ensure positive cash follow wherever he can, will serve him very well. Listed in Figure 11.2 are measures that supervisors may take or matters they may assist in that have a bearing on the all-important matter of ensuring positive cash flow.

1. Being aware of the importance of properly structuring the Schedule of Values of the Project. The Schedule of Values is more fully discussed in Chapter 18, "Ongoing Operations."

2. Being aware of, and considering in the planning and management of the project, the dates on which periodic progress payments are due to be submitted to the architect (if the supervisor's company is a prime contractor on the project) or to the general contractor (if the supervisor's company is a subcontractor).
Assisting the project manager or company-level management in any manner possible with assuring that the payment request submittal is timely, accurate, and as complete as possible.

3. Taking care in the ordering of materials for the project. As has been noted, it is very important for the supervisor to assure that adequate materials are on hand on the project for all of the work to be performed. However, an overabundance of materials leaves a surplus, which must be returned (often at less than full price credit) when the work is completed. Further, these surplus materials are themselves an additional expense in having been purchased and then not used. Additionally, these materials must be handled and stored and protected on the job site. Moreover, there is risk these materials may become damaged or degraded before they can be returned to the supplier.

4. Being aware of suppliers' return policies and returning extra materials in a timely manner. When extra materials left over from the project must be returned to the suppliers for credit, typically the supplier will accept for return, and will only issue credit for, materials within a specified time following the original purchase. Additionally, suppliers will only issue credit for materials that are returned in "as new" and salable condition.
Therefore, the supervisor must assure that materials that are to be returned are protected from damage and from degradation by the elements.

5. Returning tools and equipment that are not being used on the job site to the company equipment yard or to the equipment rental agency.

6. Taking the appropriate management measures to assure quality control and quality management in all of the work performed on the project.
As will be further discussed in Chapter 18, "Ongoing Operations," as well as in other sections of this book, performing quality work enhances the company's as well as the supervisor's reputation, and helps ensure a satisfied client.
Moreover, it is much easier to bill for work as earned value in periodic payment requests, and to receive timely payment for, work that is of unquestioned quality.
Additionally, if quality control has been the constant watchword throughout the performance of a project, the punchlist process can be completed much more expeditiously.
The punchlist is discussed in other sections of this book and further reference to this process is provided below.

7. Managing the project closeout and punchlist process.
As the project nears completion, the supervisor and all of the craft workers should be especially mindful of completing all of the work as defined in the contract, as well as to satisfying the needs of the owner.
Extensive work that needs to be performed at this time of the project, or major corrections that need to be made, are very costly at this time.
If the supervisor has been attentive to the importance of quality management throughout the duration of the project, as discussed elsewhere in this book, this time can be rendered much less stressful and much less costly.
Moreover, when the supervisor manages the punchlist and closeout procedures in a timely way, his company can submit its Final Application for Payment sooner, and can be paid the retainage that has been withheld throughout the duration of the project, and can receive payment of the balance of the contract amount at an earlier time.

Figure 11.2   Measures for the Supervisor to Take in Assuring Positive Cash Flow

MANAGING TIME ON A CONSTRUCTION PROJECT

Time is another resource that must be properly managed. Construction contracts almost invariably contain a provision that states that time is of the essence in the completion of the construction project. The things that the supervisor does, and the decisions he makes, play a major role in completing the project on time.

It is very important for the supervisor to manage in such a way as to complete the project by the completion date stipulated in the contract. Completing the project on time is always one of the project objectives, as was discussed earlier in this book. In addition, managing time and, thus, ensuring timely project completion is one of the hallmarks by which the reputation, as well as the effectiveness and the success of the supervisor, will be measured.

Construction schedules are the supervisor's primary project-planning tools and are used to manage time on a construction project. Schedules can be defined simply as, "A graphic representation of the plan for performing a construction project."

Different types of schedules are utilized for managing construction projects. Schedules that the supervisor must be familiar with include network schedules, bar charts, and short-interval schedules. Each of these schedules, as well as a number of important aspects of the construction scheduling process, are discussed more fully in Chapter 14.

SUMMARY

The supervisor should clearly understand that he has the responsibility for stewardship and management of the construction company's investment in the construction project in the form of the resources that the company commits to the project. Manpower, tools, equipment, materials, the construction site, dollars, and time, all are resources whose management is entrusted to the supervisor. Effectively managing these resources is at once an enormous responsibility and an immense challenge, as well as a tremendous opportunity. This chapter has provided a number of tools that will assist the supervisor in being successful. The supervisor who can meet the challenges and who can take advantage of the opportunity while accomplishing the objectives enumerated for every project is one who not only has proven his or her mettle but also is one who can join the ranks of respected and successful supervisors.

Learning Activities

1. Tools inventory

   If you and your company are not already doing so, delegate to one of the craft workers the task of making an inventory sheet for all of the tools and supplies that are in your company's gang box on your current project.

   (continued)

Have the worker make note of any tools that are damaged and in need of repair or maintenance, and to note any tools that may be missing accessories or attachments.

Periodically make an inventory, or have the designated worker do so, to ensure that all of the tools, equipment, and supplies that should be in the gang box are in fact present.

2. Reinforcing the cost and value of tools and supplies, and the need for their care and protection

After the inventory of the gang box has been made as described above, make a copy of the inventory sheet.

On the list, make a price determination, with the help of your project manager and your contacts at a supply house if necessary, of the cost of each item in the gang box. Generate a total value.

In one of your toolbox talk meetings, use tool cost, tool value, and tool security as a theme for the meeting. State for your crew the total value of all of the tools and supplies in the gang box. Expect looks of surprise, and perhaps expressions of disbelief. Share a copy of the inventory sheet and price determinations with your crew, in order to reinforce the value of tools, and to make the point that tool care and tool security are important.

3. Ensuring positive cash flow

Make a copy of Figure 11.2, "Measures for the Supervisor to Take in Assuring Positive Cash Flow," or write out the measures on a sheet of paper. Think about other measures you might add to the list of ways to help assure positive cash flow.

Write an "action item list" of measures you can take on your current project, to help assure positive cash flow.

CHAPTER **12**

MANAGING COSTS: UNDERSTANDING THE ESTIMATE

INTRODUCTION

One of the keys to the success of any construction company is its ability to control costs. This is a fundamental truism for any construction contractor. It is imperative for the construction supervisor to have a firm understanding of construction costs and their derivation. In addition, the supervisor must have a clear understanding of the cost reporting, cost accounting, cost control system that is being utilized by the contractor. This chapter will begin developing an understanding of construction costs, by describing the manner in which costs first become a component of a construction project, which is at the time when the cost estimate for the project is prepared. The next chapter will define and describe the project cost control system that is typically utilized by a construction company.

THE ESTIMATE

A construction project is built in accordance with the terms of a construction contract. The construction contract will, in turn, almost always contain provisions that define the cost of the project to the owner. Under the terms of the construction contract, the contractor is typically required to ensure for the owner what the owner's maximum cost of the project will be for the contractor's fulfilling all of the requirements set forth in the contract documents. While there are construction projects that are performed under contracts that do not contain such a provision, most contracts do contain this requirement; therefore, discussion in this chapter will proceed on this basis.

It follows, then, that contractors must have a method for determining what they believe their cost of performing the construction project will be. This is the only way they can guarantee that the contract requirements will be performed for a stipulated number of dollars, as required by the construction contract. This determination of the amount of money for which the contract requirements can be fulfilled is made by the contractor's estimator and the estimating department. This is where construction costs are first envisioned and defined.

Whether the estimating department consists of one individual or a large number of people, the objective is invariably the same: to produce the best possible determination of the contractor's cost of performing a construction project as the project is defined and described in the contract documents. Accordingly, the contractor's estimate is often referred to as the cost estimate.

The cost estimate can be defined as the estimator's best prediction of what the costs of performing a project will be, given the time and other resources available. Estimators take the construction documents and, through the application of their skills, produce this prediction, the cost estimate. This chapter will acquaint the supervisor with the estimating process.

TYPES OF ESTIMATES

Contractors produce different types of estimates for different purposes. For example, contractors sometimes produce estimates that are called *factor estimates*. Such estimates are intended to provide an approximation of construction costs on the basis of units, such as dollars per square foot, per cubic foot, per linear foot, per cubic yard, per fixture, or per some other unit of measure. Estimates of this kind are also referred to as parameter estimates, or as parametric estimates.

These factor or parametric estimates may be used to determine order-of-magnitude approximations of construction costs. Additionally, they may be used in feasibility studies. However, such estimates are seldom sufficiently accurate to be utilized to determine contract prices.

The type of estimate to be discussed further in this chapter is referred to as the detailed estimate. Detailed estimates provide the best estimate of construction costs. Detailed estimates are also the most time-consuming and the most costly to prepare.

Detailed estimates are used to prepare proposals that are submitted by a prime contractor to an owner, or in the case of a subcontractor, to prepare proposals for submittal to a prime contractor. Detailed estimates are also employed by prime contractors as well as by subcontractors, to prepare prices for change orders during the course of a construction project.

The fundamental purpose of a detailed estimate is to determine the number of dollars that a prime contractor or subcontractor is willing to enter into a proposal for the purpose of being selected to receive the award of a construction contract. The usual expectation is that if the contractor's proposal is selected, the contractor will enter into a contract, and that the amount on the contract will be the same as the amount on the proposal. Therefore, in a very real sense, the amount of dollars that a contractor's estimator derives in the estimate becomes the amount of money for which the contractor will be expected to perform the construction contract.

TRUISMS REGARDING THE ESTIMATE

The Supervisor's Role

It is important for the supervisor to understand the estimating process for a number of reasons. First, it is in the estimate that the costs of performing the work on the construction project are first determined. As you will see, the supervisor will be working with these cost figures throughout the duration of every project he or she performs.

Additionally, the supervisor, in an indirect way, plays a very important role in the estimating process. When the supervisor manages a project during the course of its construction, he provides coded craft labor input regarding the labor costs on the project to the construction company home office. This information is compiled, and then is stored in a historical cost database that the company maintains. As you will see, the estimator utilizes this historical cost information in a meaningful way in the formulation of construction cost estimates. In a very real sense, then, the supervisor provides input that determines the estimated cost of future projects that the company will perform.

The estimate is extremely important in terms of its basic function as we have described it. It is also important for the supervisor to recognize that estimate is also part of a larger system in the contractor's operations. This system is described as the cost determination (estimating), cost reporting, cost accounting, and cost control system that the contractor employs. This system operates as a continuous cycle within the construction company, as illustrated in Figure 12.1.

Estimating as a Career Option

Supervisors should also recognize that as their careers advance they might wish to become a project manager, or perhaps to join the construction office staff in the estimating department. In either of these roles, understanding the construction estimating process will prove to be an extremely valuable asset. Many contractors believe that the best estimators are those who have previously managed work on construction projects. In addition, most contractors consider their estimators to be among their most valuable assets. For a great many constructors, estimating is a very attractive career option.

**CONTRACTOR'S PROJECT COST
ACCOUNTING SYSTEM**

- **SYSTEM OF COST CODES ESTABLISHED**
- **INPUT TO THE SYSTEM VIA CODED LABOR TIME CARDS**
- **PAYROLL PREPARES CRAFT LABOR CHECKS**
- **PROJECT COST ACCOUNTING TO COST CODE CATEGORIES**
- **COST REPORTS PREPARED THROUGHOUT PROJECT**
- **MANAGEMENT COMPARISON TO PROJECT BUDGET**
- **MANAGEMENT ANALYSIS**
- **MANAGEMENT DECISION MAKING**

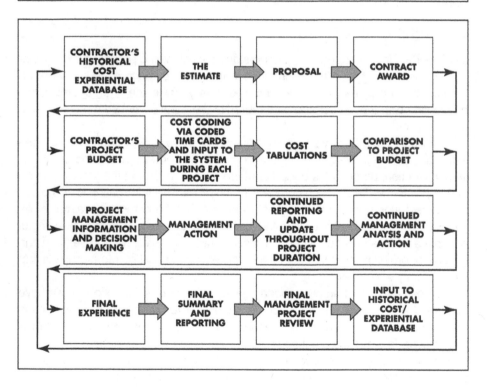

Figure 12.1 Cycle of Estimating, Project Cost Accounting, Project Cost Control, and Historical Database Information

PREPARING THE ESTIMATE

The remainder of this chapter will outline, for the supervisor's understanding, the process by which detailed estimates are prepared. While the fundamentals of the estimating process are consistent and do not change, it should also be recognized that every estimator has his or her unique approach to the estimating process, as well as his or her own subtle variations in procedure. In addition, the policies and management decisions of the construction company will determine some aspects regarding the preparation of the estimate. It has been said, therefore, that estimating is partly art and partly science.

One way for the supervisor to envision the estimating process is to consider that the estimator builds the project in his or her mind. That is, the estimator envisions each activity to be performed in constructing the project. Then, using the tools and techniques described in this chapter, the estimator formulates the best prediction of the cost of performing each activity. The estimated project cost is the summation of the estimated costs of performing each activity in the construction of the project.

THE DETAILED ESTIMATE

A prime contractor's detailed estimate consists of six components. For a subcontractor, there will be five components, unless the subcontractor is planning to utilize sub-subcontractors to perform some of the work on the project. In this case, there will also be six components in the subcontractor's detailed estimate.

The six components of a prime contractor's detailed cost estimate are: materials, labor, equipment, subcontractor quotations, indirect costs, and markup. The estimators will determine their best prediction of the project cost for each of these components. The cost estimate, that is, the amount to be entered onto the proposal, will be the total of these six components. Figure 12.2 illustrates this process in graphic form.

Prior to beginning the preparation of the detailed estimate, the contractor's estimating and management staff will make a determination of which items of work on the project the contractor will plan to perform with its own craft labor forces, called self-performed work, and which items of work the company plans to perform by subcontracting. For those items of work that the company plans to perform with its own craft labor, the materials, labor, and equipment costs will be estimated by the contractor. For those items that the company plans to perform by subcontracting, subcontractors' proposals will be sought and will be incorporated

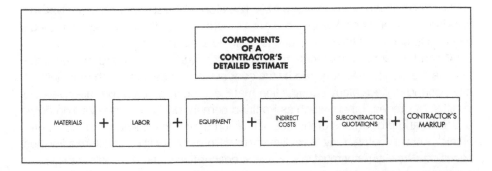

Figure 12.2 Components of a Detailed Estimate

into the estimate. Whether lump sum or unit price subcontractor proposals are taken from the subcontractors, each subcontractor's proposal is expected to include the subcontractor's consideration of all materials, labor, equipment, indirect cost, and markup for the performance of the subcontractor's portion of the work.

Additionally, prior to beginning the detailed estimating process, the estimating staff will carefully evaluate the project duration. This is the amount of time stated in the contract documents, in which the contractor must complete all of the requirements of the contract documents. Time is a very important provision of the construction contract. The project duration is a very important consideration in the estimating process.

In order to predict construction costs, the estimator must prepare at least a preliminary construction schedule, which is the contractor's plan, on a timeline, for completing the project within the specified duration. If the project's duration permits, the contractor will most likely plan to perform the work on a five days per week, eight hours per day schedule. However, as indicated by the duration and the amount of work to be performed in the project, the contractor may find it necessary to plan to work on weekends, and/or to work overtime, and perhaps even to work in consecutive shifts in order to complete the project on time. These determinations will have a profound effect on the construction costs and must, therefore, be reflected in the estimate.

PREPARING THE DETAILED ESTIMATE

Estimating Materials Costs

Quantity Takeoff

The estimator's first step in estimating materials is to identify each material that will be required in the performance of the work of constructing the project. This information will be determined from the estimator's analysis of the drawings and the specifications for the project. Most of the materials that will be required in a building construction project are not called out in the contract documents in the form of a list or a table; rather, the drawings and specifications show and describe what the finished construction product will be. Therefore, the estimator must analyze the drawings and the specifications in order to identify all of the materials that will be required to construct the project and to fulfill the contract requirements.

When the requisite materials have been identified, the estimator will next determine the quality level required by the contract documents for each material that has been identified. Many construction materials, such as plywood, are available in different grades; others, such as Portland cement concrete, have varying compositions and proportions of ingredients, which define the quality of the concrete; others, such as metals, may have different metallurgical qualities or different finishes specified. The estimator must infer from the description of the materials in the specifications, and from information that is provided on the drawings, all of the pertinent elements of quality for each material that will be used in constructing the project.

The estimator's next step will be to determine the quantity of each material that will be required for the project. This step in the estimating process is often referred to as the quantity takeoff, or QTO. Each material must be quantified in the unit of measure in which it is produced and marketed: cubic yards of cast in place concrete, square yards of carpet, sheets of plywood, bank cubic yards of excavated soil, squares of roofing, and so forth.

Some materials, such as concrete, excavated soil and fill soil, and so forth are quantified by calculation, based on the dimensions in the drawings. Others, of identical kind and description, such as door units, lighting fixtures, windows, and the like are determined by counting the number of identical items as indicated in different parts of the project, as shown in the drawings. These are referred to as count items in the estimate.

Following the quantification of each material in its appropriate unit of measure, the estimator will determine the amount of each material that will actually need to be purchased in order to complete the work. This is called determining a waste or overage allowance. For some materials, such as framing lumber, for example, the estimator must allow for the fact that stock lengths of lumber will be cut by the carpenters to the exact lengths required in the work in order to produce the pieces to be installed in the project. Thus, more lumber will need to be purchased than the precise quantity that has been calculated. For other materials, such as electrical switchgear, or air-handling units for the mechanical system, the waste or overage factor will be expected to be zero. The amount of waste or overage factor to be applied by the estimator will come from the estimator's experience and judgment.

Materials Pricing

After determining the quantity and quality of each material that will need to be purchased, the estimator will next determine prices for each of the materials. The estimator will contact materials suppliers, distributors, or manufacturers and will solicit price quotations for each of the materials to be purchased. The estimator will furnish information to the materials suppliers regarding the exact description of each material, the quality level in which the material is needed, and the quantity of the material to be purchased.

Materials suppliers will provide price quotations in response to the estimator's request, and the estimator will determine the best price for inclusion in the estimate for each material. The estimator will compile this information for each material and will then determine the dollar amount that the company will expect to pay for the material when the project is performed.

While this process appears straightforward and direct, there is actually a considerable degree of uncertainty and, therefore, risk, in this determination. Some materials prices are stable, while others fluctuate. Time will elapse between the taking of the price quotation from the supplier and the point in the future when the material is actually purchased for the project. Materials prices will often change during that time.

Sometimes materials suppliers will guarantee materials price quotations for a period of time following their initial price quotation, but often not. Therefore, the price that the estimator includes in the estimate at the time of its preparation must reflect, as nearly as possible, the price for which he or she expects the company to purchase the material on that date in the future when the material is actually purchased for the project.

Additionally, the estimator must include in the materials price determination, details such applicable taxes, shipping fees, delivery charges, and so forth. In addition, materials suppliers will often include the designation fob in their price quotations. The characters fob translate literally to "freight on board," or to "free on board." The fob designation in a price quotation will include some location or destination, as stipulated by the vendor (e.g., "fob my loading dock, Kansas City, Kansas."). This location is the designated place where the supplier will have the product located for the contractor when the material is purchased. This location might be the supplier's or the manufacturer's location, or a warehouse, or a port facility. Or, it could be the contractor's job site. This is also the location to which the supplier assumes responsibility, and the point at which the contractor assumes responsibility for the material or product at the price that has been quoted.

When all of the elements of the cost of each material have been finalized, all of the prices for all of the materials needed for the project are totaled by the estimator. This total becomes the materials cost subtotal in the estimate.

Estimating Labor Cost

As noted earlier, the estimator will calculate labor costs for the craft labor that the contractor plans to self-perform, that is, to perform with craft workers on his payroll. Having decided what parts of the work will be self-performed and what parts of the work will be performed by subcontracting, the estimating staff will determine the labor cost of the contractor's craft labor forces performing work on the project.

While there are several approaches to the process of labor estimation, many estimators believe that the best approach for the estimator to employ is for the estimator to "build the job in his mind," and to estimate labor cost in doing so. In the course of building the project in his mind, the estimator will first determine all of the activities that must be performed in order to complete the work. Activities are elements of work that are identifiable and quantifiable, and that consume resources.

Next, the estimator will envision the number and the classification—how many journeymen, apprentices, crew leaders, and so forth—of craft workers necessary to perform each activity. This is referred to in estimating terminology as making the work breakdown structure, or WBS.

The next crucial step is determining, literally estimating, the number of man-hours necessary for the crew and each of its craft workers to complete each activity that has been identified. This component of the labor estimate is both vitally important and, at the same time, filled with uncertainty.

This determination of craft labor hours is vitally important because this is where labor man-hours are quantified. Obviously, this determination will in turn form the basis for the quantification of labor cost on the project. On building construction projects, labor cost is typically the largest single component of the total cost of a construction project.

At the same time that the quantification of craft labor man-hours and, therefore, labor cost forms the basis for the largest component of the cost of a project, this determination is also the most uncertain, and therefore the most risky, component of the estimate. This is true because estimating labor man-hours is based literally on estimating the productivity of the craft labor on the project. It is one of the ironies, and also one of the elements of difficulty associated with cost estimating, that labor is both the largest component of cost within an estimate and is at the same time the most difficult and, therefore, the most risky to estimate component of construction cost.

There are a great many factors that influence what the labor productivity will be on a construction project, and many of these factors are difficult to predict in advance. Additionally, their impact on labor productivity is difficult to quantify. A few examples will illustrate this point.

Weather has a profound influence on labor productivity. Hot and cold weather, high humidity, wind, and precipitation all have an influence on the rate at which craft labor can perform the work. While general weather trends can usually be predicted, the daily weather and its variability, as well as sudden and unforeseen weather events, such as storms or precipitation, certainly affect productivity but cannot be predicted with accuracy.

Additionally, factors such as geographic location of the project and the composition of the construction team—owner, architect, general contractor personnel, subcontractors, building officials, and so forth—have a bearing on the craft labor productivity on a project. Likewise, the composition and management abilities of the contractor's team—the supervisor, superintendent, project manager, and office staff—have an influence. All of these various factors exert an influence on how craft labor does its work and on how productive the craft workers will be.

While other factors could be listed that impact labor productivity, it should now be clear that the estimator faces a daunting task in his or her endeavor to quantify and to determine the cost of craft labor and its performance of the construction work on a project. Nonetheless, the estimator must make the best prediction possible, given the time and other resources available, of the cost of labor to perform the project.

To accomplish this task, there are several different methodologies that may be employed. Both of the methods described here make extensive use of the historical cost information that the contractor's cost estimating, cost accounting, cost control system has developed. This is, in fact, the very reason for the contractor's compiling and maintaining this file of information.

The historical cost database is a tabulation of the contractor's costs of performing work, by project and by activity, on past projects. This information is generated as each project is performed, and is stored in a systematic manner in the historical cost database. This database is one of the contractor's most important and most closely-guarded assets.

By one estimating philosophy, estimators "build the job in their mind" and estimate construction costs as though this database of information did not exist. Estimators proceed with estimating labor as we have described, envisioning activities to be performed, determining the work breakdown structure, and estimating craft labor man-hours for the performance of the work. Periodically during the development of the estimate, or sometimes when the labor estimate is complete, estimators will then check the values derived in their estimates against labor man-hours for the same activities on similar or comparable projects that have been completed on the past, stored in the historical information database. This comparative analysis allows estimators to determine their comfort level with the derived values in this estimate as a function of how these values compare to the contractor's experience on similar activities and projects in the past. Estimators may or may not modify their labor estimates as derived for the project they are currently estimating following their comparison of their current numbers with costs in the historical information database.

A second method for estimating labor is the converse of the first. In this approach, estimators consult the historical information first, and based on the activities to be performed, then extract from the historical information what the number of man-hours was in the performance of these activities on past projects. The estimate being prepared currently is then structured accordingly. Finally, the estimators adjust the man-hours or production rates in the current estimate in such a way as to reflect their best determination of what the man-hours or production rates will be on the project currently being estimated.

Certainly individual estimators, as well as different construction company owners, have their opinions regarding which method of estimating labor yields the best results. In the final analysis, however, the conclusion is the same—the estimators must utilize their skill and judgment, and the best tools available, including the historical information database, to estimate with the greatest degree of certainty attainable, the cost of the craft labor in performing the construction work on the project.

Following the determination of all the activities to be performed in completing the work on the project, and the estimation of the number of man-hours of craft work necessary to complete each activity, the estimator will develop totals for labor hours, totaled by craft and by skill level. The total number of master carpenter hours, journeyman carpenter hours, carpenter apprentice hours, ironworker hours, cement finisher hours, and so forth will be tallied.

The labor cost to be included in the estimate will then be determined by multiplying the craft labor man-hours estimated by the rate (dollars per hour) that that the craft and skill classification will be paid on the project. This product is called the *direct labor* dollars.

The total of labor dollars on the project will include the determination of both *direct labor* and *indirect labor* dollars. This determination is illustrated in Figure 12.3. Indirect labor is also frequently referred to as *labor burden*.

Indirect labor dollars are defined as dollars that the contractor must pay, expressed as a fraction or percentage in several different categories, for each direct

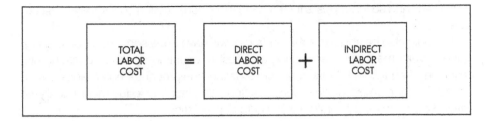

Figure 12.3   Total Labor Dollars Determination for the Estimate

labor dollar of labor cost. Examples include: federal withholding taxes, Social Security taxes, workers' compensation insurance premiums, unemployment insurance premiums, union fringe benefits, employee medical care, paid vacation time, and so on. Each of these elements of indirect labor cost is typically defined as a percentage of direct labor dollars. Most contractors will tabulate the total of all of the indirect labor components to be included for each craft and skill level, and will then apply that percentage as a multiplier to the direct labor for that craft and skill level. While the amount will vary by location and by contractor, the indirect labor is frequently in the range of 40% to 50% of the direct labor amount.

For the estimator's purposes, the labor cost for a project is determined by multiplying the number of craft labor man-hours that have been estimated times dollars per hour of total labor cost (sum of direct labor and indirect labor) in order to determine the estimated labor cost for the project. This labor cost is then entered as a subtotal into the cost estimate.

It should also be noted that because of the importance of the labor cost determination, this component of the estimate is almost always subjected to several levels of review on the part of the contractor. The estimating staff will typically make the initial determination, followed by a review on the part of the head of the estimating department or by the chief estimator. In some companies, company management may conduct a final review of the labor cost determined by the head estimator prior to finalizing the proposal.

Estimating Equipment Cost

The estimation of equipment cost begins with the estimator determining what items of equipment will be necessary for the performance of the work. Equipment will be defined, in this sense, as machinery that is necessary for the contractor's performance of the work.

Several elements of distinction should be noted here. First, there is sometimes confusion regarding the term *equipment* and its definition as machinery necessary for the performance of the work on a project, as contrasted with equipment that is to be installed in the project, such as heating, ventilating, and air conditioning equipment. Machinery necessary for the performance of the work may include cranes, backhoes, materials lifts, temporary elevators, track hoes, earthmoving equipment, and so forth. This is the component of construction cost to be estimated here.

It should also be noted that the contractor will have done some decision making in the development of his estimating and cost accounting systems regarding his definition of equipment. Most contractors define equipment as noted above, with the additional element of definition to include the fact that equipment usually is defined as a capital expenditure, subject to depreciation.

In most contractors' organizations, the concept of capital equipment is what distinguishes equipment from other similar elements necessary for the performance of the work, such as tools. While a pan scraper earthmoving machine would certainly be a capitalized piece of equipment, sledge hammers and other hand tools, as well as circular saws, electric drills, extension cords, and the like, would be most commonly be classified as tools.

Sometimes tax laws govern the determination of the classifications of capital equipment as contrasted with tools. When legislation does not provide the distinction, the contractor will make internal company definitions of equipment and tools classifications. Sometimes standard accounting practice will also indicate how these classifications are structured.

For the supervisor, it is important to know that, while both equipment and tools as described thus far have costs associated with their purchase and use, the matter of where and how the cost is tabulated and estimated is not as important for the supervisor's understanding as is the realization that every component of cost, no matter its name or its classification in an accounting sense, is a component of cost and must be included in the contractor's estimate. Many contractors utilize the expression, "If I need this item in order to perform the work, and if I do not have its cost included in the estimate, then I will be furnishing it to the project free." This is hardly a good business procedure. Therefore, every effort is made by the estimator to include a cost in the estimate for everything that will be needed in order to perform the work on the project.

In determining what elements of equipment will be needed to perform the work, the estimator must carefully assess the exact characteristics that the equipment must have in order to effectively perform the work on the project being estimated. Factors such as lifting capacity, height or depth of reach, production rate, mobility, and a great many other factors determine the suitability of the equipment to meet the needs of the work to be done.

If the contractor owns the equipment with the characteristics needed, the estimator will proceed with determining its cost to the project by calculating the contractor's ownership and operating costs for each piece of equipment and ascribing those costs to the project based on the amount of time the machine will be needed on the project. If the contractor does not own a piece of equipment whose need has been identified for a project, he may consider purchase, or he may rent or lease the necessary equipment. If a purchase decision is made, the estimator will proceed with determining ownership and operating costs.

If the contractor does not own the needed equipment and decides not to purchase it, then the equipment will be rented or leased for the project. While there certainly are differences between equipment rental and lease agreements, for purposes of understanding the estimating process, the process of estimating the costs of equipment rental or leasing will be considered to be the same.

In this case, having determined the necessary characteristics for the machine to be rented or leased, the estimator will contact equipment rental or leasing agencies in order to obtain price quotations. These quotations are usually stated by rental or lease agencies as a *rate*—dollars per hour, per day, or per week, and so on. The estimator will determine the amount of time the equipment will be needed on the project and will multiply by the best rate quoted to determine the cost of the rented equipment for the project.

The estimator must add to the quoted equipment rate his or her consideration of the cost of other typical elements of the rental or lease agreement such as delivery and pickup charges, setup fees, loss and damage waivers, and so forth. Additionally, operating and maintenance costs, as well as the cost of the equipment operator, must usually be determined and included as additional elements of the cost of the rented equipment. The estimator will tabulate all of these elements of the cost for the rented machine and will include these costs in the estimate.

For equipment that the contractor already owns that is needed for a project, the estimator will determine the *rate*, also known as the *internal rate*. The internal rate is the sum of the contractor's ownership and operating cost for the machine, and it is expressed as dollars per unit of time—dollars per hour, per day, per week, and so forth. Ownership and operating costs for equipment that the contractor owns will have been determined by the cost accounting system that the contractor has developed for his business.

Ownership costs can be defined as costs that continue to accrue, whether the machine is in use or not. Examples include: financing costs for buying the machine (or sinking fund contributions for replacing the machine at the end of its useful life) and depreciation. Operating costs, as their name indicates, include costs associated with the operation of the machine. Examples include: fuel, lube, hydraulic fluid, scheduled maintenance, repairs, and so forth.

To determine ownership and operating costs, most contractors utilize their historical cost database. Machine costs are tabulated over a period of time and are stored in the historical cost database; they are then converted to the internal rate.

The estimator will determine how long the piece of equipment is needed on the project whose cost he is estimating and will then multiply by the internal rate to determine the cost of utilizing that machine on the project. In the future, as the project is constructed, revenue received for the performance of the work will be allocated, on an internal accounting basis in the contractor's office, to pay for the ownership and operating costs of the contractor's owning this machine and furnishing it for use on this project. Additionally, as the project is constructed, the costs of the machine performing work on the project will be tabulated and will be stored in the historical information database.

The estimator will also determine the cost of the machine operator in the project cost estimate. The operator's hourly rate will be multiplied by the time he or she will spend operating the machine on the project, in order to determine dollars of operator cost for the estimate.

In a procedural note, the supervisor should know that, although equipment operator costs are estimated as noted, these costs are usually tabulated in the labor cost section of the estimate. Although it is logical to assume that the equipment cost should include equipment operator cost, the inclusion of operator costs in the labor cost section of the estimate simply facilitates calculations of indirect labor costs, which are based on labor payroll dollars.

Subcontractor Proposals

For those items of work on the project that contractors do not plan to perform with their own craft labor forces, they will plan to award subcontracts. Most contractors will maintain a file of subcontractors in a variety of specialty trades whom they know and have worked with in the past.

The estimator will inform subcontractors in trades whose work is included on this project that the contractor is planning to submit a proposal for the project, and will invite subcontractors to prepare and submit proposals for their part of the work to the contractor in advance of bid day. In addition, it is common for subcontractors who have learned of the existence of the project being proposed to inquire of a number of general contractors whether they are planning to submit a proposal, and whether they would accept a subcontractor proposal.

Each subcontractor will follow an estimating process parallel to the one being employed by the general contractor. The subcontractor will analyze the contract documents, primarily the drawings and specifications, for those elements of the project that pertain to the subcontractor's trade. The activities or items of work to be performed by the subcontractor are referred to as the scope of work, often referred to simply as scope. The subcontractor's estimator will estimate materials, labor, equipment, indirect cost, and markup, in order to arrive at a price to be entered onto a subcontract proposal for the subcontractor's defined scope of work.

The subcontractor will prepare a subcontract proposal, which includes a scope statement, a listing of items of work included in the proposal, and items of work excluded from the proposal, along with the proposal amount, which is usually in lump sum. The scope statement will provide definition for the work the subcontractor will perform in descriptive format, and often will include a listing of Construction Specifications Institute (CSI) division numbers of the proposed work.

Subcontractors will submit their proposals to one or more general contractors, in time for the general contractor to include the subcontractor's proposal in the estimate prior to the time the proposal is due to the owner. General contractors will receive all of the subcontractor proposals for each category of the work on the project that they do not plan to perform with their own craft labor forces. They will analyze each proposal, and will compare the proposals for each category of work, considering completeness of scope versus price for each subcontract proposal.

The general contractor will then select the proposal in each category of work that is the most complete and most competitive, and will include that proposal price in his own estimate as the estimated price for that component of the work on the project. In this manner, the proposals of the plumbers; electricians; heating, ventilating, and air conditioning specialists; roofers; and all of the other subcontractors who will perform the work that the general contractor is not planning to perform with their craft labor will be entered into the general contractor's estimate and will become a component of his proposal price.

Subsequently, when the general contractor's proposal has been accepted by the owner and the prime contract has been formalized, the general contractor will notify those subcontractors whose proposals have been accepted. A subcontract agreement, commonly referred to as the subcontract, will be entered between the general contractor and each of the subcontractors who have been selected. The subcontract agreement will contain the scope of work and the subcontract price in accord with the proposal that the subcontractor submitted at the time of making the estimate.

Estimating Indirect Cost

Indirect costs are also referred to as overhead costs. Indirect costs can be defined as costs that a contractor incurs and dollars that the contractor must pay for elements that do not contribute in a direct way to the completion of a project in the sense in which materials and labor and equipment do. Indirect costs are often referred to as costs for items that support the work being performed on the project.

Indirect costs are subdivided into two categories: project overhead and office overhead. Project overhead costs may also be referred to as general conditions costs or as project indirect. Office overhead costs may also be referred to as company overhead or as general overhead.

Examples of project overhead costs include: the job site office building, as well as equipment and utilities for the job site office; the job site fencing and gates; temporary lighting; temporary water; temporary heat; break areas for craft workers; drinking water; temporary parking; crushed stone travel ways on the job site; toilet facilities for workers; scaffolding; ladders; toolboxes (gang boxes); hand tools; and so on.

Usually the salaries of project supervision, foreman, and superintendent, are included in project overhead, and sometimes the project manager's salary is included here as well. These project overhead costs for a project may be determined by direct estimation. More typically, however, these costs are extracted from the contractor's historical cost database and are then included in the estimate.

A contractor's office overhead includes the cost of owning or leasing the contractor's home office; salaries of the home office staff, such as the company president, estimators, cost accountants, reception, and clerical workers; utilities and equipment for the home office; and so forth. Some companies include the salaries of project managers in the category of general overhead.

To facilitate estimating general overhead, the construction company typically tabulates the costs that have been defined as general overhead over a period such as a year. These costs are then apportioned to each project that the company estimates on a *pro rata* basis. Some companies apportion the cost of office overhead to each project being estimated based on the number of projects the company usually performs in a calendar year. Other firms pro rate the anticipated total cost of the project being estimated as a fraction of the company's volume in a year and then include that fraction or percentage of yearly company overhead costs in the cost estimate for the project currently being estimated.

For the supervisor, it is not as important to understand the exact method of apportionment of company overhead costs, as it is to recognize what the components of these costs are and to know that the company must have some reasonable method of apportioning these costs and including them in the cost estimate for each project that the company will perform. After the company has been awarded a construction contract, as the work on the project is performed and revenue is generated, some of that revenue is utilized to pay the costs of the company's general overhead.

Markup

Markup, also referred to as margin in some construction companies, refers to the number of dollars included in the estimate that the company plans to keep for itself as return for performing the work after all of the costs associated with the performance of the project have been paid. The markup determination is typically the last component to be determined when the estimate is finalized.

While some may refer to this component of the estimate as "profit," others argue that the term *profit* refers to dollars that the company has *earned* after all project costs have been paid. At the time the estimate is being prepared, consideration is given to how many dollars the company would like to have earned or anticipates having earned, after all costs have been paid. This is the rationale for the usage of the term *markup* rather than profit for this component of the estimate.

The amount of the markup to be included in the estimate for the project is typically determined by the company's executive management or by the chief estimator. The decision is momentous.

In a business sense, the company should expect to earn sufficient markup to justify the business enterprise and the effort and the risk of performing construction projects. In the competitive environment of the construction industry, however, the company cannot overprice its expected markup to the extent that it then is not sufficiently competitive to be awarded the contract.

Additionally, the markup determination is tempered by a number of other considerations. These include the amount of confidence that management or the chief estimator have in the other components of the estimate. This thought is based in the realization that if the company is awarded the construction contract, and if some component of construction cost exceeds the estimated amount when the work is performed, those additional costs can be recovered in only two ways. Either the construction of some other element of the project must be completed at a cost less than the estimated amount in order to compensate for the overage, or the cost overrun must be compensated by reduction in the amount of the markup the company will receive.

Other factors that influence the markup determination include: composition of the project team, the owner and architect; who the other contractors are who are competing for the contract award (which the company may or may not know at the time of preparation of the estimate); the company's assessment of how urgently it needs a contract award for a project in order to maintain its operations or to keep its key people employed. Also to be considered are the time of year and expected weather during the time the construction will be performed; market conditions relative to the availability of craft labor, materials, and subcontractors needed to construct the project; and numerous other elements.

The single concept that seems to best describe in summary fashion the factors that determine the amount of markup to be included in the estimate is *risk*. The greater the risk the contractor perceives in the project in all of its aspects, the greater the amount of markup he will include in the estimate in order to compensate for the risk.

FINALIZING AND SUBMITTING THE ESTIMATE

When a total has been determined for each of the six components of the estimate, a summation of the six subtotals is made. This sum and the amount of money it represents is the contractor's estimated dollar amount for the performance of the work on the project in such a manner as to comply with the requirements of the contract documents, and this is the amount that will be submitted on the general contractor's proposal to the owner.

The contractor will double-check all of the components of his estimate for completeness and accuracy and will finalize the proposal before signing the proposal and submitting it to the owner. This process is referred to as "closing out" the estimate.

The proposal is completed in exact compliance with the stipulations set forth in the Instructions to Bidders, and it is then submitted to the owner. The general contractor's proposal will be analyzed by the owner following its submittal, in competition with the proposals of the other contractors who are seeking the award of the contract for the construction of the project.

A prime contractor will be selected by the owner to receive the contract award and will sign the agreement, which formalizes the contract between the owner and the prime contractor or general contractor. The general contractor will then proceed with completing subcontract agreements and materials purchase orders.

Subsequently, the contractor will receive a letter of intent or a notice to proceed from the owner. These documents will authorize the contractor to occupy the site and to begin work, and they will denote the beginning of the project duration. The contractor will then immediately occupy the construction site and will commence work on the construction of the project.

During the performance of the work in the contract, the supervisor will perform his function of managing the work of the craft labor crew and coordinating the work of the subcontractors on the project. In addition, during the construction of the project the supervisor will contribute in a very significant way to the contractor's cost determination, cost reporting, cost accounting, cost control system. These functions will be described and defined in the next chapter.

SUMMARY

The costs associated with the construction of a project are first defined and determined in the contractor's estimating process. This chapter has described the process by which the estimate is prepared. It is important for the supervisor to understand all aspects of construction costs, including the cycle that was presented in Figure 12.1. Additionally, it is important for the supervisor to understand the elements of the estimating process and the process of preparing a detailed estimate for a construction project.

The costs that compose the estimate are the basis for the determination of which contractor will receive the contract award. As the next chapter will illustrate, after the contract is awarded, the actual costs of performing the work are tabulated as the project is performed and are closely compared to the estimated costs in the form of cost reports that are prepared by the contractor on a regular basis throughout the performance of the work on the project. The supervisor is a key participant in this process.

Learning Activities

1. Learning more about costs and about the estimating process

   Try to arrange to spend some time with an estimator in your company. Perhaps you can meet for lunch on occasion.

   Visit with the estimator to learn more about how he or she goes about the process of estimating.

   Ask about the specific methods he or she uses to quantify materials, and to estimate equipment, indirect costs, and labor.

   Ask about how the estimator utilizes the historical cost information, or the production rates and productivity factors derived from historical cost, in preparing the estimate.

   Ask the estimator how important it is for construction labor costs to be accurately reported in the field for inclusion in the historical cost information database.

2. One of the longest-running debates among construction company managers is whether it is better for estimators to be people who have worked extensively in the field before becoming estimators or not.

Sit down with an estimator, or project manager, or construction company executive whom you know, and discuss this matter so that you will have the benefit of their viewpoint.

3. Try to arrange to attend a bid opening, or to be an observer at a presentation in conjunction with a competitive sealed proposal where your company is submitting a proposal.

CHAPTER 13

MANAGING COSTS— ELEMENTS OF COST CONTROL

INTRODUCTION

I t is essential for the supervisor to have a firm understanding of construction costs, as well as a clear understanding of the cost reporting, cost accounting, and cost control system utilized by the contractor. This system provides a critically important function in the contractor's operations, and since the supervisor is a direct participant in its operation, the supervisor absolutely *must* understand every aspect of this system.

THE SYSTEM IS A CYCLE

For the supervisor, it is best to view the important matter of cost control not only as a vitally important function unto itself but also as a component of a larger operational system within the construction company. While details will vary somewhat from one construction company to another, most contractors have established a system whose functional components operate in the form of a continuous cycle. Figure 13.1 illustrates such a system.

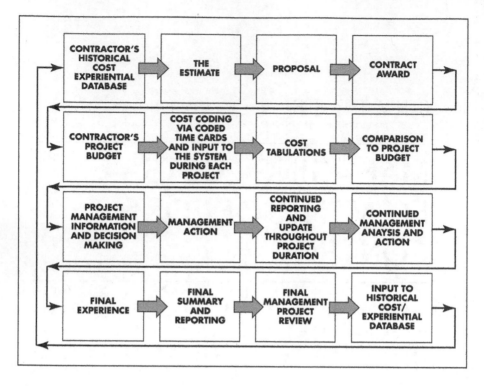

Figure 13.1 Project Cost Reporting, Cost Accounting, Cost Control System

Central to the process, and fundamental for supervisors' understanding, is the *project cost accounting system*. The key features of this system, which is at the heart of the cycle, is illustrated in Figure 13.2. The basic operational component of this system is a series of "cost codes," which have been developed by the contractor to describe the activities that the company performs during the course of completing

CONTRACTOR'S PROJECT COST ACCOUNTING SYSTEM

- **SYSTEM OF COST CODES ESTABLISHED**
- **INPUT TO THE SYSTEM VIA CODED LABOR TIME CARDS**
- **PAYROLL PREPARES CRAFT LABOR CHECKS**
- **PROJECT COST ACCOUNTING TO COST CODE CATEGORIES**
- **COST REPORTS PREPARED THROUGHOUT PROJECT**
- **MANAGEMENT COMPARISON TO PROJECT BUDGET**
- **MANAGEMENT ANALYSIS**
- **MANAGEMENT DECISION MAKING**

Figure 13.2 Central Elements of a Project Cost Accounting, Reporting, and Control System

its work on a construction project. Each activity has an assigned cost code number and a name. Typical examples of cost codes might be: "3117 Wall Forms," "16124 Conduit in Framed Walls," and "9256 5/8" Fire Code Drywall on Ceilings."

The contractor will prepare such a system of cost codes that describe the activities in the work to be performed. Some contractors develop their cost code numbers and activity descriptions themselves, in-house. Others utilize the Construction Specifications Institute's (CSI) naming and numbering structure. Still others utilize a system developed by a contractors' professional association. Whatever their derivation, once these cost code numbers and activity names have been developed by the contractor, they will be utilized on every project that the contractor performs.

Typically, once a contractor has developed and operationalized this set of cost codes, they will remain unchanged in terms of their basic structure and composition. At the same time, most cost-coding systems are established with sufficient inherent flexibility so as to accommodate changing technology and its effects on the work the contractor does. Clean room construction, fiber-optic cable installation, and photovoltaic glazing are examples of fairly recent technological developments whose construction activities were necessarily incorporated by contractors into their existing cost-coding systems.

The supervisor will utilize this cost-coding system on all of the work that he manages. He will be provided with the set of cost codes that the contractor utilizes in his system. As the construction work in the field is performed, the supervisor is expected to monitor the work, and on a daily basis is expected to complete a "coded time card" for each person under his supervision who worked on the project during that day. The supervisor will enter onto each construction craft worker's time card, the number of hours which that person worked in performing various activities during the day, accompanied by the cost code number for each of those activities from the contractor's cost-coding system.

Figure 13.3 illustrates a typical time card for a construction worker as prepared by the construction supervisor for a typical five-day work week. Note that, for each day's work by Alfred during the week, the total number of hours worked that day is recorded on the time card. Additionally, and importantly, each activity on which this craft worker performed labor during each day is recorded on the time card by the activity cost code number for that activity. Each cost code number is accompanied by the number of hours that the worker spent working on that activity during the day.

The supervisor will prepare a time card in similar fashion for each person in the supervisor's crew, for every work day throughout the duration of the project. This is referred to as "cost coding the work." It is a critically important supervisory function.

Two points merit special emphasis here. First, it is very important for the supervisor to accurately record the cost codes for all of the activities on which each worker performed labor during each work day, by using the correct cost code number for each activity. If the supervisor fails to accurately cost code all of the construction labor, then the consequence is that inaccurate information is provided to the company's cost accounting and cost control system, and to the historical information database, which were discussed in Chapter 12.

JRRS CONSTRUCTION COMPANY
WEEKLY TIME CARD

EMPLOYEE NAME						ALFRED KLEIN
COMPANY EMPLOYEE IDENTIFICATION NUMBER						3224
CLASSIFICATION: J6					RATE:	

DAY AND DATE		JOB NUMBER	COST CODE	REGULAR HOURS	OVERTIME HOURS	NOTES
MONDAY	8-16	2251	6410	4.0	0	Helper: Billy
		2251	6265	2.0	0	Helper: Billy
		2251	6110	2.0	0	Helper: Billy
TUESDAY	8-17	2251	6110	8.0	0	
WEDNESDAY	8-18	2251	6110	8.0	0	
THURSDAY	8-19	2251	6190	8.0	0	Helper: John
FRIDAY	8-20	2251	6152	6.0	0	Helper: John
		2251	6190	2.0	0	Helper: John
SATURDAY				0	0	
SUNDAY				0	0	
WEEKLY TOTAL:				40.0	0	

Figure 13.3 Typical Craft Labor Time Card

Additionally, it is very important that this cost coding be completed no later than the end of each work day. The supervisor must avoid the temptation to wait until some later time, or until the time when time cards are due for submittal to the company office for payroll purposes, to enter each worker's cost codes on the time card. The accuracy of cost coding is sure to suffer if this practice is followed, and when labor costs are not accurately coded by the supervisor, a great number of problems surely will result, as will be pointed out in discussions that follow.

The supervisor enters cost codes and hours worked on each activity on the time cards for every craft worker on the project for every work day throughout the duration of the project. Once per week, or more often in some companies, these time cards are submitted to the contractor's office. The payroll department in the company office will collate the hours of work performed during the week by each construction worker and will use this information to prepare a payroll check for each craft worker, usually on a weekly basis.

Additionally, the cost-coded information from each time card for each craft worker on the project will be tabulated into the cost-reporting system for the project. This system utilizes the project budget as a basis, accrues all of the costs associated with each construction activity, and compares cost to date against progress to date with project budget values. This important information is then reported by means of a series of periodic cost reports throughout the duration of the project. This process will be further discussed in the section that follows.

THE PROJECT BUDGET AND THE PROJECT LABOR BUDGET

The project budget can be defined as a document that is prepared by the construction contractor for each project that the contractor performs. The project budget is an internal document, which is prepared for the contractor's own use in tabulating and controlling his project costs on each project he constructs.

The contractor will incorporate into the project budget each element of construction cost on the project for which he wishes to monitor costs for the purpose of exercising management control during the construction of the project. While various elements of cost may be included in project budgets, the element of cost that is almost always included is construction labor cost.

As noted previously, on most construction projects, labor costs constitute the largest single component and also the largest percentage of the total project cost. Additionally, labor costs are highly variable, and they vary in response to a large number of factors. The productivity of individual craft workers, and of construction crews, varies on different projects and in different environments. In addition, labor costs are difficult to estimate, so there is uncertainty with regard to labor cost when construction estimates are prepared.

For all of these reasons, labor costs are a source of uncertainty and risk for the contractor on construction projects. Therefore, the construction contractor, as well as the supervisor, must make every effort to continuously monitor and to control labor costs throughout the duration of every project that the contractor performs. The management team absolutely must know on a regular basis throughout the life of the project whether labor costs are in line with the costs that were estimated and budgeted for the project. For this reason, labor costs are regularly tabulated and reported, so that the contractor's management team can analyze the costs as they are incurred and can take management action if labor costs exceed the project labor budget amount. The supervisor plays a vital role in this entire process.

The project budget is the basis for the tabulations and cost reports as noted above. Following the formalizing of the construction contract and signing the agreement, and before commencing work on the project, the contractor will prepare the project budget, which, as has been noted, almost always includes the project labor budget. The project budget is the primary tool that will be utilized by the contractor's management team, in the office as well as in the field, for monitoring and controlling project costs throughout the life of the project.

The project budget consists of a listing of the all of the activities that the contractor will perform in completing the work on the construction project. As noted previously, activities are defined as elements of work that can be identified and quantified and that consume resources. All of the activities that the contractor must complete, from start to finish on the project, are listed on the project budget. These activities are listed in roughly the same order in which they will be performed during the course of the work, and in an amount of detail to suit the management needs of the contractor.

Along with each activity is listed the amount of money, the number of labor man-hours, or the number of production units that the contractor has determined are required for the work that must be completed on the project. In a literal sense, the contractor quantifies on the project budget the amount of resources budgeted for the completion of each activity. These amounts are referred to as the budgeted amount for each activity in the project budget.

Additionally, the project budget will include the amount of each activity to be performed on the project. These quantities will be based in those determined during the preparation of the cost estimate that the contractor formulated in the course of bidding or negotiating for the contract award.

The activities, their quantities, and their budgeted amounts are usually listed in columnar form on the project budget. Two illustrative examples of project budgets are shown in Figure 13.4 and Figure 13.5.

Throughout the life of the construction project, the project budget, and the periodic project cost reports for which the project budget forms the basis, will be used by all levels of the contractor's management team in order to monitor labor costs on the project. This information provides a continuous status report with regard to project costs for the contractor. The contractor simply must know,

Phase	Division	Cost Code	Description	Original Budget	Uncommitted Costs
3 digit phase must exist in project	2 digit division must exist in project	8 digit cost code must exist in project	Budget code description	Original budget amount	Uncommitted costs amount
100	01	01111000	General Condition	$2,680,101.00	$2,680,101.00
150	01	01912000	Sub-Guard	$618,548.00	$618,548.00
150	01	01901000	General Liability Insurance	$504,214.00	$504,214.00
150	01	01911000	P&P Bond	$440,576.00	$440,576.00
150	01	01902000	Builders Risk Insurance	$122,382.00	$122,382.00
100	10	10000000	General Trades	$2,242,642.00	$2,242,642.00
200	01	01226000	LEED/BIM Coordination	$75,000.00	$75,000.00
200	01	01222000	Owner Office Equipment	$5,000.00	$5,000.00
200	01	01204000	Owner Office Utilities	$14,000.00	$14,000.00
200	01	01201500	Owner Office Trailer	$21,000.00	$21,000.00
200	01	01590500	Temp Utilities	$939,000.00	$939,000.00
200	02	02300000	Site Work and Paving	$2,325,331.00	$2,325,331.00
200	02	02925000	Landscape and Irrigation	$558,905.00	$558,905.00
200	03	03000000	Key Concrete	$5,560,000.00	$5,560,000.00
200	04	04000000	Masonry	$1,571,430.00	$1,571,430.00
200	05	05100100	Misc. Metals	$2,521,602.00	$2,521,602.00
200	06	06200000	Millwork	$485,568.00	$485,568.00
200	07	07500000	Roofing and Sheet Metal	$1,519,341.00	$1,519,341.00
200	07	07100000	Waterproofing	$364,382.00	$364,382.00
200	08	08000000	Glass and Glazing	$2,592,524.00	$2,592,524.00
200	09	09000000	Drywall	$3,180,799.00	$3,180,799.00
200	09	09300000	Ceramic Tile and Stone	$812,660.00	$812,660.00
200	09	09600000	Flooring	$673,416.00	$673,416.00
200	09	09500000	Painting	$490,421.00	$490,421.00
200	12	12050000	Furnishing	$323,100.00	$323,100.00
200	14	14000000	Elevators	$716,295.00	$716,295.00
200	15	15000000	HVAC/Plumbing	$7,065,008.00	$7,065,008.00
200	15	15050000	Fire Protection	$661,325.00	$661,325.00
200	16	16000000	Electrical	$6,517,765.00	$6,517,765.00
800	23	01902000	Construction Contingency	$1,649,461.00	$1,649,461.00
900	26	01901000	Construction Phase Fee	$1,701,065.00	$1,701,065.00
			Total:	$48,952,861.00	$48,952,861.00

Figure 13.4 Sample Project Budget
Courtesy of Skanska USA Building Inc.

| | **PROJECT BUDGET** | | | | | | |
| | **SHINING START OFFICE TOWER** | | | | | | |
COST CODE	ITEM NAME	MATERIALS	LABOR	SUBCONTRACT	EQUIPMENT	MISC. BUDGETED COST	TOTAL
1000	**GENERAL CONDITIONS**						
1001	BONDS					96770	96770
1002	EXCESS UMBRELLA					26800	26800
1003	BUILDER'S RISK					24900	24900
1004	OWNER PROTECTION LIABILITY					13000	13000
1005	TRAVEL & LODGING					47000	47000
1006	SUBSISTENCE					19000	19000
1007	AGC DUES					23240	23240
1008	PROJECT MANAGER		69500				69500
1009	PROJECT SUPERINTENDENT		115800				115800
1010	ASSISTANT SUPERINTENDENT		62000				62000
1011	FOREMAN		59000				59000
1012	ADMINISTATIVE ASSISTANT		0				0
1013	TIMEKEEPER		26000				26000
1014	FIELD ENGINEER		7000			7000	14000
1015	JOB CLEANUP—REGULAR		7000			14000	21000
1016	JOB CLEANUP—FINAL		1550			2000	3550
1017	TEMPORARY OFFICE		2200			14000	16200
1018	TEMPORARY OFFICE STORAGE					4800	4800
1019	TEMPORARY TOILETS					5600	5600
1024	TEMPORARY ROADS & DRAINS		1500			3600	5100
1025	TEMPORARY BARRICADES		600			2200	2800
1026	TELEPHONE					5600	5600
1027	TEMPORARY ELECTRICAL					6000	6000
1028	TEMPORARY WATER					2800	2800
1029	TEMPORARY HEAT & AC					1900	1900
1030	ICE AND CUPS					1800	1800
1031	TEMPORARY FENCING					7400	7400
1032	FIRE EXTINGUISHERS					1100	1100
1033	JOB SIGNS	600				1700	2300
1034	JOB PHOTOS					2500	2500
1035	MOVE IN AND MOVE OUT					22000	22000
1037	CONCRETE TESTING					3900	3900
1061	CONCRETE VIBRATORS				4700		4700
1062	CONCRETE FINISHING MACHINE				5800		5800
1063	AIR COMPRESSOR & PNEUMATIC TOOLS				4900		4900
1064	WELDING MACHINE & SUPPLIES				5600	1900	7500
1065	TAMPERS				1800		1800
1066	WATER PUMPS				3900		3900
1078	CRANE				38000		38000
1079	DROTT—CHERRY PICKER				26000		26000
1080	SKYLIFT				21500		21500
1081	HAND TOOLS				3450		3450
1082	POWER TOOLS					6200	6200
1091	GAS AND OIL					8100	8100
1092	MAINTENANCE AND REPAIR		1400			28000	42000
9310	HAULING—CRANES				38000		38000
1094	PAYROLL TAXES & INSURANCE					269886	269886
1095	GENERAL CONTRACTOR'S SALES TAXES					61098	61098
	TOTALS GENERAL CONDITIONS CATEGORY						**1256194**

Figure 13.5 Sample Project Budget
Courtesy of Bryan Construction

on a regular recurring basis, what the status is on the project with respect to costs incurred as compared to costs that were estimated and budgeted. Additionally, and importantly, this information is utilized as the basis for management decision making and management action throughout the conduct of the work on the project.

LABOR COST REPORTS

The contractor's cost control system will include a mechanism for the preparation, analysis, and use of periodic cost reports. Most contractors prepare cost reports on a weekly basis throughout the duration of each project. It should be noted that the information that appears in the labor cost reports is derived directly from the coded labor information that the supervisor has noted on each craft worker's time card.

Project cost reports utilize the project budget as a basis with some additional columns added. Again, the project activities, their quantities, and their budgeted amounts are usually listed in columnar form on the cost report form. A series of additional columns are also included whose typical titles are: "cost this period," "cost to date," "percentage complete," and "variance from budget." Additionally, sometimes a column is included whose title is "projected cost to complete." An illustrative example of a contractor's cost report is shown in Figure 13.6. Another example of a contractor's cost report form is illustrated in Figure 13.7.

As the work on the project is performed, at the end of each reporting period (usually weekly, as previously noted), labor costs for each activity on which work was performed during the reporting period are tabulated in the appropriate column on the cost report form. From this information, a mathematical determination is made, of total costs to date for that activity, and those values are entered onto the cost report form.

At the same time a determination is made, of the current percentage complete on the project for each activity. This percentage complete may be determined on the job site by counting, measurement, calculation, or estimation.

A calculation is performed of the percentage complete of the work on each activity compared to the percentage of the project budget amount that has been expended for that activity. The result is entered onto the cost report as "variance from budget." A notation is also made on the cost report form to indicate whether a positive (+) variance means the work is costing more than the budgeted amount at this time or that a positive (+) variance indicates the activity is costing less than the budgeted amount and is, therefore, "ahead of budget."

For example, if it is determined that the work on a certain activity is 60% complete, and 40% of the budgeted amount has been expended, it can be seen that the work is "ahead of budget" for this activity. Similarly, if the cost-reporting system indicates that the work on an activity is 20% complete, and 50% of the project budget amount has been expended, clearly the work on that activity is "over budget," and management attention is warranted.

Additionally, cost reports often include a calculation of "projected cost to complete" for the activities. This is simply a mathematical calculation of what the cost to complete will be for each activity, on the current trend, and assuming a linear progression. This projection is sometime helpful in depicting what the final outcome may be for the work on each activity or for calling attention to the variance from budget for an activity.

Thus, it can be seen that the cost report provides a summary view of the status of each activity as of each cost report date, as compared to the project budget.

JRRS CONSTRUCTION COMPANY

PERIODIC LABOR COST REPORT

PROJECT 2021:	SHINING STAR OFFICE BUILDING
PROJECT START DATE:	MM-DD-YY
REPORT DATE:	MM-DD-YY
REPORT TYPE:	LABOR OPERATIONS REPORT
DATA DATE THROUGH:	MM-DD-YY

COST CODE	ACTIVITY DESCRIPTION	LABOR BUDGET	LABOR COST THIS PERIOD	LABOR COST TO DATE	PERCENT COMPLETE	PERCENT COMPLETE TIMES BUDGET	VARIANCE UNDER BUDGET (-)	VARIANCE OVER BUDGET (+)	PROJECTED COST TO COMPLETE
1010	Project Supervisor	70000	7000	21000	30	7000	0	0	70000
1050	Field Engineer: Layout, Line & Grade	12000	1100	3900	40	3000	- 900		9750
2200	Install & Maintain Silt Fence	3000	300	3240	90	2700		+ 540	3600
2210	Fine Grade Beneath Spread Footings	4000	490	3100	100	4000	- 900		3100
3140	Wall Forming	12190	1230	8120	70	8533	- 413		11560
3180	Deck Forming, Pan Slabs	26850	1920	17240	60	16110	- 510		28733
3175	Edge Forming, Pan Slabs	36280	2800	13190	40	14512	- 1322		32975
3250	Keyway & Waterstop	3100	350	2920	100	3100	- 180		2920
3255	Bulkheads & Blockouts	2800	280	940	20	560		+ 380	4700
3315	Place & Finish Concrete Flatwork	28900	1260	5520	20	5780	- 260		27,600
3332	Rub Finish Architectural Concrete	14300	3000	4950	30	4290		+ 660	16,500
3370	Cure & Protect Concrete	6850	900	3210	40	2740		+ 470	8025
3215	Install Spread Footing Reinforcing & Chairs & Bolsters	16300	0	15280	100	16300	- 1020		15,280
5420	Steel Stud Partition Framing	48105	3800	11210	20	9621		+ 1589	56050
12350	Install Base Cabinets	13235	0	0	0	0	NA	NA	NA
1560	Job Cleanup	9980	700	2670	30	2994	- 324		8900
	ETC. TO BOTTOM LINE TOTALS IN EACH COLUMN	TOTAL	TOTAL	TOTAL	TOTAL	TOTAL	TOTAL VARIANCE SHOWN UNDER BUDGET	TOTAL VARIANCE SHOWN OVER BUDGET	PROJECTED COST TO COMPLETE, LINEAR PROGRESSION ON CURRENT TREND

Figure 13.6   Periodic Labor Cost Report

The cost report provides powerful information for the supervisor, as well as for all of the other members of the contractor's management team. This information can be analyzed and discussed by the members of the team, and it can be used as the basis for management decision making.

For example, sometimes a change in crew size or a modification in crew composition may be indicated. Sometimes the need for developing an alternative work method, or the need for utilizing equipment, additional equipment, or different equipment, may be determined. In summary, cost reports are commonly used as the basis for management analysis and management decision making as the work is performed. Further, cost reports and the management action they initiate are

Anticipated Cost Report

Detailed

Big Manufacturing Hdqtrs - Project #12345

Budget Code/ Description	A Original Budget	B Approved Revs	C=A+B Approved Budget	D Pending Revs	E=C+D Pending Budget	F Approx Revs	G=E+F Project'd Budget	H Contracts Executed	I Changes Executed	J Contracts Pending	K Changes Pending	L In Scope	M Out of Scope	N In Scope	O Out of Scope	P Uncommit Costs	Q=H thru P Project'd Costs	R=C-Q Approved Budget vs. Projected Cost	S=G-Q Projected Budget vs. Projected
Phase 100 General Conditions																			
100-01.01111000.5010 General Conditions	$2,680,101	$ 0	$2,680,101	$ 0	$2,680,101	$ 0	$2,680,101	$685,353	$0	$0	$0	$0	$0	$0	$0	$1,994,748	$2,680,101	$ 0	$ 0
100 General Conditions Subtotal:	$2,680,101	$ 0	$2,680,101	$ 0	$2,680,101	$ 0	$2,680,101	$685,353	$0	$0	$0	$0	$0	$0	$0	$1,994,748	$2,680,101	$ 0	$ 0
Phase 150 Reimbursable General Conditions																			
150-01.01901000.5040 General Liability Insurance	$504,214	$1,502	$505,718	$2,975	$508,691	$0	$508,691	$504,214	$0	$0	$0	$0	$4,467	$0	$10	$0	$508,691	($2,975)	$0
150-01.01902000.5040 Builders Risk Insurance	$2,680,101	$364	$122,746	$721	$123,467	$0	$123,467	$82,841	$0	$0	$0	$0	$1,083	$0	$2	$39,541	$123,467	($720)	$0
150-01.01911000.5040 Payment & Performance Bond	$2,680,101	$1,312	$441,888	$2,599	$444,487	$0	$444,487	$415,365	$0	$0	$0	$0	$3,903	$0	$8	$25,211	$444,487	($2,599)	$0
150-01.01912000.5040 Sub-Guard	$2,680,101	$2,187	$620,735	$4,333	$625,068	$0	$625,068	$643,290	$0	$0	$0	$0	$6,506	($24,742)	$14	$0	$625,068	($4,333)	$0
150 Reimbursable General Conditions Subtotal:	$2,680,101	$5,365	$1,691,085	$10,628	$1,701,713	$0	$1,701,713	$1,645,710	$0	$0	$0	$0	$15,959	($24,742)	$34	$64,752	$1,701,713	($10,628)	$0
Phase 200 Cost of Work																			
200-01.01201500.5031 Owner Office Trailer	$21,000	$0	$21,000	$0	$21,000	$0	$21,000	$3,970	$0	$0	$0	$0	$0	$0	$0	$17,030	$21,000	$0	$0
200-01.01204000.5031 Owner Office Utilities	$14,000	$0	$14,000	$0	$14,000	$0	$14,000	$0	$0	$0	$0	$0	$0	$0	$0	$14,000	$14,000	$0	$0
200-01.01220000.5031 Owner Office Equipment	$5,000	$0	$5,000	$0	$5,000	$0	$5,000	$0	$0	$0	$0	$0	$0	$0	$0	$5,000	$5,000	$0	$0

Figure 13.7 Contractor's Cost Report Form
Courtesy of Skanska USA Building Inc.

Big Manufacturing Headquarters: Project Number 12345

Anticipated Cost Report — Detailed

Column legend: A, B under **Approved Budget** (A = Original Budget, B = Approved Revs); C = A+B Approved Budget; D = Pending Revs (under **Pending Budget**); E = C+D Pending Budget; F = Approx Revs (under **Projected Budget**); G = E+F Project'd Budget; H, I, J, K under **Committed Costs** (H = Contracts Executed, I = Changes Executed, J = Contracts Pending, K = Changes Pending); L, M under **Pending Cost Events** (L = In Scope, M = Out of Scope); N, O under **Approx Cost Events** (N = In Scope, O = Out of Scope); P = Uncommit Costs; Q = H thru P Project'd Costs; R = C-Q Approved Budget vs. Projected Cost; S = G-Q Projected Budget vs. Projected.

Budget Code / Description	A	B	C=A+B	D	E=C+D	F	G=E+F	H	I	J	K	L	M	N	O	P	Q=H thru P	R=C-Q	S=G-Q
200-01.01224000.5031 LEED/BIM Coordination	75,000	0	75,000	0	75,000	0	75,000	12,774	0	0	0	0	0	0	0	62,226	75,000	0	0
200-01.01590500.5031 Temporary Utilities Consumption Allowance	939,000	0	939,000	0	939,000	0	939,000	0	0	0	0	0	0	0	0	939,000	939,000	0	0
200-02.02300000.5020 Brazos Paving	2,325,331	19,145	2,344,476	275,620	2,620,096	32,207	2,652,303	2,325,331	0	0	99,609	0	214,863	0	12,500	43,529	2,695,832	(351,356)	(43,529)
200-02.02925000.5020 Landscape and Irrigation	558,905	0	558,905	41,900	600,805	0	600,805	0	0	0	0	0	41,900	0	0	558,905	600,805	(41,900)	0
200-03.03000000.5020 Keystone Concrete	5,560,000	127,682	5,687,682	0	5,687,682	69,540	5,757,222	5,560,000	0	0	127,682	0	0	5,000	64,540	81,700	5,838,922	(151,240)	(81,700)
200-04.04000000.5020 RW Pfeffer Masonry	1,571,430	0	1,571,430	0	1,571,430	2,500	1,573,930	1,571,430	0	0	0	0	0	0	2,500	0	1,573,930	2,500	0
200-05.05100100.5020 Mynex	2,521,602	0	2,521,602	0	2,521,602	5,000	2,526,602	2,421,102	0	0	0	0	0	5,000	5,000	117,601	2,543,703	22,101	(17,101)
200-06.06200000.5020 MGC	485,568	0	485,568	0	485,568	7,500	493,068	485,568	0	0	0	0	0	0	7,500	0	493,068	7,500	0
200-07.07100000.5020 Alpha Insulation and Waterproofing	364,382	0	364,382	0	364,382	0	364,382	364,382	0	0	0	0	0	0	0	0	364,382	0	0
200-07.07500000.5020 Rain King	1,519,341	0	1,519,341	0	1,519,341	0	1,519,341	1,519,341	0	0	0	0	0	0	0	0	1,519,341	0	0
200-08.08000000.5020 Floyd's Glass	2,592,524	0	2,592,524	0	2,592,524	0	2,592,524	2,592,524	0	0	0	0	0	0	0	0	2,592,524	0	0
200-09.09000000.5020 Drywall Systems, Inc.	3,180,799	(709)	3,180,090	0	3,180,090	16,000	3,196,090	3,180,800	0	0	0	0	(709)	1,000	15,000	0	3,196,091	(16,001)	(0)
200-09.09300000.5020 Ceramic Tile and Stone	812,660	0	812,660	0	812,660	500	813,160	812,660	0	0	0	0	0	0	500	0	813,160	(500)	0
200-09.09500000.5020 Radian Point Contractors	490,421	(335)	490,086	0	490,086	500	490,586	490,421	0	0	0	0	(335)	0	500	0	490,586	(500)	0
200-09.09600000.5020 Flooring	673,416	0	673,416	0	673,416	0	673,416	673,416	0	0	0	0	0	0	0	0	673,416	0	0
200-12.12050000.5020 Furnishing	323,100	0	323,100	0	323,100	0	323,100	0	0	0	0	0	0	0	0	323,100	323,100	0	0
200-14.14000000.5020 ThyssenKrupp Elevator	716,295	0	716,295	0	716,295	0	716,295	716,295	0	0	0	0	0	0	0	0	716,295	0	0

Figure 13.7 (Continued)

Big Manufacturing Headquarters: Project Number 12345

Budget Code/ Description	A Original Budget	B Approved Revs	C=A+B Approved Budget	D Pending Revs	E=C+D Pending Budget	F Approx Revs	G=E+F Project'd Budget	H Contracts Executed	I Changes Executed	J Contracts Pending	K Changes Pending	L In Scope	M Out of Scope	N In Scope	O Out of Scope	P Uncommit Costs	Q-H thru P Project'd Costs	R-C-Q Approved Budget vs. Projected Cost	S-C-Q Projected Budget vs. Projected
200-15.15000000.5020 Lockridge Priest	$ 7,065,008	$ 0	$ 7,065,008	$ 0	$ 7,065,008	($ 14,500)	$ 7,050,508	$ 7,065,008	$ 0	$ 0	$ 0	$ 0	$ 0	$ 5,500	($ 20,000)	$ 13,073	$ 7,063,581	$ 1,427	($ 13,073)
200-15.15050000.5020 Standard Automatic Fire Protection	$ 681,325	$ 0	$ 681,325	$ 0	$ 661,325	($ 5,000)	$ 656,325	$ 805,475	$ 0	$ 0	$ 0	$ 0	$ 0	($ 5,000)	$ 0	$ 0	$ 800,475	$ 139,150	($ 144,150)
200-16.16000000.5020 Britt Rice Electric	$ 6,517,765	$ 0	$ 6,517,765	$ 203,601	$ 6,721,368	$ 31,000	$ 6,752,366	$ 6,517,765	$ 0	$ 0	$ 0	$ 0	$ 203,601	$ 13,000	$ 18,000	$ 125,000	$ 6,877,366	$ 359,601	$ 125,000
200 Cost of Work Subtotal:	$38,993,872	$ 145,783	$ 39,139,655	$ 521,121	$ 39,660,776	$ 145,247	$39,806,023	$37,118,262	$ 0	$ 0	$ 227,291	$ 0	$ 459,320	$ 20,000	$ 105,540	$ 2,300,164	$40,230,577	($ 1,090,922)	$ 424,554
Phase 800 Contingencies																			
800-23.01902000.5040 Construction Contingency	$ 1,649,461	($ 159,643)	$ 1,489,818	($ 232,342)	$ 1,257,476	($ 167,247)	$ 1,090,229	$ 0	$ 0	$ 0	$ 0	$ 0	$ 0	$ 5,000	($ 3,000)	$ 1,088,229	$ 1,090,229	$ 399,589	$ 0
800 Contingencies Subtotal:	$ 1,649,461	($ 159,643)	$ 1,489,818	($ 232,342)	$ 1,257,476	($ 167,247)	$ 1,090,229	$ 0	$ 0	$ 0	$ 0	$ 0	$ 0	$ 5,000	($ 3,000)	$ 1,088,229	$ 1,090,229	$ 399,589	$ 0
Phase 900 Fee																			
900-26.01901000.4400 Construction Phase Fee	$ 1,701,065	$ 8,495	$ 7,709,560	$ 17,085	$ 1,726,645	$ 0	$ 1,726,645	$ 1,701,064	$ 0	$ 0	$ 0	$ 0	$ 25,580	$ 0	$ 0	$ 0	$ 1,726,645	$ 17,085	$ 0
900 Fee Subtotal:	$ 1,701,065	$ 8,495	$ 7,709,560	$ 17,085	$ 1,726,645	$ 0	$ 1,726,645	$ 1,701,064	$ 0	$ 0	$ 0	$ 0	$ 25,580	$ 0	$ 0	$ 0	$ 1,726,645	$ 17,085	$ 0
Phase SP1 Self Perform Work																			
SP1-10.10000000.5020 General Trades	$ 2,242,642	$ 0	$ 2,242,642	$ 0	$ 2,242,642	$ 4,500	$ 2,247,142	$ 2,242,642	$ 0	$ 0	$ 0	$ 0	$ 0	$ 0	$ 4,500	$ 0	$ 2,247,142	$ 4,500	$ 0
SP1 Self Perform Work Subtotal:	$ 2,242,642	$ 0	$ 2,242,642	$ 0	$ 2,242,642	$ 4,500	$ 2,247,142	$ 2,242,642	$ 0	$ 0	$ 0	$ 0	$ 0	$ 0	$ 4,500	$ 0	$ 2,247,142	$ 4,500	$ 0
Grand Totals	$48,952,861	$ 0	$ 48,952,861	$ 316,492	$49,269,353	($ 17,500)	$49,251,853	$43,393,030	$ 0	$ 0	$ 227,291	$ 0	$ 500,859	$ 258	$ 107,074	$ 5,447,894	$49,676,407	($ 723,546)	($ 424,554)

Figure 13.7 (Continued)

232

the primary control tool utilized by the contractor to ensure that the work on the construction project will be completed at a cost equal to or less than the project budget.

This process of taking coded labor information from the supervisor's time cards and incorporating that critically important information into periodic project cost reports, and then utilizing that information to determine project labor status relative to the project budget continues throughout the duration of the project. This is the management control mechanism that most contractors employ in order to track their costs and to manage their costs on the construction contracts that they perform.

It should again be noted that the information that appears in the labor cost reports is derived directly from the coded labor information that the supervisor has entered onto each craft worker's time card, when the supervisor submits the time card to the company office. The importance of the matter of accurately coding and accurately reporting the craft labor hours expended in the performance of each activity on the project should now be clear to the supervisor.

THE CYCLE CONTINUES

As you have seen, when the supervisor submits craft workers' coded time cards to the company office, the office staff uses this information in order to prepare payroll checks for all of the construction craft workers. Additionally, the coded labor that is indicated on the time cards that the supervisor has prepared is tabulated and incorporated into periodic project cost reports.

There is yet another very important application of the coded labor information that has been submitted by the supervisor. This application is likewise a vitally important component of the contractor's operations.

The coded labor information that the supervisor provides is also compiled into a database of historical cost information that the contractor maintains. This historical cost is considered by most contractors to be one of their most important assets, for it represents, quite literally, the company's history in terms of its costs of performing work (and its labor costs in particular), on all of the projects that the company performs. As was discussed in the previous chapter, this information is regularly used by the contractor's estimator(s) in preparing the cost estimates for future projects for which the company will submit proposals.

Generating Historical Cost Information

Following the utilization of the supervisor's coded time cards for payroll and for cost report preparation, company office staff will take the time cards and will tabulate the coded costs into the company's historical cost information database. This historical cost information file is typically structured with the same activity names and numbers that the contractor's system uses for labor coding, project budgeting, and cost reporting.

The historical cost database will compile the history of the contractor's costs of completing each activity for all of the projects that the contractor performs. It is considered to be some of the most valuable and closely guarded information that the contractor has.

Typically, information is recorded in the company's historical cost file in such a way that the historical cost for performing each activity can be sorted by project, and also by an average for all of the projects that the contractor has performed. As was discussed in Chapter 12, this information is then used by the contractor's estimating department in determining the costs that are estimated for future projects.

Thus, it can be seen that the coded labor man-hours that the supervisor enters onto each craft worker's time card prior to each payroll period form the basis for the labor cost totals that are maintained in the contractor's historical cost information database. So, in a very literal sense, the labor budget, which the supervisor sees as a component of the project budget, and as a component of the cost reports for the projects that the supervisor is currently managing, are a function of the information that the supervisors (and others like them in the company) have submitted to the system when they "cost coded" the time cards that they submitted. And, as is true of any system, the accuracy and, therefore, the value of the information that the system provides is a function of the accuracy of the information that is input to the system by the supervisors.

Earlier in this chapter, emphasis was placed on the critical importance of supervisors coding craft labor on the projects that they manage in a timely and accurate fashion. The reason for this should now be apparent to the supervisor.

PROJECT REVIEW

A management practice that is utilized to great benefit by a number of construction firms is referred to as "project review." This activity typically is conducted following final completion of all of the work on the project, and after final payment and demobilization. Many contractors believe that project review, and the information and the company records that result, are extremely valuable resources for all managers, including supervisors.

This process entails assembling the management team for the project on one final occasion and, in open and honest fashion, assessing the successes and shortcomings in the performance and management of the project in its entirety. Included in this review are the final project cost numbers as compared to the project budget values, as well as any other information that the management team believes may be of value.

The information generated from this project review can be entered into the company's institutional memory in terms of best practices, or in terms of matters to be aware of or cautious of on future projects. The information gathered by this process may also be used to modify the information in the company's historical cost information database.

This process is illustrated in the cycle of construction cost reporting, analysis, and so forth, which was depicted in Figure 13.1. The significance and the value to the contractor of conducting project reviews are further discussed in Chapter 19. The concept of project review is noted here inasmuch as almost invariably a component of project review centers upon analyzing the reasons why construction costs in the performance of the work were greater than or were less than those that were anticipated when the project was estimated. The practice of conducting project reviews and of documenting what was learned from the reviews is held by many contractors to be among their most valuable management functions. The supervisor should know that there is a wealth of valuable information to be learned by studying these project reviews.

SUMMARY

The project budgeting, cost reporting, cost control, and historical cost information processes are extremely important components of the contractor's management operations. The supervisor must fully understand these processes and must recognize the vitally important role that he or she plays in the effectiveness of these processes.

Learning Activities

1. Ask whether it is possible for you to see some of your company's historical cost information files, so that you can better understand how the labor cost information that is cost coded by the supervisor in the field becomes part of the company's cost files and how this information is used in the preparation of estimates for future work.

   See how the man-hours of labor or dollars of labor cost incurred in the work on various jobs become production rates or productivity factors that can become information to be used by the estimator.

2. Make arrangements to talk with the estimator in your company, or with a member of the estimating department. See how the estimator approaches the matter of determining (predicting) labor costs for a project being estimated. See how production rates and productivity factors are determined, and how they are used in pricing labor during the preparation of the estimate.

CHAPTER 14

MANAGING TIME ON A CONSTRUCTION PROJECT

INTRODUCTION

Time is an important element in the performance of almost every construction project. Owners typically make a statement in the contract documents for a project that prescribes that time is of the essence in the performance of the construction contract. The General Conditions of the Contract for Construction, AIA Document 201–2007 (Consensus Document) contains the following provision: "Time limits stated in the contract documents are of the essence of the Contract" (Article 8.2.1).

Therefore, completing all of the requirements set forth in the contract documents within the time limit established by the owner becomes a provision of the construction contract. Should the contractor fail to complete the project on time, the result will be some very unpleasant consequences, which are also specified in the construction contract.

These take the form of liquidated damages, which are defined as a dollar amount set forth in the contract documents that the owner will assess the contractor for each day past the stipulated project completion date that the contractor requires to fulfill all of the requirements of the contract. Note that if the contractor is late—not having the project completed within the specified duration—the project and must still be completed all requirements set forth in the contract documents fulfilled. However, the contractor must pay the owner the amount specified in the liquidated damages clause of the contract for each day past the specified completion date that is required to satisfactorily complete all of the requirements of the contract.

Liquidated damages amounts can vary from hundreds to several thousands of dollars per day.

Supervisors must be aware of the exact provisions that define the completion time for every project on which they manage. *They must manage the project in such a way as to ensure completion on time.* The construction schedule is the primary management tool with which project time is managed. The more thoroughly supervisors understand the process of scheduling, and the more completely they understand the construction schedules that result from this process, the more effective their management will be.

SCHEDULING DEFINED

Construction scheduling can be defined as: "Defining and placing in sequence a series of events and determining a timeline for the accomplishment of each event in order that an overall timeline can be planned." In the language of scheduling, the series of events that are to be defined and placed in sequence are referred to as *activities*.

Activities are defined by the contractor as the steps that must be taken, the things that must be done, in performing all of the work on the project, in order to complete the entire project. In a scheduling sense, activities must be discrete, that is, each activity must have a unique identity. Activities must be quantifiable, in terms of their identity and also in terms of a specific amount of time, usually expressed in days, that is allotted for their accomplishment. Activities also consume resources—materials, manpower, dollars, or time, or all of the above.

Contractors will define the activities that are to be performed for a project in an amount of detail to suit their needs with regard to the particular schedule being prepared. There are different types of schedules prepared for different purposes and, thus, there are different amounts of detail included relative to the definition of activities in the several different kinds of schedules.

The scheduling definition above notes that a timeline is to be determined for each activity. This timeline for the completion of each activity is referred to as the duration of that activity. Activity durations are usually stated in days. By the logic being developed, the activities that have been defined must be completed within a duration as determined, so that the overall timeline for the completion of the project can be realized.

The scheduling definition and discussion in the paragraph above note that an overall timeline is to be planned. This overall timeline for the project is referred to as the project duration. The project duration is typically established by the owner and the architect, and is stated in the contract documents, and is usually stated in days.

In fact, there will typically be an entire section of the General Conditions of the Contract for Construction, devoted to "Time." Time is defined in Article 8 of AIA Document 201–2007, General Conditions of the Contract for Construction.

The contract documents also state that the beginning of the time allotted for completion of the project will be defined in a very definite manner. Beginning on this date, the contactor will have the stipulated number of days, the project duration, within which to complete all of the requirements of the project as set forth in the contract documents. Thus, the owner and the contractor, as well as everyone else associated with the project, can determine the date in the future when the project must be completed.

As will be demonstrated as the logic of scheduling is further developed, construction schedules are portrayed as bar chart diagrams or as network diagrams. Therefore, the schedule can also be defined in simplified terms as, "A graphic representation of the contractor's plan for the performance of the project."

THE SCHEDULE AS A PLANNING TOOL

Scheduling in the management of a construction project fits within the context of the management function of planning. In Chapter 2, planning was defined as one of the five fundamental functions of management. Planning was defined as "setting goals and objectives, and determining specific elements to be accomplished, in order to ensure their fulfillment." Further, it was noted that planning is "determining how to get from where we are now to where we want to be." It was noted, additionally, that planning is performed in both the short term and the long term.

If the definition of scheduling is overlaid with the definition of planning, it can be seen that there is an exact correlation between the two. The schedule can be said to be the tool with which contractors accomplish planning, both in the long term and in the short term. Additionally, the schedule is the communication tool with which contractors convey their plans for the performance and completion of the project, to all concerned.

IMPORTANCE OF THE SCHEDULE

The contractor's schedule is vitally important to the owner of the project. For this reason, the owner will usually require the contractor to provide to the owner a construction schedule that has been prepared in a format that is satisfactory to the owner. The owner needs to know that the contractor has a plan, that the plan is reasonable, and that the plan demonstrates that the contractor will complete the project within the project duration that has been specified by the owner. There are a number of additional reasons why the owner will require the contractor to provide a schedule and to keep it updated throughout the life of the project. These additional reasons will be discussed as the topic of scheduling is further developed in this chapter.

The contractor has a strong management need to prepare a personal schedule, as well. In developing the schedule, the contractor will demonstrate, and will reflect in the planning, that the project can be completed within the number of days specified by the owner as the duration. Additionally, the contractor needs to know, and needs to plan, whether completing the project within the specified duration will entail a five-day, 40-hour per week work week, or whether weekend shifts, overtime work, or perhaps double shifts will be required. By graphically plotting all of the activities with their durations and their relationships with one another on a timeline, the contractor can make these important determinations.

In fact, as was noted in Chapter 12, there is a correlation in this regard between the contractor's estimating and scheduling functions. As the estimate is prepared, a schedule, or at least a preliminary schedule, will need to be prepared in parallel.

People Who Place Reliance on a Contractor's Schedule	
Owner	Insurance Company
Designer, Architect, or Engineer	Bonding Company
General Contractor	City Building Officials
Subcontractors	Fabricators
All Levels of Management Within General Contractor's and Subcontractors' Management Teams: Home Office, Project Manager, Superintendent, Supervisor	Material Suppliers
Financier, Banker	

Figure 14.1 People Who Place Reliance on a Contractor's Schedule

As discussed above, the determinations of whether the work is to be performed in regularly scheduled work hours or if overtime, weekend work, or double-shift work are required in order to complete the project within the allotted duration will have a profound effect upon the estimated cost of the work.

After the contract has been awarded and the work on the project begins, the contractor will need to utilize the schedule as a control tool to monitor the actual progress as compared to planned progress on the project. The contractor absolutely must know, on a continual basis throughout the life of the project, what the status is relative to completion of the planned project activities on a timeline.

Additionally, modern scheduling methods allow the contractor to identify certain activities in the work whose completion on time is critical to completing the project on time. The contractor can then set priorities in such a way as to ensure timely completion of these activities. The Critical Path Method (CPM), which will be examined later in this chapter, as well as the Last Planner Process and the Planning Reliability determinations as discussed in other chapters of this book, allow the contractor to focus efforts and resources on those activities that have a priority with regard to maintaining the project on the planned timeline.

The schedule that is prepared by the contractor is valuable to a number of people in addition to the owner, the architect, and the contractor. As indicated in Figure 14.1, a great many of the people involved with a construction project make use of, and place their reliance on, the contractor's construction schedule.

TYPES OF SCHEDULES

Construction schedules can be divided into two categories, in terms of the graphical method that is used to portray the scheduling information: bar charts and network schedules. As will be demonstrated, each of these graphical methods can be used to depict project scheduling information in a variety of different formats, so as to meet the needs of those who are involved with the project, including the supervisor. The scheduling graphics can be used to portray the overall project schedule, or any segment of a project schedule, the critical activities within a schedule, a short-interval schedule, or any number of additional elements of the project and its planned timeline.

The Bar Chart

The bar chart schedule, also referred to as a Gantt chart, was invented during the early 1900s. It has been used in construction scheduling for many years. It remains one of the most valuable, and certainly one of the most effective, construction scheduling tools available today.

The bar chart is constructed by listing the activities to be referenced in a column on the left side of the schedule. Usually, these activities are listed in the sequential order in which they are to be performed on the project. At the top of the bar chart, the headings for a series of additional columns contain the timeline for the schedule, usually in days or weeks. To depict the planned schedule, a bar is drawn to the right of each activity to delineate the time frame in which the start and the completion of that activity are planned. Figure 14.2 illustrates a bar chart schedule for a small project. Figure 14.3 shows a bar chart for a larger project, a courthouse building.

One of the key advantages of the bar chart is its intuitive nature. It is easily understood and can readily be interpreted by people at all levels on a construction project, from the owner to the craft worker.

An additional advantage that the bar chart offers is its versatility. Activities can be defined in greater or lesser amounts of detail, to suit the needs of the manager. Likewise, the timeline can be varied and can be stated in units of hours, quarter-days or half-days, days, weeks, months, and so on. The entire project schedule can be depicted, or any portion of the overall schedule can be shown on the bar chart. Additionally, a short-interval schedule, which is an extremely valuable planning tool for the supervisor to use and which will be discussed later in this chapter, can be prepared by use of the bar chart.

In addition, bar charts are easily constructed. They may be drawn by hand on paper or even on a scrap of plywood or drywall on the job site. Or, they may be computer-generated, in the form of a spreadsheet or as the output of a computer scheduling software. Colors, shading, or cross-hatching can easily be added to the bar chart schedule, in order to depict and emphasize critical activities (to be defined later in this chapter) or to indicate actual progress versus planned progress.

Additionally, bar charts can be derived from, and can be related to, the network schedule for a project. For example, after the overall project schedule has been determined in network schedule format by the use of computer scheduling software, a bar chart can be extracted by the software to depict any aspect of the schedule, such as activities within a certain window of time, critical activities, activities by craft, activities involving usage of a certain material, and so forth.

Along with all of their many benefits, however, bar charts do have some limitations. For example, bar charts cannot portray the dependency relationships between activities. Dependencies and dependency relationships are concepts to be defined

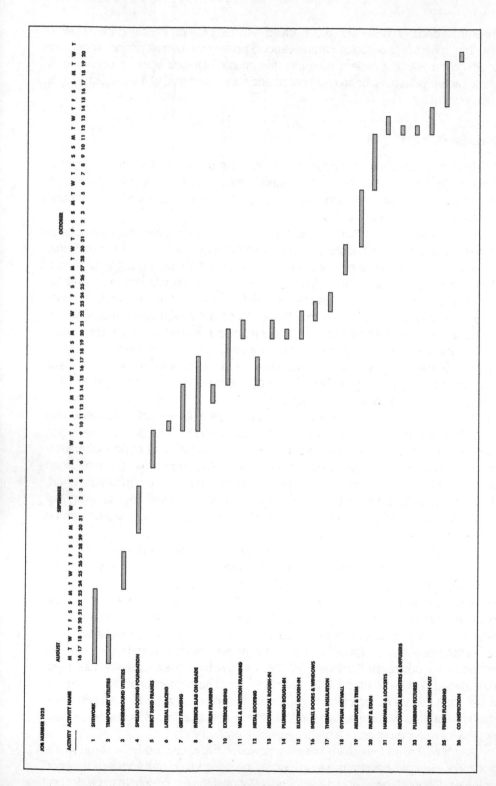

Figure 14.2 Bar Chart Schedule to Depict All of the Activities on a Small Project

Act ID	Description	Orig Dur	Early Start	Early Finish	Total Float	%	Phase - pions b	
PROJECT MILESTONES								
00005	County Approval of Site Pkg GMP	0	20JAN09		-1d	0	AME	
00010	Notice to Proceed	0	21JAN09		-1d	0	AME	
00030	Mobilize	10	21JAN09	03FEB09	-1d	0	AME	
00015	Construction	423	04FEB09	12APR10	0	0	AME	
00040	Turnover New Drive & East Parking to Owner	0		03MAR09*	0	0	AME	
00050	Turnover West Parking to Owner	0		11MAR09	274d	0	AME	
SITEWORK								
10010	Install Construction Entries	2	22JAN09	23JAN09	-1d	0	SITE	
10020	Site Clearing	10	26JAN09	06FEB09	-1d	0	SITE	
10030	Install Base of Skanska Office	2	26JAN09	27JAN09	0	0	SITE	
10350	Setup Office Trailers	5	28JAN09	03FEB09	0	0	SITE	CMA R
10040	Install Erosion Control	2	02FEB09	03FEB09	0	0	SITE	
10050	Cut for Concrete Paving - East	3	02FEB09	04FEB09	-1d	0	SITE	
10060	Demo Existing Parking - East	3	02FEB09	04FEB09	4d	0	SITE	
10070	Install Storm Drain Utilities	15	03FEB09	25FEB09	-1d	0	SITE	
10110	Cut for Concrete Paving - West	3	03FEB09	09FEB09	276d	0	SITE	
10140	Cut for Concrete Paving - North	3	10FEB09	12FEB09	278c	0	SITE	
10080	Install Base for Paving - East	3	12FEB09	18FEB09	-1d	0	SITE	
10090	Form, Rebar, Place Concrete Paving - East	5	19FEB09	25FEB09	269c	0	SITE	PAVE
10120	Install Base for Paving - West	5	19FEB09	25FEB09	0	0	SITE	PAVE
10100	Cure Concrete Paving - East	7	26FEB09	04MAR09	-1d	0	SITE	PAVE
10130	Form, Rebar, Place Concrete Paving - West	5	26FEB09	04MAR09	272d	0	SITE	PAVE
10160	Install Base for Paving - North	5	26FEB09	04MAR09	269d	0	SITE	PAVE
10370	Install Sanitary Utilities - Sta. 0+00 to SSMH1	8	26FEB09	09MAR09	276d	0	SITE	UTIL
10330	Stripe Paving - East	2	03MAR09	04MAR09	0	0	SITE	PAVE
10150	Cure Concrete Paving - West	7	05MAR09	11MAR09	386d	0	SITE	PAVE
10180	Form, Rebar, Place Concrete Paving - North	5	05MAR09	11MAR09	269d	0	SITE	PAVE
10190	Demo Existing Paving & Walls - North	10	05MAR09	18MAR09	123d	0	SITE	PAVE
10170	Stripe Paving - West	2	09MAR09	10MAR09	275c	0	SITE	PAVE
10200	Cure Concrete Paving - North	7	12MAR09	18MAR09	380d	0	SITE	PAVE
10210	Cap Abandoned Utilities - Lot C	5	12MAR09	18MAR09	123d	0	SITE	UTIL
Foundation								
Basement								
10220	Excavate Building Pad	3	19MAR09	23MAR09	123d	0	FNDN	SITE
10230	Excavate Basement	5	24MAR09	30MAR09	123d	0	FNDN	SITE
10240	Drill Piers at Basement	6	31MAR09	07APR09	123d	0	FNDN	CONC
10250	Form,Rebar,Place Basement Walls - West Section	10	06APR09	17APR09	123d	0	FNDN	CONC
10480	Install Bldg Pad Select Fill - West Section	7	06APR09	14APR09	250d	0	FNDN	SITE
10490	Drill Piers at Level 1 - Grid Ln 1 to 3	4	10APR09	15APR09	249c	0	FNDN	CONC
10400	Form,Rebar,Place Basement Walls - Middle Section	15	16APR09	05MAY09	123d	0	FNDN	CONC

Big Courthouse Building

Number/Version 01/GMP1
Data date 20JAN09
Run date 12AUG10
Page number 1A
© Primavera Systems, Inc.

Legend:
■ Early bar
■ Progress bar
■ Critical bar
■ Summary bar
◆ Start milestone point
◆ Finish milestone point

Figure 14.3 Bar Chart Schedule for a Courthouse Building
Courtesy of Skanska USA Building Inc.

Figure 14.3 (Continued)

Act ID	Description	Orig Dur	Early Start	Early Finish	Total Float	%	Phase	span b
10260	Cure Basement Walls - West Section	3	18APR09	20APR09	313c	0	FNDN	CONC
10270	Strip Basement Wall Forms - West Section	2	21APR09	22APR09	222c	0	FNDN	CONC
10280	Below Grade Waterproofing on Walls - West Sect.	2	23APR09	24APR09	222c	0	FNDN	CONC
10390	Form/Rebar/Place Basement Walls - East Section	10	01MAY09	14MAY09	123c	0	FNDN	CONC
10410	Cure Basement Walls - Middle Section	3	06MAY09	08MAY09	204c	0	FNDN	CONC
10420	Strip Basement Wall Forms - Middle Section	2	11MAY09	12MAY09	204c	0	FNDN	CONC
10440	Below Grade Waterproofing on Walls - Middle Sect	2	13MAY09	14MAY09	204c	0	FNDN	CONC
10380	Cure Basement Walls - East Section	3	15MAY09	19MAY09	172c	0	FNDN	CONC
10420	Strip Basement Wall Forms - East Section	2	18MAY09	19MAY09	123c	0	FNDN	CONC
10340	Below Grade Waterproofing on Walls - East Sect.	2	20MAY09	21MAY09	123c	0	FNDN	WTR F
Column Line AA to GG								
10360	Install Bldg Pad Select Fill - Grid Ln AA to HH	3	31MAR09	03APR09	238c	0	FNDN	SITE
10320	Drill Piers at Level 1 - Grid Ln AA to HH	4	08APR09	16APR09	236c	50	FNDN	CONC
00060	Install Underground Plumbings AA to HH	5	10APR09	16APR09	242c	0	FNDN	PLUM
00100	Excavate Grade Beams 50% of AA to HH	3	10APR09	13APR09	236c	0	FNDN	CONC
00120	Place Rebar Grade Beams 50% of AA to HH	4	14APR09	17APR09	236c	0	FNDN	CONC
00140	Excavate Grade Beams 100% of AA to HH	2	14APR09	15APR09	238c	0	FNDN	CONC
00130	Pour Grade Beams 50% of AA to HH	1	20APR09	20APR09	240c	0	FNDN	CONC
00160	Place Rebar Grade Beams 100% of AA to HH	4	20APR09	22APR09	236c	0	FNDN	CONC
00170	Pour Grade Beams 100% of AA to HH	1	24APR09	24APR09	236c	0	FNDN	CONC
00260	Form/Poly/Rebar for SOG AA to HH	5	27APR09	01MAY09	236c	0	FNDN	CONC
00110	Place SOG AA to HH	1	04MAY09	04MAY09	236c	0	FNDN	CONC
Column Line HH to C3								
00280	Backfill West Basement Walls	3	27APR09	29APR09	222c	0	FNDN	SITE
00150	Install Select Fill HH to C3	3	30APR09	04MAY09	222c	0	FNDN	SITE
00180	Drill Piers HH to C3	1	05MAY09	05MAY09	222c	0	FNDN	CONC
00190	Install Underground Plumbings HH to C3	5	05MAY09	11MAY09	225c	0	FNDN	PLUM
00200	Excavate Grade Beams 50% of HH to C3	2	06MAY09	07MAY09	222c	0	FNDN	CONC
00210	Place Rebar Grade Beams 50% of HH to C3	2	08MAY09	11MAY09	222c	0	FNDN	CONC
00220	Excavate Grade Beams 100% of HH to C3	2	08MAY09	11MAY09	222c	0	FNDN	CONC
00230	Pour Grade Beams 50% of HH to C3	2	12MAY09	13MAY09	224c	0	FNDN	CONC
00240	Place Rebar Grade Beams 100% of HH to C3	2	12MAY09	13MAY09	222c	0	FNDN	CONC
00250	Pour Grade Beams 100% of HH to C3	1	14MAY09	14MAY09	222c	0	FNDN	CONC
00260	Form/Poly/Rebar for SOG HH to C3	5	15MAY09	21MAY09	222c	0	FNDN	CONC
00270	Place SOG HH to C3	1	22MAY09	22MAY09	222c	0	FNDN	CONC
Column Line C3 to G								
00290	Backfill Basement Walls Middle Section	3	15MAY09	19MAY09	204c	0	FNDN	SITE
00300	Install Select Fill C3 to G	3	20MAY09	26MAY09	204c	0	FNDN	SITE
00310	Drill Piers C3 to G	1	26MAY09	26MAY09	204c	0	FNDN	CONC
00320	Install Underground Plumbings C3 to G	5	26MAY09	01JUN09	211c	0	FNDN	PLUM
00330	Excavate Grade Beams 50% of C3 to G	2	27MAY09	28MAY09	204c	0	FNDN	CONC
00340	Place Rebar Grade Beams 50% of C3 to G	4	29MAY09	03JUN09	204c	0	FNDN	CONC

Number/Version	01/GMP1
Data date	20AUG09
Run date	12AUG10
Page number	2A
© Primavera Systems, Inc.	

Big Courthouse Building

Legend:
- Early bar
- Progress bar
- Critical bar
- Summary bar
- Start milestone point
- Finish milestone poin

244

Schedule table — Big Courthouse Building

Act ID	Description	Orig Dur	Early Start	Early Finish	Total Float	%	Phase	
00350	Excavate Grade Beams 100% of C3 to G	2	29MAY09	01JUN09	204d	0	FNDN	CONC
00360	Pour Grade Beams 50% of C3 to G	1	04JUN09	04JUN09	209d	0	FNDN	CONC
00370	Place Rebar Grade Beams 100% of C3 to G	4	04JUN09	09JUN09	204d	0	FNDN	CONC
00380	Pour Grade Beams 100% of C3 to G	1	10JUN09	10JUN09	204d	0	FNDN	CONC
00390	Form/Poly/Rebar for SOG C3 to G	5	11JUN09	17JUN09	204d	0	FNDN	CONC
00400	Place SOG C3 to G	1	18JUN09	18JUN09	204d	0	FNDN	CONC
Column Line G to L								
00410	Backfill Basement Walls East Section	3	22MAY09	27MAY09	123d	0	FNDN	SITE
00420	Install Select Fill G to L	3	28MAY09	01JUN09	123d	0	FNDN	SITE
00430	Drill Piers G to L	2	02JUN09	03JUN09	123d	0	FNDN	CONC
00440	Install Underground Plumbing G to L	5	02JUN09	08JUN09	131d	0	FNDN	PLUM
00450	Excavate Grade Beams 50% of G to L	2	04JUN09	05JUN09	123d	0	FNDN	CONC
00460	Place Rebar Grade Beams 50% of G to L	4	08JUN09	11JUN09	123d	0	FNDN	CONC
00470	Excavate Grade Beams 100% of G to L	1	08JUN09	09JUN09	125d	0	FNDN	CONC
00480	Pour Grade Beams 50% of G to L	1	12JUN09	12JUN09	127d	0	FNDN	CONC
00490	Place Rebar Grade Beams 100% of G to L	4	12JUN09	17JUN09	123d	0	FNDN	CONC
00500	Pour Grade Beams 100% of G to L	1	18JUN09	18JUN09	123d	0	FNDN	CONC
00510	Form/Poly/Rebar for SOG G to L	5	19JUN09	25JUN09	123d	0	FNDN	CONC
00520	Place SOG G to L	1	26JUN09	26JUN09	123d	0	FNDN	CONC
Structure								
00530	Erect Columns and Beams Level 1	7	29JUN09	08JUL09	123d	0	STRC	
00540	Plumb/Detail Level 1	3	09JUL09	13JUL09	123d	0	STRC	
00545	Install Edge Angle/Detail Level 2	3	14JUL09	16JUL09	123d	0	STRC	
00550	Install Deck Level 2	5	17JUL09	23JUL09	123d	0	STRC	
00560	Install Beams Roof Level	5	20JUL09	24JUL09	123d	0	STRC	
00570	Install Joists at Low Roof	10	21JUL09	03AUG09	166d	0	STRC	
00620	Install Pan Stairs	10	24JUL09	06AUG09	166d	0	STRC	
00710	Lay 4 courses of CMU Level 1	2	24JUL09	27JUL09	156d	0	STRC	
00600	Install Trusses at High Roof	10	27JUL09	07AUG09	166d	0	STRC	
00720	Install Door Frames at CMU Level 1	3	27JUL09	28JUL09	156d	0	STRC	
00730	Rough-In MEP at CMU Walls Level 1	3	29JUL09	31JUL09	156d	0	STRC	
00740	Top out CMU walls Level 1	3	29JUL09	05AUG09	156d	0	STRC	
00580	Install Roof Opening Frames at Low Roof	2	04AUG09	05AUG09	123d	0	STRC	
00590	Install Low Roof Deck	3	06AUG09	12AUG09	144d	0	STRC	
00640	Rebar/Place and Finish Slab on Deck AA to A	3	06AUG09	10AUG09	166d	0	STRC	
00750	Shore/Form/Rebar Mezzanine Level	2	06AUG09	10AUG09	166d	0	STRC	
00630	Place Pan Stair Fill	2	07AUG09	07AUG09	166d	0	STRC	
00760	MEP Rough in at Mezzanine Level	5	07AUG09	11AUG09	166d	0	STRC	
00610	Install Deck at High Roof	5	10AUG09	14AUG09	166d	0	STRC	
00650	Rebar/Place and Finish Slab on Deck A to F	1	11AUG09	13AUG09	150d	0	STRC	
00770	Place and Finish Mezzanine Slab Level 1	1	12AUG09	12AUG09	166d	0	STRC	

Big Courthouse Building

Legend:
- Early bar
- Progress bar
- Critical bar
- Summary bar
- Start milestone point
- Finish milestone point

Number/Version	01/GMP1
Data date	26JUN09
Run date	12AUG10
Page number	3A

© Primavera Systems, Inc.

Figure 14.3 *(Continued)*

245

Figure 14.3 (Continued)

Act ID	Description	Orig Dur	Early Start	Early Finish	Total Float	%	Prior b
00660	Rebar/Place and Finish Slab on Deck F to L	3	14AUG09	18AUG09	162d	0	STRC
00670	Lay 4 Courses Int CMU Walls Level 2	2	14AUG09	17AUG09	150d	0	STRC
00700	Install Detention Frames & Embeds Level 2	2	17AUG09	18AUG09	150d	0	STRC
00680	Rough-in MEP at CMU Walls Level 2	2	19AUG09	20AUG09	150d	0	STRC
00690	Top-out Int CMU Walls Level 2	3	21AUG09	25AUG09	150d	0	STRC
00780	Shore/Form/Rebar Mezzanine Level 2	3	26AUG09	28AUG09	152d	0	STRC
00810	Layout Partitions at/conc Topping Slabs Level 2	1	26AUG09	26AUG09	150d	0	STRC
00790	MEP Rough In at Mezzanine Level 2	3	27AUG09	31AUG09	152d	0	STRC
00820	Install Styrofoam/Form/Rebar Topping Slab Lvl 2	5	27AUG09	02SEP09	150d	0	STRC
00800	Place and Finish Mezzanine Slab Level 2	1	01SEP09	01SEP09	152d	0	STRC
00830	Place and Finish Topping Slabs Level 2	1	03SEP09	03SEP09	150d	0	STRC

Exterior Finishes

Act ID	Description	Orig Dur	Early Start	Early Finish	Total Float	%	Prior b
00840	Install Perimeter Studs Level 1 West Elevation	3	11AUG09	13AUG09	123d	0	EXT
00850	Install Sheathing Level 1 West Elevation	3	12AUG09	14AUG09	164d	0	EXT
00940	Install Perimeter Studs Level 2 West Elevation	3	14AUG09	18AUG09	123d	0	EXT
00930	Install Sheathing Level 2 West Elevation	3	17AUG09	19AUG09	123d	0	EXT
00860	Install Tyvek West Elevation	2	20AUG09	21AUG09	123d	0	EXT
00870	Install Stone West Elevation	10	24AUG09	04SEP09	123d	0	EXT
00920	Install Scaffold for Stucco West Elevation	2	08SEP09	09SEP09	123d	0	EXT
00880	Install Metal Lath at West Elevation	5	10SEP09	16SEP09	123d	0	EXT
00890	Install Stucco Scratch Coat West Elevation	4	16SEP09	17SEP09	123d	0	EXT
00900	Install Stucco Brown Coat West Elevation	3	18SEP09	22SEP09	123d	0	EXT
00910	Install Stucco Finish Coat West Elevation	3	23SEP09	25SEP09	123d	0	EXT
00950	Remove Stucco Scaffold West Elevation	2	28SEP09	29SEP09	123d	0	EXT
00960	Install Storefront Glass and Glazing West Elev.	10	30SEP09	13OCT09	123d	0	EXT

Big Courthouse Building

Number/Version	01/GMP1
Data date	20JAN09
Run date	12AUG10
Page number	4A
© Primavera Systems, Inc.	

Legend:
- Early bar
- Progress bar
- Critical bar
- Summary bar
- Start milestone point
- Finish milestone point

246

later in this chapter. One can determine from bar charts which activities are to be conducted in parallel and which activities are planned to be conducted in sequence. However, dependency relationships as we will define them shortly cannot be determined from a bar chart.

Additionally, bar charts cannot be used to make network calculations (also to be defined later in this chapter), nor to determine the critical path. Bar charts can be used to display which activities are critical after this determination has been made by the Critical Path Method in a network schedule. However, bar charts cannot be used to make the determination of which activities are critical and which are not.

In addition, bar charts do not lend themselves well to depicting the other values that result from network calculations. Early Start, Early Finish, Late Start, Late Finish, and Float cannot be determined on a bar chart schedule, and neither does the bar chart usually depict these values.

While the bar chart does have some limitations, the supervisor should recognize that it is most assuredly one of the most basic, and at the same time one of the most powerful, scheduling tools available to the supervisor. In fact, at the foreman and crew leader level of supervision, the bar chart is perhaps the most effective scheduling tool the supervisor can utilize.

Later in this chapter short-interval schedules will be discussed more fully. A definition of the short-interval schedule will follow and then will be reinforced in the forthcoming discussion. The short-interval schedule can be defined as a schedule that breaks down the overall project schedule into short look-ahead intervals such as a few days, or a week, or perhaps two or three weeks. It is constructed in a bar chart format. To emphasize: the short-interval schedule is one of the most powerful, and one of the most valuable scheduling tools and communication tools that the supervisor can employ. Its use by the supervisor is very highly recommended.

Network Schedules

Network schedules are the basis for most project schedules today. Network schedules are at their best when they are utilized to provide a graphic representation of the plan for the completion of the entire project, along with the critical path and other important planning information. Modern computer softwares that perform "computerized scheduling" use network schedules and the Critical Path Method as their basis of operation.

Network schedules may also be referred to as critical path schedules, as PERT (Program Evaluation and Review Techniques) schedules, or as precedence schedules. The terms PERT schedules and precedence schedules actually refer to the origins and evolution of network schedules. However, these terms are frequently used synonymously in referring to network schedules today.

The determination and use of the critical path is based in network scheduling. The Critical Path Method will be more fully discussed later in this chapter. The Critical Path Method uses the network schedule and a set of calculations, based on information that the contractor inputs to the schedule, to determine a great deal of extremely valuable scheduling information. Included in these determinations is the critical path, as well as the other values that were referred to earlier in this chapter: Early Start, Early Finish, Late Start, Late Finish, and Float. All of these values have important meaning in the planning and scheduling of a construction project. The more fully supervisors understand the Critical Path Method, and the matter of network scheduling in general, the more effective they can be in their management of construction projects.

As noted earlier, network schedules are most commonly used to depict the overall schedule for an entire construction project. Network schedules show the activities to be performed, along with the duration of each activity. They also depict the dependency relationships between the activities, and they denote which activities are planned to be conducted in sequence and which are planned to be conducted in parallel. In addition, they convey the other important information noted above for each activity in the project: Early Start, Early Finish, Late Start, Late Finish, and Float.

There are two forms of network schedules, Activity on Arrow (AOA) and Activity on Node (AON). Activity on Arrow schedules were the first network schedules to be developed, and were the first to utilize the Critical Path Method. They served a valuable function historically and, for a time, were considered the state-of-the-art scheduling methodology. Activity on Arrow schedules have been replaced by their modern evolution, Activity on Node network schedules.

Reference to Activity on Arrow Network schedules is included here in order that the supervisor may understand the historical significance of this form of schedule in the evolution of the Critical Path Method. Additionally, since Activity on Arrow schedules were in use for quite a number of years, and since it is possible that the supervisor may encounter an Activity on Arrow schedule when the documentation for a remodeling or renovation or restoration project is encountered, these schedules are given brief mention here.

Activity on Arrow (AOA) Network Schedules

Activity on Arrow schedules, as their name implies, depict the names of all project activities on an arrow. A circular figure called a *node* is used at the head and at the tail of each arrow. The arrows point in the direction of the flow or sequence of the work. Nodes graphically depict the beginning and the end of an activity, and as noted above, are placed at the head and tail of each arrow. Nodes serve as logical connectors to link activities in sequence or in parallel. Numerical values, with a unique number for each node, are typically assigned to the nodes, and activities are identified by reference to the node values at the tail and head of each arrow. Figure 14.4 shows an activity as it would appear in an Activity on Arrow network schedule. This activity would be referred to as "Activity 37-38 Excavate Footings," referencing the node numbers at the beginning and end of the activity along with the name of the activity.

Activity on Arrow network schedules depict the planned flow of the work for all of the activities defined for the project. Dependency relationships are shown, as well as parallel and sequential relationships between activities.

As was noted earlier, the Activity on Arrow network schedule was developed along with the Critical Path Method. The critical path for the project is typically indicated on the network schedule in a red line and/or with a bold line, indicating the pathway through the network that connects critical activities. In the Activity on

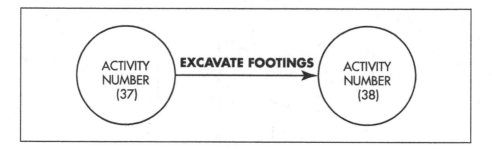

Figure 14.4 Activity on Arrow Depiction with Nodes at Each End of the Arrow

Arrow network scheduling method, network calculation values (to be discussed later in this chapter) were determined and were typically shown on a table that accompanied the graphic schedule.

An Activity on Arrow network schedule is read beginning at the top left and reading toward the right and downward. As a person reads the schedule in this fashion, he or she can see the planned flow of the work in the activities on the construction project. Figure 14.5 depicts an Activity on Arrow network schedule for a small construction project. Depiction of the critical path is purposely omitted from this schedule. As indicated above, the critical path will be defined and illustrated in a network schedule later in this chapter.

Activity on Node (AON) Network Schedules

Activity on Node network schedules are an evolution from the original Activity on Arrow network schedules. This scheduling method uses the same logic of depicting the flow of the work by showing activities and their dependency relationships, with the activities arranged in sequential or parallel order. As the name indicates, in this network scheduling method, the activities for a project are indicated on nodes. Arrows connect the nodes and point in the direction of the flow of the work. The Critical Path Method for making network calculations and for determining the critical path is the same as it was in the original Activity on Arrow network schedules. An Activity on Node network schedule is read beginning at the top left and reading to the right and downward.

The nodes in the AON scheduling method consist of squares that contain the activity name. Additionally, the squares are subdivided into smaller squares and rectangles, where additional scheduling information is conveyed. A typical node in an Activity on Node scheduling network is depicted in Figure 14.6.

It should be noted that the node contains the name of the activity and also its duration. An identifying number for the activity may also be included within the node. Other values, determined from network calculations, are indicated in smaller boxes within the node: Early Start, Early Finish, Late Start, Late Finish, and Float. Later in this chapter we will discuss the derivation and meaning of this additional information, which is shown on each node.

As noted earlier, arrows connect the nodes and show the flow of the work. The arrangement of the nodes and arrows also depicts the dependency relationships between the activities (to be defined later in this chapter). Additionally, the

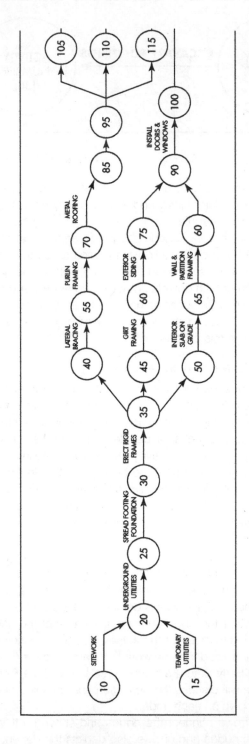

Figure 14.5 Activity on Arrow Network Schedule for a Small Project

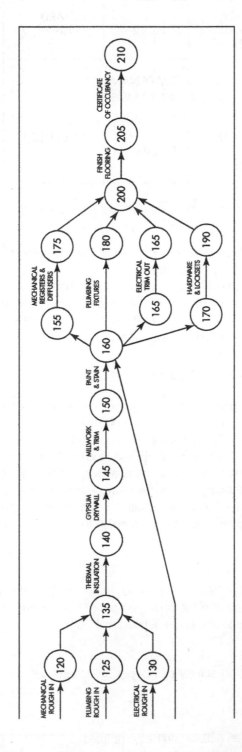

Figure 14.5 (Continued)

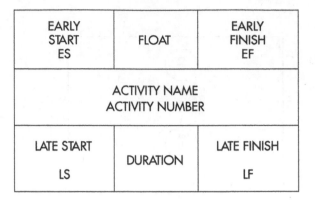

Figure 14.6 Node in an Activity on Node Network Schedule

arrangement of the arrows and nodes indicates whether the plan is for the activities to be conducted in sequence or in parallel.

Figure 14.7 depicts an Activity on Node network schedule for a small construction project. This is the same project whose schedule was depicted in an Activity on Arrow schedule in Figure 14.5, so that the similarities and differences between AOA and AON network schedules can be seen.

THE CRITICAL PATH METHOD (CPM)

The Critical Path Method uses the logic portrayed in network schedules, along with a series of calculations to determine some very important scheduling information. The methodology employed in the Critical Path Method will determine for each activity, its Early Start, Early Finish, Late Start, Late Finish, and Float. Additionally, the method will determine whether each activity is a critical activity or not. When critical activities have been identified, a pathway through the network schedule can be shown, which is called the critical path. All of this information is significant for the supervisor, as well as for all others who are involved in managing the construction project.

Definitions

As noted above, all of the information derived from the Critical Path Method has value to those associated with the management of the construction project. The values that were referred to above are defined as follows.

Early Start (ES):

The earliest that an activity can begin (in accord with this plan, and this schedule logic).

Early Finish (EF):

The earliest that an activity can be completed.

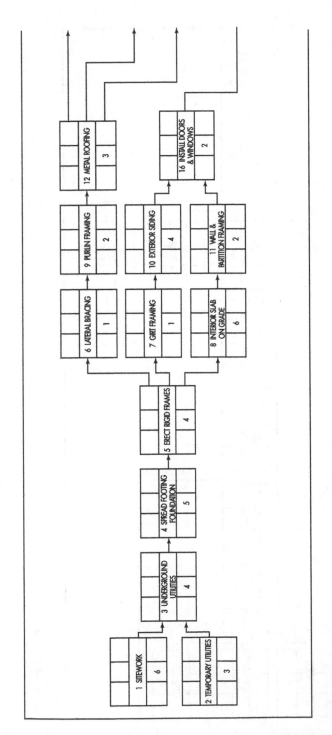

Figure 14.7 Activity on Node Network Schedule for a Small Project

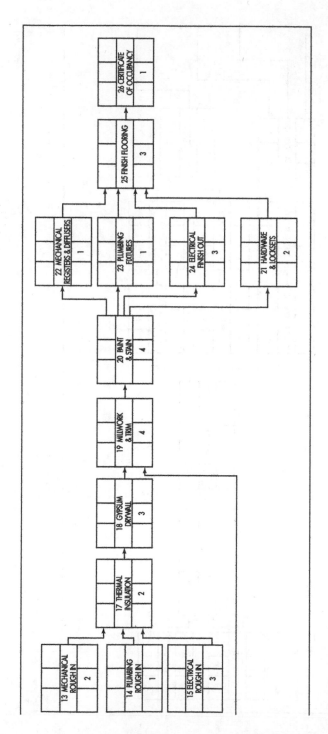

Figure 14.7 *(Continued)*

Late Start (LS):

The latest that an activity can begin.

Late Finish (LF):

The latest that an activity can be completed.

Float:

The amount of time that the completion of an activity can be delayed without impacting project duration and project completion date.

Critical Activity:

An activity which has zero float.

Critical activities must be completed on time, or the project completion date will be delayed by an amount of time equal to the delay in the completion of the critical activity (unless management actions are taken to save time elsewhere in the project).

Critical Path:

The pathway through a network schedule which connects critical activities.

The critical path is the longest pathway through the network schedule that will complete the project on time.

Network Calculations

The Critical Path Method entails a series of calculations, which are performed after the network schedule has been developed and after the relationships between activities have been determined. These calculations are referred to as network calculations, and they are the Forward Pass and the Backward Pass. Both the Forward Pass and the Backward Pass follow a discrete set of rules, meaning that each is always performed in exactly the same way, in accordance with a set of rules that are inherent in the process that is called the Critical Path Method.

The Forward Pass and Backward Pass calculations are used to determine the values Early Start, Early Finish, Late Start, and Late Finish. After these determinations have been made, other very important information can be derived, including the earliest project completion date, the amount of float for each activity, critical activities, and the critical path through the network schedule.

While supervisors are not likely to be involved in making these network calculations, at least not early in their supervisory career, it is important for them to understand how they are performed, and to realize that this is the basis for determination of much of the scheduling information that they work with on a daily basis. Later in this chapter a series of examples will demonstrate how these calculations are performed.

DEVELOPMENT OF A NETWORK SCHEDULE AND THE CRITICAL PATH

Network schedules are developed in a step-by-step process, which will be summarized here. It is important to understand that, while network schedules are often referred to as "computerized schedules," the important steps in the development of a network schedule are performed by people. Thought and judgment and analysis are applied to the development of the various components of the scheduling process and the relationships among the activities. These things the computer cannot do.

Supervisors should also realize that while they may not be involved in making the network schedules for the project, it is nevertheless important to know that the network schedules that they receive for the project are derived by a method that follows a process, and a set of consistent rules. A basic understanding of the process and rules will enhance supervisors' understanding of network schedules and will also allow them to discuss network schedules with the project manager and others who make daily use of these schedules. Further, supervisors should realize that as their career advances, they will be increasingly involved with network schedules and their preparation and application.

The steps in the development of a network schedule are summarized here:

1. **List all of the activities that are to be performed on the project.**

    All of the activities to be performed on the project are listed. They are determined and set forth in accord with the definition of activities as discussed earlier, and they are listed in an amount of detail so as to be most useful as the schedule is consulted throughout the project.

2. **Determine the duration of each activity.**

    Activity durations are determined for all activities. Durations are almost always measured in days on a network schedule. The durations are determined by the scheduler through his insight and analysis regarding the time that will be required to complete each activity, as well as through his analysis of the estimate for the project.

    The estimate will contain activity names as determined by the estimator, along with estimated production units (units of production for the activity per hour or per day). The scheduler can transform this information into days allotted for the completion of each activity, that is, the duration of each activity, in making the network schedule.

3. **Determine the sequence in which the activities are to be performed.**

    The order in which the activities will be performed is next determined by the scheduler. Some activities must be performed in a certain order, or sequence, as defined by the nature of the activity. For example, a cast-in-place concrete column must be formed, then reinforced, then the concrete placed, then cured—these steps and this sequence are inherent in the nature of the work with regard to this particular material and its use in a column. However, the person making the schedule must determine the sequence and relationship in which this activity is to be performed relative to the other activities in the project.

4. **Determine dependency relationships between activities**

Some activities are dependent upon other activities for their performance and completion. This means that some activities cannot be performed before another activity or activities that precede them have been completed or are underway.

For example, the activity, "Tape and Float Gypsum Drywall" is dependent upon the completion of a preceding activity, "Install Gypsum Drywall." Similarly the activity, "Texture Gypsum Drywall" is dependent upon the completion of the activity, "Tape and Float Drywall." The activity that occurs earlier in this relationship is referred to as the predecessor or predecessor activity, while the one that follows is referred to as the successor or successor activity.

Network schedules can depict different kinds of dependency relationships between predecessor and successor activities. However, the dependency relationship that is utilized most often is one that is referred to as a "Finish-Start" relationship. This means that the start of the second activity is dependent upon the completion (finish) of the first activity.

The activity "Place Concrete" cannot be performed until the predecessor activity "Install Reinforcing Steel" has been completed. "Place Concrete" is, therefore, in a Finish-Start relationship with "Install Reinforcing Steel." The Finish-Start relationship between activities is depicted in an Activity on Node network schedule as shown in Figure 14.8.

5. **Determine lag times to be used.**

An accompaniment to the scheduler's use of Finish-Start relationships is the use of lag time, which is also referred to as lag. Lag times are used when the predecessor activity does not need to be finished in its entirety before the successor activity can begin. For example, the activity, "Install Conduit in Framed Walls" would logically be defined in a Finish-Start relationship with a predecessor activity, "Frame Walls."

However, the scheduler's judgment might well determine that *all* of the wall framing does not need to be completed before *any* of the conduit installation can begin. Analysis may indicate that on this project, two days after wall framing has begun, conduit installation can be started. The rationale is,

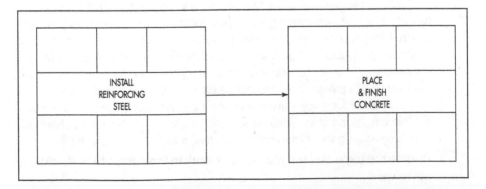

Figure 14.8 Depiction of Finish-Start Relationship between Activities in an AON Network Schedule

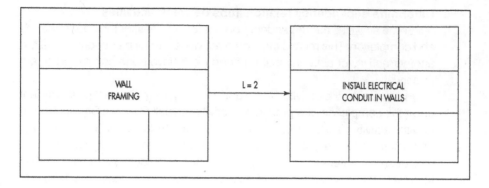

Figure 14.9   Finish-Start Relationship between Activities, with a Two-Day Lag

that sufficient wall framing will have taken place after two days to make ready for the beginning of conduit installation, and that as wall framing continues conduit installation will also continue, but conduit installation will never catch up with wall framing.

Lag time is graphically depicted in a Finish-Start relationship between activities as indicated in Figure 14.9.

6. **Determine which activities are to be performed in sequence and which are to be performed in parallel.**

The plan for the flow of the work is determined and set forth by the scheduler, in a process of arranging activities in sequential and parallel fashion from the beginning to the end of the project. Some activities are performed in sequence, one after another, often because there is a dependency relationship between them. This is illustrated in Figure 14.8, where the activity "Place and Finish Concrete" would follow in sequence after the activity "Install Reinforcing Steel."

Activities can also be planned so as to be performed in parallel with other activities, meaning that they are performed simultaneously with those other activities with which they have no dependency relationship.

7. **Plot to a logic diagram.**

Many schedulers will depict the planned logic and flow of the work on a flow chart prior to representing the logic in the computer scheduling software, which is Step 8 in the process. Often schedulers apply the names of activities to sticky-note paper and then arrange the sticky notes on a surface such as a whiteboard in order to develop and portray the scheduling logic and the planned flow of the work, and to define parallel and sequential activities.

This allows for easy visualization of the scheduler's plan for the work. Additionally, as the plan is visualized and analyzed, it can be readily changed, simply by moving the sticky notes (activities) into new arrangements.

8. **Portray the logic and the plan for the work in the computer scheduling software.**

Schedulers utilize a variety of different computer software for network scheduling. The three programs that are most commonly employed are

Microsoft Project, Prima Vera, and Sure-Trak. All of these software applications are extremely powerful programs, which can develop and portray
network schedules with amazing speed.

After activity dependency relationships and sequential relationships have
been determined by the scheduler, the logic that the scheduler has developed is entered into the computer. Each of the computer programs has a
protocol whereby the scheduler will input the activity names, activity numbers, durations, and dependency and sequence relationships, including lag.
It is interesting to note that some schedulers are able to omit Step 7 above
and are able to proceed directly from Steps 5 and 6 to this step.

9. **Perform the Network Calculations.**

Network calculations and the procedures by which they are performed
are defined in the Critical Path Method. As noted earlier, these calculations
follow a discrete set of rules, so that they are always performed in exactly the
same way. The calculations take the form of a Forward Pass and a Backward
Pass and are used to determine the scheduling values in a network schedule:
Early Start, Late Start, Early Finish, and Late Finish for each activity. From
these values, other important scheduling information is determined such as
the Float for each activity, critical activities, the critical path for the project,
and the earliest completion time for the project.

Prior to the development of computer scheduling programs, these Forward Pass and Backward Pass calculations were performed manually, a very
tedious and time-consuming process. Today, scheduling programs perform
these network calculations very quickly and unerringly.

10. **Make the Forward Pass.**

The Forward Pass involves the application of a set of rules and procedures that are inherent in the Critical Path Method and is done exactly the
same way each time. The Forward Pass determines the Early Start and Early
Finish times for each activity in the project. This process also determines
the earliest possible project completion time, that is, the minimum project
duration, according to this plan for the work.

The rules for performing the Forward Pass are as follows:

1. The ES of each activity is entered onto the node in the top-left box within
   the node.

2. The Early Start (ES) of the first activity in the project is defined to be 0.

3. The Early Start of each of the other activities in the network is the Early Start
   of the previous activity (predecessor), plus the duration of the previous
   activity.

$$ES2 = ES1 + Duration1$$

4. Where the network branches, the Early Start of the activity at the end of
   the branch (where the network reconvenes) is the *highest* value derived
   from calculating the Early Start along each leg of the branch.

5. Calculate the ES of each activity in the network until the end of the Network
   Schedule has been reached.

Figure 14.10 shows an AON schedule with the Early Start (ES) values determined for each activity.

6. Calculate the Early Finish (EF) of each activity in the network.

The Early Finish of each activity is its Early Start, plus its duration.

$$EF_n = ES_n + D_n$$

The EF value for each activity is entered onto the node in the top right box within the node.

The Early Finish value for the last activity in the Network Schedule is the earliest possible project completion time (according to the plan represented in the Network Schedule).

Figure 14.11 depicts the network schedule with the Early Finish (EF) values determined for each activity.

11. **Make the Backward Pass.**

The Backward Pass also utilizes a set of rules and procedures that are inherent in the Critical Path Method so that it is performed exactly the same way each time.

The Backward Pass determines the Late Start (LS) and Late Finish (LF) times for each activity in the network.

The rules for performing the Backward Pass are as follows:

1. Forward Pass calculations must be completed for the network before the Backward Pass can begin.

The Backward Pass begins at the last activity in the network.

2. Determine the Late Start (LS) for the last activity in the network.

The LS of the last activity in the network is defined as being equal to the Early Start of that activity.

3. The LS value is entered in the lower left box within the node for each activity.

4. Work backward through the network.

5. The Late Start of the current activity (the activity whose LS has yet to be determined), is the Late Start of the following activity (successor) minus the duration of the current activity.

$$LS10 = LS11 - D10$$

6. When the network branches, the Late Start of the activity at the end of the branch where the network reconvenes is the *lowest* value derived from determining the Late Start of each activity along each leg of the branch.

7. If the network calculations have been performed correctly, when the Backward Pass has been completed and the first activity in the network is reached, the Late Start of the first activity in the network schedule will be zero, the same as the Early Start value for that activity.

Figure 14.12 depicts the network schedule with the Late Start (LS) values entered.

8. Determine the Late Finish (LF) for each activity.
    The Late Finish for each activity is its Late Start plus its duration.

$$LF_n = LS_n + D_n$$

9. The Late Finish value is entered in the lower-right box within the node for each activity.
    Figure 14.13 shows the network schedule with the Late Finish (LF) values entered.

12. Determine Float for each activity.

Float was defined earlier as the amount of time that the start, finish, or duration of an activity can be extended without an adverse effect on project duration and project completion date.

The float for each activity can be determined after the Forward Pass and Backward Pass are complete. The float (start float) for each activity is the difference between the Late Start and the Early Start for that activity.

$$Float_n = LS_n - Es_n$$

While the concept of float has many implications for the owner, and for the general contractor and the subcontractors on a project, knowing the definition of Float, and its significance with regard to the critical path (to be discussed in the paragraphs which follow), will be of tremendous value to the supervisor.

13. Identify critical activities.

When the Float for each activity has been determined, critical activities can be identified. Critical activities are those whose Float is zero.

This means that if an activity is a critical activity, any delay in the completion of that activity will delay the project duration and, therefore, project completion by an amount equal to the delay in the completion of the critical activity (unless management action is taken). For this reason, supervisors and other managers place a high priority on identifying critical activities and then ensuring that they manage the work in such a way that these critical activities are completed on time.

Figure 14.14 shows the network schedule with the critical activities determined. Critical activities are those whose float is zero.

14. Plot the critical path.

The critical path is the pathway through a network schedule that connects critical activities. On a typical network schedule, the critical path is indicated with red arrows and often with the arrows emphasized in boldface, so that the critical path is easy to see.

Figure 14.15 shows the network schedule with the critical path indicated.

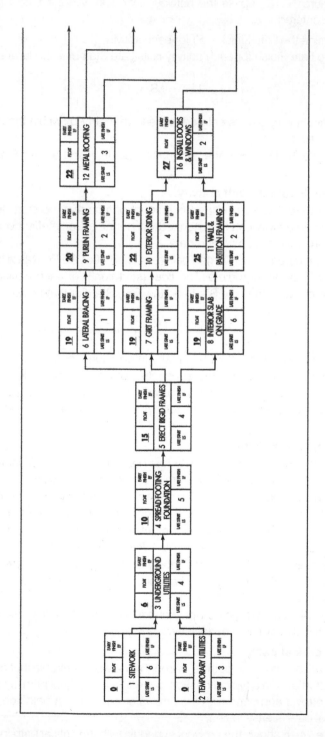

Figure 14.10 Network Schedule with Early Start (ES) Values Calculated

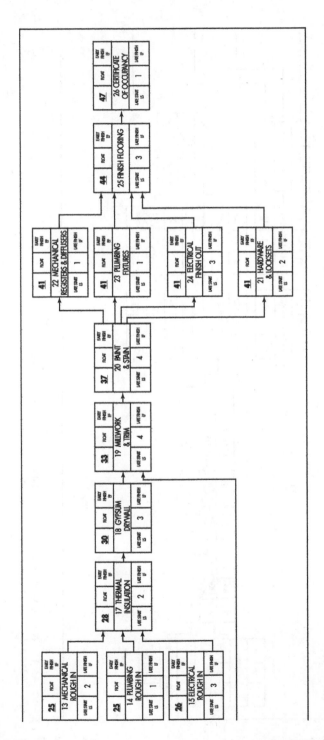

Figure 14.10 (Continued)

263

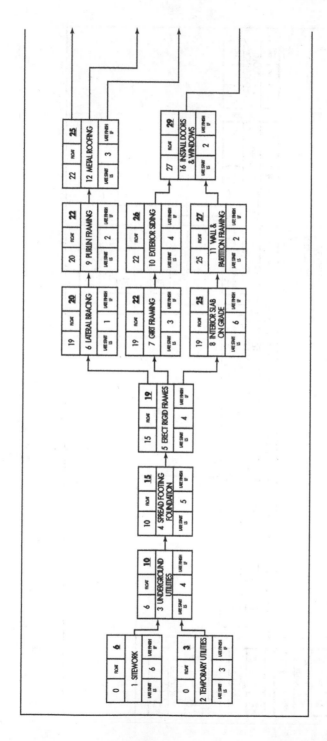

Figure 14.11 Network Schedule with Early Finish (EF) Values Calculated

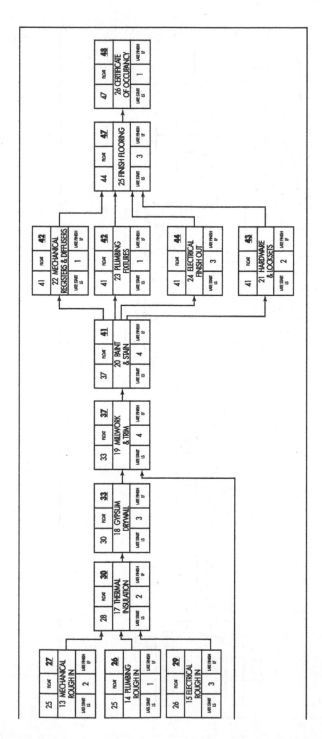

Figure 14.11 *(Continued)*

265

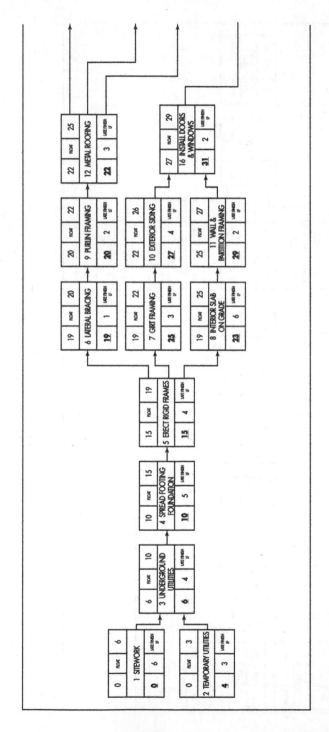

Figure 14.12 Network Schedule with Late Start (LS) Values Calculated

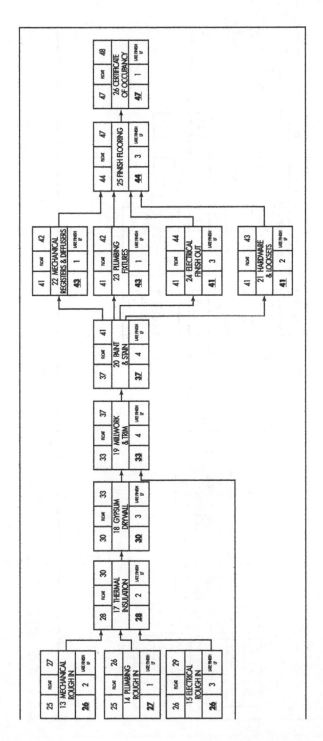

Figure 14.12 *(Continued)*

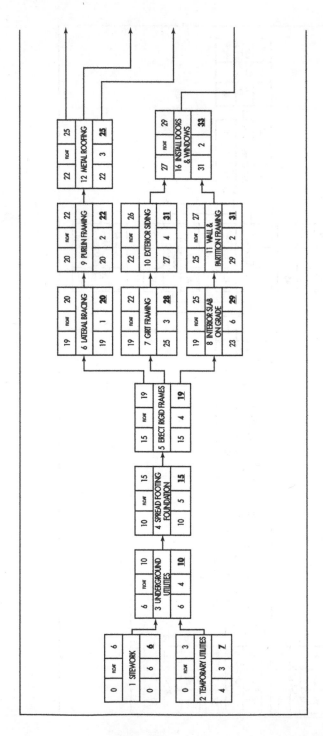

Figure 14.13 Network Schedule with Late Finish (LF) Values Calculated

268

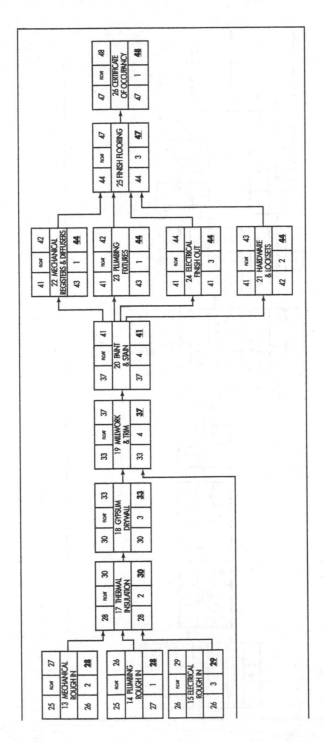

Figure 14.13 (Continued)

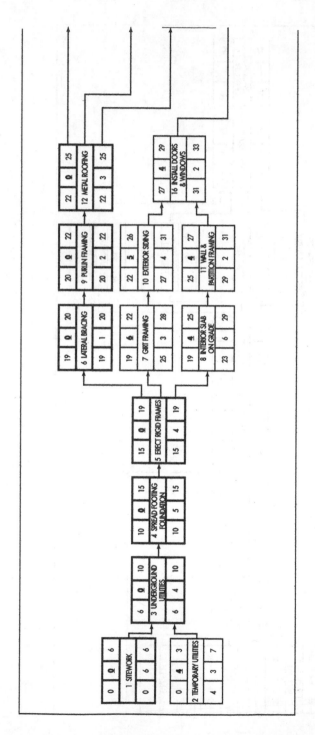

Figure 14.14 Network Schedule with Critical Activities Indicated

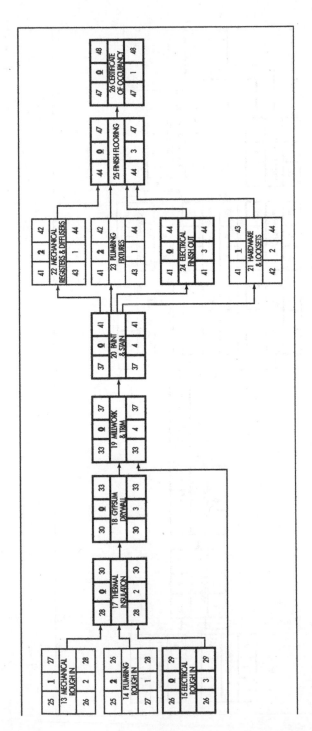

Figure 14.14 *(Continued)*

271

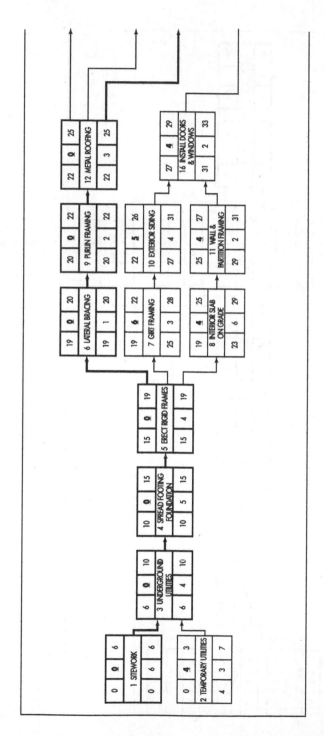

Figure 14.15 Network Schedule with Critical Path Indicated

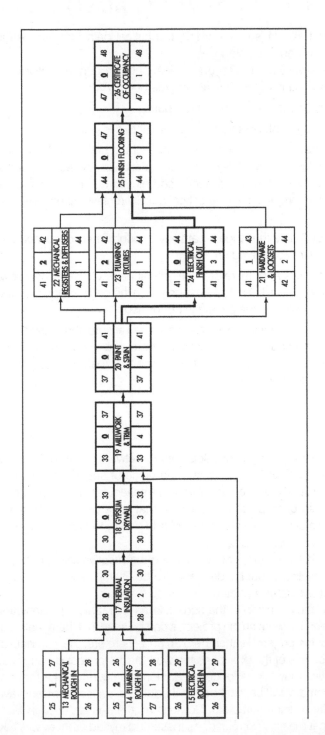

Figure 14.15 (*Continued*)

BASIC TRUISMS WITH REGARD TO THE CRITICAL PATH

There are some basic truisms regarding the critical path that are useful for the supervisor to know and understand:

1. There will always be at least one critical path through the network schedule by the application of the Critical Path Method (CPM).
2. There may be more than one critical path.
3. The critical path links in sequence, those activities which are *critical activities*.
4. The significance of critical activities, and of the critical path, lies in the fact that, by definition, any delay in the completion of any activity that is on the critical path will delay the project's completion by an amount equal to the delay in the completion of the critical activity (unless management action is taken).
5. The critical path is the longest pathway through a network schedule that will get the project completed on time.
6. Critical activities and, therefore, the critical path will likely change when the schedule is updated to reflect changes from the original plan that have taken place as the work on the project is performed.
7. Therefore, it is imperative that supervisors ensure that they are keeping abreast of schedule updates, working from and communicating the latest schedule revisions, and continuing to place their focus on completing critical activities on time, throughout the project.

NETWORK SCHEDULE UPDATES

This chapter has discussed the development of network schedules and the application of the Critical Path Method. The importance of supervisors understanding the terminology involved in the use of network schedules has been emphasized. Emphasis has also been placed on the fact that supervisors' understanding the derivation and use of network schedules will add greatly to their effectiveness in the management of construction projects.

There is an additional component regarding the effective use of network schedules that the supervisor should understand. To be useful, network schedules must be regularly and accurately updated.

If we refer to the definition of the schedule as a graphic representation of the plan for performing the work on the project, it can be seen that there really is not ever *one* schedule for the project. Rather the plan and, therefore, the schedule changes throughout the course of the project as the work on the project is performed.

Some activities may take a longer or a shorter amount of time for completion than planned; some activities must be rescheduled because of material availability or the availability of subcontractors. Sometimes there is an error in the original planning logic that must be corrected. As these and myriad other events take place on a construction project, the plan changes; and when the plan changes, its graphic representation, the schedule, also must change.

When changes are to be reflected in the schedule, schedule updates are prepared. Some contractors perform schedule updates on an as-needed basis. Other contractors update project schedules monthly, while others may do so more often. The point is that the latest schedule revision should reflect the most current plan for the performance of the work.

When schedules are updated, activity relationships may change. Activities that were previously in parallel may now be in sequence, or *vice versa*. Additionally, activity durations may be changed when the schedule is updated. When activity relationships or durations change, and when network calculations are then performed, the schedule obviously changes. Additionally, the critical path may change. Activities that were formerly not on the critical path may become critical activities and, therefore, will be on the critical path following a schedule update, and *vice versa*.

Most network schedules have, or should have, a "Date of Last Revision" indicated on the schedule itself. Supervisors should form the habit, anytime they are consulting the schedule or anytime they are communicating verbally or in writing with regard to any scheduling matter, of ensuring that they have checked the Date of Last Revision and ensuring additionally that the schedule they are referring to is the most current version.

In an additional note, the supervisor should realize that there is sometimes frustration and misunderstanding in planning the work, with regard to schedule updates. It is not uncommon for the supervisor to have identified critical activities from the schedule as recommended previously, and to have his planning directed toward the timely completion of those activities, only to have those priorities changed when the schedule is updated. The activity that the supervisor had been planning for the craft workers to be working on may no longer be critical on the updated schedule, and another activity, which previously had float, is now a critical activity and, therefore, must be given priority in order to ensure that it is completed on time.

Some have argued, in fact, that network schedules are useless, inasmuch as they are represented to be a planning tool, and then when the schedules are updated, the plan and its priorities change, causing misunderstanding and confusion along with the necessity of rearranging the plan for the work. The supervisor who understands the basis for planning and for network scheduling, soon realizes that the inverse is actually true. If the schedule is indeed a graphic representation of the plan for the work, then it follows logically that, when events change on the project, a new plan is in order and, therefore, the schedule must be updated to represent the new plan.

SHORT-INTERVAL SCHEDULES

As was emphasized in the previous discussion, network schedules are extremely useful for portraying the overall project schedule. However, they contain so much information, and on such a large scale, that they are not very useful for daily or weekly planning of the work.

One of the most effective planning tools the supervisor can employ is the short-interval schedule. Short-interval schedules break down the overall project schedule, the network schedule, into manageable time frames and activities to be performed.

JRRS CONSTRUCTION										
PROJECT 2021: SHINING STAR OFFICE BUILDING										
ACE OWNER										
CERNA & GARZA ARCHITECTS										
WEEK OF: AUGUST 2 - 6										
CREW 1: CARPENTERS JESSIE, CHARLES; HELPERS LOUIS, SAMMY										
	MON		TUES		WED		THURS		FRI	
	AM	PM	AM	PM	AM	PM	AM	PM	AM	PM
Plywood Decking. Pan Slab Forms, Level 2. North Bay	▓									
Edge Forms & Blockouts		▓	▓	▓						
Form Tendon Anchorage Pockets					▓	▓				
Brickledge						▓				
Align & Brace Edge Forms							▓			
Layout for Metal Stud Partition Framing. Room 107. South Bay								▓		
Metal Stud Partition Framing. Room 107. South Bay										▓

Figure 14.16   Five-Day Short-Interval Schedule

Short-interval schedules are almost invariably prepared and represented in the form of a bar chart.

Short-interval schedules can be four-week, three-week, or one-week look-ahead schedules. On very urgent projects that have a short turnaround time, such as might be encountered in shutdown work in a factory or industrial facility, short-interval schedules may be prepared to plan the work at hourly, or even shorter, intervals. The key point for supervisors to realize is that the short-interval schedule is a powerful tool that they can use to meet the planning needs at hand for short-term planning, whatever that short term is determined to be.

Many managers believe that the best short-interval schedule for a supervisor to utilize on a typical project is a five-day look-ahead, with each work day divided into half-day intervals. Figure 14.16 provides an example of such a schedule. Figure 14.17 depicts a four week short interval schedule where the supervisor is planning for manpower (craft labor).

A schedule of this type provides a great deal of useful planning information, both for the supervisor and also for the craft workers. Everyone can see at a glance what the plan is, for each day or segment of a day, as well as for the entire week. Additionally, with this plan in hand, the supervisor and the workers can be thinking about and planning for the materials and tools that will be needed to complete the work.

In addition to its use as a planning tool, such a schedule has also been shown to be a very effective communication tool. Construction craft workers often are heard

4-WEEK MANPOWER SCHEDULE
Big Manufacturing Hdqtrs. - Proj#: 12345

Activity	WEEK 2 8/10 Mon	8/11 Tue	8/12 Wed	8/13 Thu	8/14 Fri	8/15 Sat	8/16 Sun	WEEK 3 8/17 Mon	8/18 Tue	8/19 Wed	8/20 Thu	8/21 Fri	8/22 Sat	8/23 Sun	WEEK 4 8/24 Mon	8/25 Tue	8/26 Wed	8/27 Thu	8/28 Fri	8/29 Sat	8/30 Sun
– Sanitary Sewer – Lines D & F																					
– 30" Storm Sewer Run-out to Inlet Boxes	4	4	4	4	4																
– 12" Equalization Line @ Cistern	3	3	3	3	3			3	3	3											
– Misc Drain Lines to Cistern (6" to 12")																		3	3		
– Site Cuts & Fills @ Bldg Areas	3	3	3	3	3																
– Grading @ Service Yard	1																				
– Approaches from Kumbuya Blvd.	6	6	6	6	6			6	6	6	6	6			6	6	6				
– Light Pole Bases	3	3	3																		
– Detention Pond Structures											5	5			5	5	5	5	5		
– Backfill @ Crawlspace Walls															2	2	2				
– Backfill @ Pile Caps – 11–16	2	2	2	2	2																
MANPOWER TOTALS	**22**	**21**	**21**	**18**	**18**			**9**	**9**	**9**	**11**	**11**			**13**	**13**	**13**	**8**	**8**		

Figure 14.17 Four Week Short Internal Schedule
Courtesy of Skanska USA Building Inc.

to complain that they are not apprised of the planning and that they frequently feel as though they do not know what is going on.

Managers who have employed short-interval schedules, and who have communicated this planning information to the craft workers, have consistently reported that workers very much appreciate knowing what the plans are. Further, when they know what the plans are, the craft workers become participants in the planning process and can provide their input with regard to materials, tools, and equipment that will be needed in performing the work, or with regard to the plan itself. This greatly facilitates and improves the planning.

Additionally, many believe that when the workers are kept informed and involved with regard to the planning, they become more motivated and more productive in their performance of the work. It is noteworthy that these findings are in exact accord with the understanding that was developed in Chapter 7.

SUMMARY

The importance of planning has been emphasized numerous times throughout this book. In this chapter, the importance of managing time on a construction project was overlaid with the planning function. Construction scheduling is the contractor's primary planning and control tool for managing time on a construction project.

Understanding construction scheduling and the various forms of construction schedules that are utilized in the construction industry will prove to be an extremely valuable asset to the supervisor throughout his career. In addition, when that understanding translates to proper use of schedules of all kinds for more effective planning and management of the work, the supervisor becomes much more effective in an overall sense and, therefore, much more successful in managing construction projects.

Learning Activities

1. Learning more about network schedules

   Arrange to get a copy of the network schedule that is in use on the project where you are now working.

   Spend some time analyzing the network schedule, to see for yourself the various components discussed in this chapter.

   See for yourself the "big picture" plan for the project as portrayed on the network schedule.

   It is likely you will conclude that while the network schedule is useful for determining and depicting the overall planning for the entire project, it is not very useful for planning day-to-day work activities.

2. Utilization of a short-interval schedule

   Prepare a five-day short-interval schedule for your crew, on the project you are working on at the present time. Use the bar chart form of schedule, and draw it by hand. If necessary, ask for assistance from your superintendent or project manager in making the schedule and in determining what should be on it.

   Use the schedule to obtain, or to be sure you have on hand, the necessary resources for the work.

   Use your schedule to set priorities, and to make assignments, and to manage the work for the next five days.

   Show your schedule to the workers in your crew. Use the short-interval schedule as a communication tool, to let your crew know what the plan is, and what lies ahead, for tomorrow and the following days.

   Let the workers know that now that they can see what the plan is for the next five days, they are welcome to contribute their thoughts with regard to tools, materials, equipment, or supplies that may be needed.

   Let the workers know too, that if they know of anything that might make this plan unworkable, or if they know of some work item that has a higher planning priority than something on the short-interval schedule, they should let you know.

CHAPTER 15

MANAGING PRODUCTION

INTRODUCTION

Managing production is at the very core of what a supervisor does. The foreman manages at the crew level. The superintendent manages at the job level, but supervisors at all levels have the responsibility to ensure that the project is completed on time, on schedule, safely, and at a high level of quality. Hence, they have the responsibility to manage the production process.

Managing production is of personal importance to a supervisor. Often, the reason craft workers become foremen is because they think that they can lead the crew and accomplish the work more effectively than other foremen they have known or worked for. Of course, becoming a foreman might not be by the initiative of the craft worker, but rather by the determination of someone higher up in management. Craft workers are often selected to be a supervisor because they demonstrate a number of traits and skills that lead to effective production, which is what management wants. Thus, whether it is their own decision or that of a manager above them, the decision to have a craft worker become a supervisor is based primarily upon the ability to produce.

The primary metric for evaluating supervisory performance is how well the supervisors manage their part of the project. Supervisors are evaluated by the managers above them, based on such criteria as how well their crew accomplishes assigned work, and how effectively work assignments move the project toward successful completion. Supervisors are also evaluated based upon how efficiently (economically) work is completed, what the quality of the work is when it is completed (eliminating rework), and how safely the work was done. All of these are measurements of production.

UNDERSTANDING PRODUCTION AND PRODUCTIVITY

Production can be defined as the process of producing a product. In construction, production is accomplished by bringing together materials, tools, and workers at a specific location so that the workers can use the tools to install the materials and produce their component of a complex project.

Productivity measures the efficiency of production. It expresses the classic relationship that exists between what is used to produce the product (inputs) and how much product is produced (outputs). This relationship can be shown as:

$$Productivity = Outputs/Inputs$$

Higher productivity means more is produced with less input. Lower productivity means that less is produced with the same input, or more input is required to produce the same product.

In construction, productivity is typically measured in terms of the number of units produced per hour of work. The hours might be in terms of individual worker hours, crew hours, or machine hours. Examples include:

- Feet of 2-inch conduit installed by an electrician in an hour
- Bricks laid by a masonry crew in an hour
- Cubic yards of soil moved by a scraper in an hour

Money may be used as the measurement of investment instead of time. However, measuring money is not as useful as measuring time because the amount of money expended per unit of output varies with a number of different factors. Labor cost depends on the labor rate (dollars per hour paid to the craft worker), which in turn varies due to a number of factors, such as where the job is located, when the job occurs, and how the labor rates were negotiated. Hours tend not to change nearly as much from one location to another and from one time frame to another.

Example

To illustrate the relative stability of time as a measurement of investment in a project as opposed to money, an electrician in Phoenix can be expected to install about the same amount of conduit in an hour as a similarly skilled electrician in Denver. However, the labor wage rate in Denver will be different from that in Phoenix, so the cost of the work will differ from one city to the other. Also, the same electrician working in Phoenix would have installed about the same amount of conduit in 2000 as she can be expected to install in Phoenix in 2010, but her wage can be expected to be considerably more in 2010, resulting in a higher cost to complete the same work.

Another impediment to using money as the basis of measuring production is that management often believes that production units expressed in terms of money are proprietary and are a closely held secret of the company. Management tends to be far less sensitive about how much production a craft worker or crew can achieve in an hour than how many labor dollars per installed unit it costs the company to accomplish work.

Supervisors, too, generally prefer measuring production in terms of hours rather than in terms of money. Hours are useful because they equate to schedule, and time is always important in construction. Additionally, from the time supervisors began as apprentices, they experienced and observed work accomplishment in terms of time, not money. Thus, basing evaluations of production on time is more natural to the supervisor, who can relate the time required to complete a certain amount of work to personal experience.

For all of these reasons, production and productivity are evaluated primarily in terms of time. Production measurements can easily be converted to monetary terms based upon local conditions, if required.

Example

To illustrate the conversion of production rates between time and money, if the electrician installed 1000 feet of 2-inch conduit in Phoenix in 2000, at a production rate of 10 feet per hour, the rate in terms of money might have been $2.50 per foot. In 2010, the production rate would be the same, but the money based rate would be $3.00 per foot because of an increase in the wage rate of the craftsman. In Denver in 2010, the money based rate would be $2.75 per foot even though the production rate would still be 10 feet per hour, for the same reason.

IMPROVING PRODUCTION

Production can be improved at many levels. The classic approach to improving production in construction starts by identifying inefficient activities through cost reports, as discussed in Chapter 13, through schedule updates, or through weekly coordination meetings, as discussed in Chapter 14. With more sophisticated companies, a problem-solving process, like the one described in Chapter 8, is then followed to determine the root cause, to identify and implement the best solution, and to evaluate the results after implementation.

All of the production improvement measures noted above are *reactive* measures, since they are not undertaken until a problem has been identified. It should be noted that it is far preferable to take *proactive* measures to improve production. These proactive improvement measures can be undertaken at several levels. The first level to consider is improving ongoing production at the activity level. Then, stepping back to the beginning of an activity, the planning process can be used to develop the design of a more productive way to execute the activity even before that activity has begun. Finally, stepping back still farther, consideration will be given to how execution of the overall construction process can be improved.

IMPROVING PRODUCTION IN AN ONGOING ACTIVITY

In manufacturing, where activity execution tends to be highly repetitive, every effort is made to design the manufacturing process to operate as efficiently as possible. After production begins, highly repetitive manufacturing processes are studied intensively in order to identify potential productivity improvements. Manufacturing engineers understand that even those production processes that are functioning very well can be improved, and with further iterations these improved processes can be further improved.

On the other hand, constructors tend not to look for process improvements, especially if a project is meeting its schedule and cost goals. Construction activities tend, by nature, to be inefficient. This is because of both the harsh environment within which construction is executed and the fact that construction activities are not highly repetitive like manufacturing activities. Therefore, it is accepted in the industry that construction activities will not be very efficient; and as a result, estimates of time and cost are based upon inefficient activities.

In construction, activities tend to be planned according to the way similar activities were done in the past. And once activities have been planned, little consideration is given to improving the process unless cost reports or schedule updates indicate that there may be a problem.

Process improvement is based on repetition. However, even though contractors repeat the same types of activities from job to job, the specific conditions of the job vary so the processes are not repetitive. Even a repetitive activity on a specific job does not repeat exactly the same way. In the relatively controlled environment after a building project has been closed in, the location changes and crews change so that repetitive type activities are not truly repetitive. For outside activities, not only do the location and crews change, but weather and working conditions change as well.

Sophisticated contractors try to capture lessons learned from past projects to improve planning and execution of activities on current and future projects, and to develop a record of best practices. However, few engage in active production improvement studies on an ongoing project, especially if the project is going well as measured by meeting or exceeding project objectives in terms of time, cost, quality, and safety.

In construction, production improvements are generally not pursued unless a problem is identified. However, as constructors can learn from the manufacturing industry, even an activity that has not been identified as a problem activity because it is behind schedule or over budget can generally be improved. A number of tools are available to help identify where opportunities lie to improve production in repetitive activities. Three such tools will be considered.

Production Analysis Tools

A *process chart* (see Figure 15.1) shows the steps in an activity and the duration of each step.

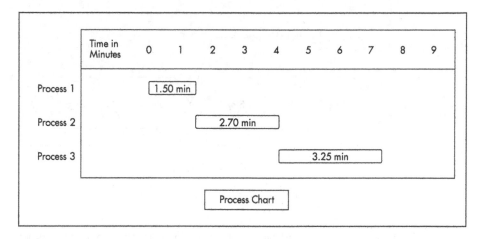

Figure 15.1   Process Chart

The process chart helps to define the steps that make up an activity and to quantify how long each step takes. After observing an activity and determining the basic steps that compose the activity, the time required for completion of each step is determined by measurement. Since, in construction, the time required to execute a step in an activity can vary significantly from cycle to cycle because of variability in the way workers perform, the accuracy of the times put into the chart can be improved significantly if time is measured for several cycles and averaged.

A *crew balance chart* (see Figure 15.2) shows the activity in terms of the work performed by each crew member in carrying out each of the steps.

Again, the data for this chart are gathered by direct observation and measurement. The crew balance chart shows clearly which crew members are contributing to the activity at any given time and, hence, how much of the time each crew mem-

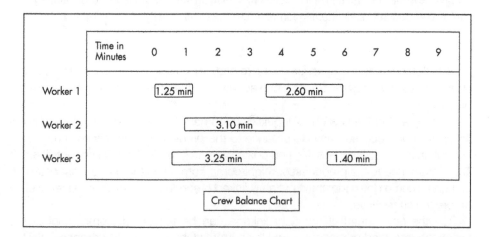

Figure 15.2   Crew Balance Chart

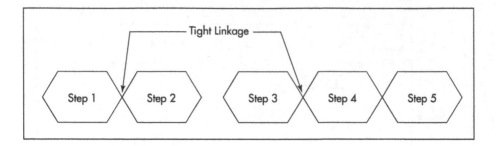

Figure 15.3 Flow Chart

ber is working. When using a crew balance chart, it is important to explain to workers that the activity on which they are working is being studied in order to improve its execution. It must be made clear to the workers that there is no intent to collect data on individuals other than to analyze *activity production* and that activity production is the focus of the study, not worker performance.

The third production analysis tool is the *flow chart* (see Figure 15.3), which represents how the steps relate to each other.

Construction is achieved by execution of a series of interrelated or linked steps. Many of the steps are *tightly linked*, meaning that the predecessor activity must be completed prior to starting (or completing) the following step, which is called the successor activity.

Example

To illustrate the predecessor-successor relationship, when considering a sequence of activities in the construction of a concrete wall, placing forms is the predecessor activity to placing concrete and stripping forms is the successor activity.

The flow chart helps supervisors understand how one step is influenced by other steps, or in terms of activities, how one activity is influenced by other activities that precede it.

In considering the flow chart, it quickly becomes clear that if one step is tightly linked to a predecessor, any disturbance to the predecessor activity will have an impact on the successor activity, and possibly on a number of successor activities. The disturbance to the predecessor might come from within the predecessor step, or it could come down the chain of tightly linked predecessors from some step well before the current step.

On the other hand, if the tight linkage can be relaxed at some point between the disturbed predecessor and the current step, then the disturbance in that predecessor will have diminished impact on the current step. This holds true until the flexibility in the linkage of steps is used up.

An example will demonstrate the use of these three tools. Here is a case study of cutting electrical cables for installation into cable trays in an industrial plant. This illustrative example is based upon a real study of an activity in a project that was considered at the time to be successful, and the cable cutting activity was considered to be going well.

The cable-cutting process involved writing labels, attaching a label to one end of the cable, pulling the cable off of a large reel to the required length, reeling up the measured cable and cutting it from the large reel, and finally attaching a label to the newly cut end. Figure 15.4 shows the flow chart for this activity.

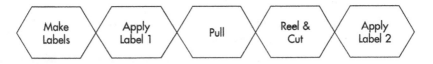

Figure 15.4   Cable-Cutting Flow Chart

There are five steps in this activity, and each is completed before the next is begun. Time measurements for the activity taken through four cycles indicated that the entire activity averaged 5.25 minutes. The resulting process chart is shown in Figure 15.5.

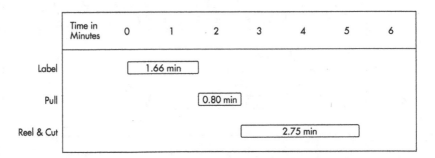

Figure 15.5   Cable-Cutting Process Chart

This activity was initially carried out by a crew of three electricians. The work accomplished by each worker is illustrated in the crew balance chart shown in Figure 15.6.

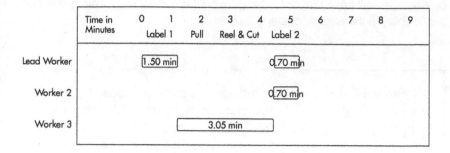

Figure 15.6   Cable-Cutting Crew Balance Chart

The crew balance chart clearly shows a significant amount of unproductive time on the part of the crew members. The conclusion can be made from this chart that one of the crew members is not essential to this operation and that the operation can be performed by two electricians instead of three.

Several options are available to improve the production. The supervisor could have two electricians pulling, reeling, and cutting instead of just the one. Or, he could reassign one of the electricians to another activity. Or, he could lay off one of the electricians. On a job the size of this industrial plant, it would probably be easy for the supervisor to reassign one of the electricians. After the reassignment, the two-person crew proceeds with the activity. Again by observing and measuring time, a revised crew balance chart is drawn (see Figure 15.7).

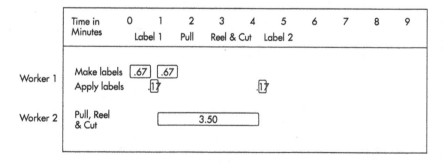

Figure 15.7   Revised Cable-Cutting Crew Balance Chart

It should be noted at this point that the initial investment of worker time per cycle, based upon a crew of three electricians working for 5.25 minutes per cycle, averaged 15.75 worker minutes. In the improved process, with two electricians working for 4.50 minutes, the improved investment of worker time was 9.00 minutes, showing a significant savings over the initial process. This large industrial project required about 2.5 million feet of cable with an average length of 100 feet. If the burdened labor rate were $40.00 per hour, the savings of 6.75 worker minutes per cable would result in saving $4.50 per cable or $112,500 for the 25,000 cables.

However, this was not the end of the improvement exercise. The revised cable-cutting crew balance chart (Figure 15.7) indicates that a significant amount of time was still being wasted by the fact that the electrician who was performing the labeling was waiting for the cable to be pulled, reeled, and cut. It was determined that there was no reason why a single electrician could not make and apply the labels, as well as pulling, reeling, and cutting cable.

A third opportunity for improvement was also identified. The step of making the labels, was analyzed, and it was determined that the labels did not have to be prepared at the time they were applied to the cable ends. Instead, they could be prepared beforehand offsite and provided in a batch to the electrician. The cables were defined by a tabular schedule on the drawings indicating the start point and end point of each cable. This is precisely the information required to make a label. So a set of labels was prepared by a clerical person in the office, using the information in the cable schedule, and then the premade labels were provided to the electrician in the field.

The resulting activity is represented by both a crew balance chart and a flow chart in Figure 15.8.

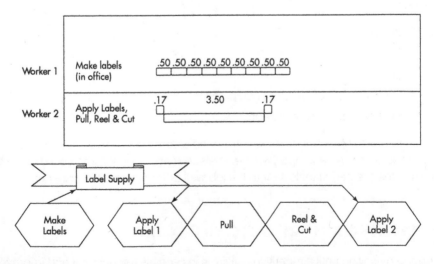

Figure 15.8   Final Cable-Cutting Crew Balance Chart and Flow Chart

The final improvements resulted in a single electrician working in the field for an average of 3.83 minutes per cycle, plus some clerical time in the office to make the labels. It was determined that the label making could be done independently of the field activity and with less costly labor. The resulting savings over the previous improvement were significant, and the overall savings resulting from the activity analysis were very significant.

Howell, G., Laufer, A., and Ballard, G. (1993). "Interaction between Subcycles: One Key to Improved Methods." *Journal of Construction Engineering and Management*, ASCE, New York, NY, 119 (4) 714–28.

Two Basic Production Improvement Principles

The illustration described in the paragraphs above demonstrates how to use the three tools introduced earlier, in order to improve production. It also provides some basic principles that can be applied in looking for improvements in ongoing activities.

First, supervisors will do well to look for and eliminate slack time, which is time when workers are not productively engaged in work. We must remember, and must reassure the workers, that the intent is not to place blame, but rather to identify where inefficiency lies so that the inefficiency can be eliminated.

Second, supervisors can look for opportunities to relax or remove the link between tightly linked steps. In the case study, the step of making labels could be separated from the field operation by inserting a *buffer* of prewritten labels between two tightly linked steps, thus separating one from the other. A buffer is something inserted in a construction operation in order to relax the interaction between, or to disconnect, related activities. Buffers fall into a variety of categories, including:

- Space
- Time
- Labor
- Tools and/or construction equipment
- Material and or installed equipment

In the example, buffers in the form of time and space were inserted between the steps of making and applying the labels. A material buffer was also used in the form of the premade labels sent to the job well before the cables were cut.

DESIGNING A MORE EFFICIENT TASK

Instead of waiting until a repetitive activity is in process to analyze it and improve it, supervisors should think in terms of planning the activity in such a way that it will be more efficient when it is performed. This can be an extremely effective technique. Such an analysis may involve a *First Run Study*, which can be defined as an explicit, detailed plan for an activity prior to starting work on that activity.

First run studies consider many aspects of the activity, such as:

- Safety
- Operation
- Sequencing and relationship to other activities
- Crew balance
- Tools and construction equipment
- Anything else that may affect this activity or related activities

First Run Study Setup

To begin the First Run Study, an appropriate activity is selected. Several characteristics are considered in selecting an appropriate activity.

An activity should be selected that is about to begin. The activity might be at the beginning of the project or when the project is about to enter into a new phase. An appropriate activity will be repetitive, with a relatively short cycle of repetition. That is, cycle duration should be in terms of minutes, rather than hours or days. The activity will involve a limited number of workers since, if possible, all workers should be involved in the study process.

When learning the First Run Study process, additional characteristics should be considered. The activity should be a standard activity and one familiar to the team that will be performing the First Run Study. It should be on a project that is considered to be successful, and it should not be a highly complex activity. After the First Run Study technique is mastered, more complex activities can be studied and projects that are in trouble can be considered. However, studying complex activities or troubled projects before mastering the First Run Study process is not a good idea, because it may be unclear whether the First Run Study or other events have actually resulted in the observed improvements, or the lack thereof.

It is also important to start with an attitude that this will be a learning process. Early studies may not be very successful, but early on, it is more important to learn the process than to have a resounding success in production improvement.

It is interesting to note that, in many cases, it has been found that novices are more successful at first run studies than highly experienced supervisors. More experienced supervisors tend to be constrained by past experience and are often not able to think creatively. The tendency for these experienced supervisors is to fall back to solutions that have resulted in success in the past, rather than seeking new and creative solutions. Less experienced supervisors often are more willing to ask questions and seek nontraditional approaches.

After the activity for the First Run Study has been selected, a study team is brought together. The First Run Study team is comprised of a broad cross-section of those participating in the activity, including the foreman, craft workers, and perhaps suppliers and workers from related activities. To get a different perspective, a participant from the office might be included, such as the project manager or the estimator who estimated the project. Although they often are not involved in day-to-day activities on the job, both estimators and project managers can frequently come up with very creative ideas.

First Run Study Plan

To start the study, a process chart is developed. This helps identify critical steps in the activity and the relationships between the steps. A brainstorming process can be used to encourage creativity and to involve all participants in the process. Particular care should be used to ensure that all critical steps in the activity are identified and clearly defined.

Once the activity process chart is completed, it should be reviewed and then checked against several criteria. The first check should be for safety. Where does the risk lie, in the performance of this activity? Are there hazards to be accounted for? What preventative measures need to be used to minimize risks and hazards?

Next, the supervisor should consider where the activity might go wrong. How might the quality of the work be impaired? What could cause delays in the various steps? What mitigations can be used to lower the risk of something going wrong? In the event one of the identified potential problems actually happens, what steps can be taken immediately to deal with the problem?

A key part of the First Run Study is to observe early cycles to learn what is working as planned, what is not working, and where a new approach might be incorporated into later cycles. During the initial planning stage, the supervisor should determine what will be observed, what data should be collected, and what key factors indicate whether the activity is improving or not. It is important to identify what is to be observed and what data are to be collected prior to the first work cycle.

First Run Study Execution

Next, the supervisor should take the workers, and the materials and equipment, to the work site and execute one cycle of the plan. The observer should watch the specific steps identified in the plan and collect the specified data. The observer will most likely see problems with the plan, but this is normal. The cycle should not be stopped, and no corrections should be made during the cycle, unless they relate to an urgent safety situation.

In addition to collecting specific data, the observer should keep in mind the following questions:

- Have the planned steps been executed? If not, why not?
- Have any unanticipated interactions between steps or with other work going on been identified?

- Were there any problems with the quality and quantity of work completed?
- Were any safety issues detected?
- Were essential resources available and accessible when required?

First Run Study Analysis and Redesign

After the cycle has been completed, the team reconvenes to evaluate the cycle. The observer reports to the team on all observations. The workers add observations of their own. Each observation is discussed. A revised plan is developed, and the next cycle of work is initiated. This process of planning, execution and observation, analysis and redesign is continued through several cycles until the team agrees that the obvious improvements have been identified and incorporated. Work then proceeds in a steady state.

Two Operational Changes

Taking a broader look at the process, two changes from the traditional approach become apparent. First, planning becomes more important than doing. Doing the work should be a result of the planning process. When given a new task, the tendency of the workers is to immediately begin working. The inclination is to start as quickly as possible, so as to be able to finish as soon as possible. When doing First Run Studies, it becomes apparent that, with proper planning, the activity can be started later but can be finished earlier and more efficiently. While this may be counterintuitive, it has been shown to be true time and time again.

Second, much of what supervisors do is to answer questions. Answering questions is reactive management. With a First Run Study, the focus shifts from the supervisor providing answers, to the supervisor asking questions. Supervisors move from a reactive to a proactive management position.

Engineering students learn very early in their academic preparation that formulating the problem (i.e., asking the right questions) is of fundamental importance to getting to a useful solution. Without asking the right questions, it may be impossible to get to a solution. Even if a solution is attained, if it is the right solution to the wrong question, it will lead down the wrong path. If it is a wrong solution resulting from an improperly formulated question, it will again lead down the wrong path, causing significant inefficiencies. On the other hand, asking the right questions on a proactive basis will lead to effective solutions.

IMPROVING CONSTRUCTION OPERATIONS

We have considered how to study an ongoing activity in order to improve production. We then moved on to explore pre-planning as a means to design a more efficient activity. We will now look at design of the construction process to improve overall operation of the project by introducing concepts of Lean Construction.

To be able to understand and analyze the complex operations on a construction project, it is important to develop a model that demonstrates how construction operations work. This will become a production model for construction. The simpler the model, the more effective it will be, as long as it provides an accurate representation of how construction works.

Many different types of models are used to represent different aspects of construction, so the concept of a model should be quite familiar to supervisors. The schedule is a time-based model of construction that shows when activities happen, how long they take, and how they relate to other activities. The budget is a cost-based model that reflects construction activities in terms of investment of resources, predominantly money. Drawings are a graphical model of the project and specifications are a verbal model of the project. Without drawings and specifications, it would be impossible to visualize the project, share information about it, and build it.

The production model of construction should correctly represent the processes of construction. It is important to think of construction as a series of processes, because processes can be learned, repeated, controlled, analyzed, and improved. A primary reason to develop the production model for construction is to be able to effectively control construction operations through better understanding.

The Breakdown of Traditional Controls

In earlier chapters, traditional tools that are utilized to control construction operations have been discussed, including scheduling and cost control systems to control time and budget. However, experienced supervisors know that, even though these are among the best tools traditionally available to move the project in the desired direction, they are frequently not very effective. This truism is illustrated by the fact that although supervisors talk about driving the project with a schedule, the fact is that the project drives the schedule. The result is that schedules need to be continually updated to realign them with the reality of where the project is at any given time.

Although it is important to use these traditional cost and schedule management tools to control the job, the supervisor should realize that there are a number of problems associated with their use. First, traditional controls focus at the project level, not at the point of execution of the work, the point that we have referred to as the workface. Schedules are generally developed and maintained by project managers, and then information is spun off to the project with more limited scope and greater focus as schedules work their way down through the trade contractors to the crews executing the work. Cost control systems are also developed at the project level, with information directed down to the crew on a limited basis, and most often only when the cost control system indicates that budgets are not being met.

These strategic management tools are valuable, providing a high-level view of activities as they relate to other elements of the project on and off the job site. This view is important since schedules and budgets are affected by many factors remote from the workface, as well as factors where the work is being performed. However, strategic tools lose their power as attempts are made to use them in the field to provide tactical direction at the activity level.

A second problem is in timing. Schedule and cost control systems use information gathered at the site, which is then fed back to a central system for periodic updates of project status. Conditions on the site are continually changing, yet schedule and cost reports reflect conditions from days or even weeks in the past and, hence, are often irrelevant or, even worse, may be inaccurate compared to the current status.

A third problem is that traditional reports show symptoms, not problems. They indicate when projects, or major parts of a project, are behind schedule or over budget, but they give no information regarding why. These reports are of little help in developing effective solutions for the problems that cause the symptoms. Moreover, the solutions that are posed will often produce side effects, which, in turn, must be addressed when identified. Hence, traditional approaches to project control, though important and somewhat useful at the overall project level, are often inadequate for tactical control at the activity level.

A new set of tools is required that will enable better understanding of the current status of the job at any time and that will highlight problems, rather than symptoms so that effective solutions can be developed and implemented on a timely basis. The goal of this new approach is to improve project outcomes, not just activity outcomes and, thereby, to improve company performance overall.

A Production Model for Construction

A new production theory for construction will be developed based upon a simple model that can provide an understanding of how construction works. This model will lead to more effective tools to control the construction process. This new model is based on activities and relationships between these activities. This concept should be familiar to the supervisor, from the discussion of network scheduling in Chapter 14. Here, activities were defined, and you learned that project schedules are built based on activities and their durations (time), and the relationships between activities. In the new model, activities are linked by a physical product or information and also by shared resources (see Figure 15.9).

Example

To illustrate the new production model, a sequence of activities required to construct a partition in an office building will be used. The sequence begins with an existing product—the floor on which the partition will be framed. Information is required, such as partition layout, the materials required to go into the partition, and the location of openings. Physical resources are acquired, such as studs and channels for framing. Other resources are also used, such as time, space, and labor. Once these resources have been assembled, a process is executed to install the studs and channels, resulting in a new product—the framed partition.

This sets up a new cycle to provide electrical rough-in for the partition. The electrician needs new information, including what electrical units go into this partition, such as switches and outlets, and where they are located. The electrician takes new resources, such as conduit and boxes, and executes a new process resulting in a new product—the roughed-in framed partition.

Finally, the drywaller comes in to hang gypsum drywall on the framed, roughed-in partition. New information is needed, including where the electrical boxes are, where the openings are, and what type of gypsum drywall and fasteners are specified. After the drywaller executes the new process, a new product is created—the drywalled partition.

The new model shows a system of linked activities. These activities are often tightly linked, which means that one must be completed before the next can start, or that resources from one process must be released before they can be used in the next process. The drywall cannot be installed before the electrical rough-in is completed. Examples of shared resources are space and time: electricians and drywallers cannot work in the same space at the same time.

This model also has another noteworthy characteristic. It contains a great deal of variability. The model is time-based, so durations for each activity must be developed. The estimate of activity durations will attempt to capture the most likely time in which the activity will be completed. However, in reality, an experienced supervisor knows that sometimes everything goes right and the activity can be done in a shorter time. There is also a likelihood that problems will occur and the duration will be exceeded. Occasionally exceptional performance is encountered in the

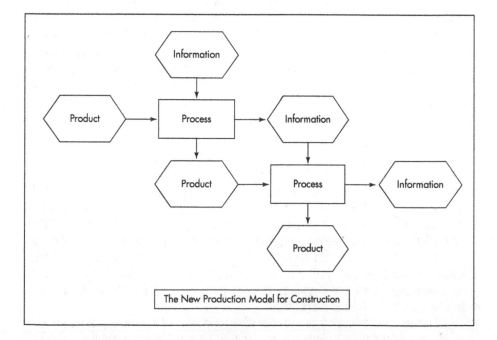

The New Production Model for Construction

Figure 15.9   The New Construction Model

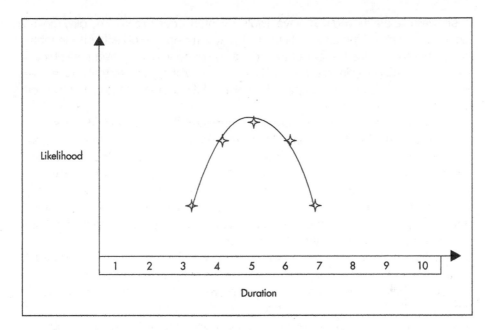

Figure 15.10 Variability of Activity Duration

execution of an activity and the actual duration is significantly less than estimated. On the other hand, sometimes multiple problems occur, and the duration is significantly extended beyond that which was estimated. Exacerbating this variability is the influence of the tight linkage of activities, so that not only is activity duration variable but also when activities occur is variable. This variability is illustrated in Figure 15.10.

A statistician would immediately recognize this type of variability as being statistical. Hence, our model of construction is characterized as a system of tightly linked activities subjected to statistical variation. Consider now what happens when this model becomes operational. The model will be represented by a game or simulation that involves moving objects through a series of stations and off the table of play. The movement is controlled by the roll of an ordinary die, of the kind which might be used in a number of dice games.

Production Model Simulation

The simulation is set up by placing a team of five players around a table. The team is given a die and a store of 100 objects. The objective of the game is to move all 100 objects around the table and off the board. As play begins, the first player rolls the die, takes that number of objects from the store and makes those objects available player two. These objects become the inventory of player two. Player two rolls the die and takes from his inventory the number of objects shown on the die, making them available to player three. If there are not enough objects in player two's inventory, all those available are moved and player two's inventory goes to zero. The play moves successively around the table until the last player moves the objects off the table. Play then moves to the first player to begin the next cycle of play. When all objects have pulled through a players position, he drops out of the game so that successive play skips him. The game is completed when the last player takes the last objects off the table.

In analyzing the play, it soon becomes apparent that only the first player can move all possible objects each time the die is rolled because he has the initial store of objects. Successive players are constrained by the number of objects in their inventory, and sometimes this is less than the die would allow them to move. To put this in terms of production, the player is resource constrained. The roll of the die represents production capacity and the number of objects moved represents production performance. Performance and capacity are always matched if there are no resource constraints, the situation represented by the first player. However, all other players, from time to time, will experience resource constraints, so they will waste capacity by not being able to move as many objects as the die allows.

Further analysis reveals that the second player is affected only by negative performance of the first player, but the third player is affected by negative performance of either the first or the second player. This cumulative impact is the result of the tight linkage between the players, and it increases as the number of players increases.

Two lessons about production in construction are learned from this model. First, the construction process is made up of a sequence of tightly linked activities so that each activity is directly affected by the previous activity and indirectly affected by a long chain of prior activities, perhaps extending all the way back to the beginning of the job. Second, when planned work is disrupted by an earlier activity, performance suffers because capacity, based on the plan, cannot be fully used and is, therefore, wasted.

Statistical variation is introduced into the model by the roll of the die. It represents the variability in estimated project activity durations, which is a key part of the planning process. If, in the simulation, we want to improve performance, one way to accomplish this is by eliminating the variability. To take this to the extreme, if the same number is placed on each of the six faces of the die, performance is balanced throughout the game. Each time the die is rolled, the exact same number of objects will be passed from player to player. Waste is eliminated. Performance is totally predictable. The game will be finished more quickly because there is no loss in performance at any point. Of course, putting the same number on each face of the die makes the game trivial. However, eliminating this variability on a construction project would be highly desirable.

Applying what we have learned to the construction process, we can see that eliminating the statistical variation in the model represents eliminating the variability in the planning process in construction. If planning is reliable, negative impacts on successive activities are eliminated. Waste is eliminated and completion is achieved sooner. The key is to *improve the reliability of planning* in the construction process.

This dice game simulation was initially set forth by Eli Goldratt in his classic book on production, *The Goal* (North River Press, 1985). The dice game simulation was extensively studied by Iris Tommelein and is summarized in her paper titled "Parade Game: Impact of Work Flow Variability on Succeeding Trade Performance" (Proceedings Sixth Annual Conference of the International Group for Lean Construction, IGLC-6, 13-15 August 1998, held in Guaruja, Brazil).

Construction Process Design

The planning process is where production is designed in construction. In manufacturing, the production process is dictated by the design of the manufacturing tools and the layout of those tools on the manufacturing floor. As a product is manufactured, it moves from tool to tool, each tool executing a process to move the product toward completion.

In construction, the successive manufacturing processes are moved through the product in a sequence of activities similar to those previously discussed in the example of building the partition. Construction process design focuses on planning of individual activities and then assembling these activities into an efficient sequence. The construction planner is a construction process designer. Just as in manufacturing, where the quality of the design of the tools and manufacturing floor have a strong influence on production efficiency, in construction, the reliability of the construction process planning plays a major role in determining whether the project will be completed on time and within the budget.

Planning in construction is now, and has been in the past, notoriously unreliable. The Lean Construction Institute has carried out many studies over the years on a variety of projects, which have indicated that planning reliability on successful jobs generally hovers a bit over 50 percent. That is, on an average successful job, construction plans are reliable about half the time. Yet, from the simulation, we concluded that planning reliability has a great impact on project success. Clearly, it is important then, for the supervisor to consider how to improve reliability in construction planning.

Improving Reliability in Construction Planning

The Last Planner Process©, developed by the Lean Construction Institute, has been used to improve reliability in construction planning. In construction, many plans are developed, expanded, and revised throughout the duration of the project, but the only plan that is implemented is the one the foreman makes when work is assigned to the crew. Thus, the foreman is the last planner in the succession of construction planners, and the foreman's plan is the one that has significant impact on the execution of the work. The plan that the foreman uses is typically formulated through the weekly planning process on most commercial construction jobs. At a weekly coordination meeting, supervisors from the various trades meet to discuss what will be accomplished during the next week, resulting in a short interval schedule for the week. The intent is that those responsible for the numerous tightly linked activities in process on the job site can coordinate with one another to sequence their activities and develop plans that will allow them to bring together the resources to move the work forward efficiently. Commitments are made at these meetings, and based upon commitments, plans are developed.

The typical driver for planning is the project network schedule that shows critical and subcritical activities that must be accomplished during the ensuing week so that the project will not be delayed. Trade supervisors are expected to commit to working on the critical activities, since, if these activities are delayed, the project will be delayed. On a schedule-driven project, other considerations required to complete the activity, such as adequate labor, resources, and equipment, become secondary. The critical path network scheduling method leads managers to conclude that work must be performed as a first priority, on those activities that are on the critical path.

Unfortunately, since the other considerations noted above do most certainly have an impact on completing activities, at the next coordination meeting, it is revealed that many of the planned activities are not completed. Based on the fact that planning is, on average, about 50% reliable, if each task is reviewed to determine whether it was completed or not, the result would be that about half the activities were not completed. The conclusion is that it is important to consider all of the factors that will affect activity completion, not just whether the activity is on the critical path, when developing the weekly production plan.

The Last Planner Process© suggests considering four criteria before making the commitment to place an activity on the short interval plan. The four criteria are:

- Definition
- Soundness
- Sequence
- Size

Definition says that the work is well enough described that it can be assigned and the completion of the activity will be well defined. For example, when reviewing planning success at the end of the period, if the sheet metal contractor had listed an activity to complete installation of ductwork in 10 specific rooms, but only 9 rooms were completed, or if 10 rooms were completed but they were not the 10 specified, that activity would have not been completed successfully.

Soundness means that all resources required to complete the activity will be available when execution of the activity takes place. This includes physical resources, such as materials and tools, as well as nonphysical resources, like information, and space.

Sequence focuses on the physical requirement to complete certain tasks prior to other tasks. In the event that several activities could take place at the same time, proper sequencing means that the choice among those activities available is the best one to move the project toward achieving its goals.

Size means that the correct amount of work is assigned so that the crew can complete it with a standard level of effort. Too much work assigned will stress the crew and begin to impact safety and quality, as well as raising the risk that all the work will not be completed. Too little work assigned will lead to inefficiency, delaying the project and negatively impacting the budget.

The Last Planner Process© begins by developing a list of activities that can be accomplished. They are well defined and sound. Activities that do not meet these two criteria are placed on a make-ready list so that focus can be placed on securing the required resources or redefining the activity until it is does meet these two criteria.

Next, sequence is considered when the activities are evaluated with regard to the project schedule and sequenced in order to ensure that critical activities will be completed first, then subcritical activities. Finally, based on labor resources available, work that is appropriately sized is assigned to the crew.

At the end of the planning period, each activity should be reviewed to determine whether the planning was successful or not. This is a clear-cut process. Either the activity was completed 100% or it was not completed; 90% completion is a planning failure.

One additional step is to follow up on unsuccessful activities in order to determine the cause of the failure. This adds a learning and accountability component to the process. It reveals where chronic failures lie, which allows steps to be taken to eliminate the problem. It also indicates who is responsible for the problems, so that those causing chronic problems can be held accountable for their disturbance to the project. Great diplomacy must be exercised in dealing with those responsible for the chronic problems, because they are often at higher levels in the project, and could even include the designer or owner of the project. The accountability aspect is not intended to assign blame, but rather is intended to identify and eliminate problems so that they do not impact later activities.

SUMMARY

In this chapter, the following key points have been presented.

- Production is the process of producing a product, whereas productivity is a measure of the efficiency of production.
- Rates of production can use either money or time as a basis of measurement, but a time basis is more stable than a money basis.
- A number of tools are available to analyze production, including:
  - The process chart
  - The crew balance chart
  - The flow chart
- To improve production, first eliminate worker slack time, then relax tight linkage between activities.
- A First Run Study is used to plan more efficient execution of an activity prior to starting that activity.

- An improved production model for construction is based on a sequence of tightly linked activities subject to statistical variation.
- Production can be improved by containing the variability in planning and wise use of buffers.

Learning Activities

1. Production study of an ongoing operation

   Identify an operation to study on an ongoing project. This could be an operation on a project on which you are working, or it could be an operation that can be studied by observing it from off the site. Some guidelines to use in selecting the operation to study include:

   - The operation should involve a small number of workers.
   - The operation should have a relatively short cycle time.
   - A well-running operation is easier to study and improve than a poorly running operation.

   Develop a report on your study, including such features as:

   - A brief description of the operation
   - Flow, process, and crew balance charts
   - Photographs of the operation
   - Suggestions for improvements

2. Implementation plan for the Last Planner Process©

   Consider an ongoing real or imaginary construction project. Develop a plan to implement the Last Planner Process© on the project. A key part of the plan will be to address how various participants will be brought into the process. In particular, what will persuade the various supervisors on the project to participate in the process? What will persuade the designer to participate? What will persuade the owner to participate?

   Develop a report describing the project and the plan.

SECTION IV

PROJECT SUPERVISION

CHAPTER **16**

PROJECT ORGANIZATION

INTRODUCTION

On the construction projects that supervisors manage they will encounter different types of contracts, and those contracts will have been formed through various types of project procurement methods. Because of these facts, and because supervisors are likely to hear the names of a variety of different contracting systems in use by those around them, a number of different terms for types of construction projects and project procurement methods, as well as various contracting methods and various different forms of contracts, will be discussed. Supervisors will do well to understand this terminology, so that they will be more conversant with the language employed in this aspect of the construction industry. In addition, supervisors should realize that they may be involved firsthand with any of the systems discussed here on the construction projects that they manage.

TYPES OF PROJECTS

Construction projects can be classified in a number of different ways. One method of classification is to define projects as being either public projects (or contracts), or private projects (or contracts). Public projects are those where the owner is some level of government, or an agency of government, whether national or state government, or school district, and so forth. Public projects may also be defined as those where the dollars that are utilized to pay for the project are tax dollars or public funds.

Public construction projects usually are closely regulated by legislation and/or by public policy. This is done in order to avoid favoritism or nepotism, and to provide everyone in the contracting community an equal opportunity to participate. For example, legislation typically requires that public construction projects be announced to all interested and qualified parties through an advertisement for bids or public notice to bidders. Similarly, statutes typically require that when bids or proposals have been received on bid day, they must be opened and read publicly. In addition, the contracts for such projects usually must be awarded by competitive bidding, and the lowest valid bidder usually must be awarded the construction contract.

Construction projects that are classified as private projects are those where the owner is an individual or a business entity, whether a sole proprietorship, a partnership, or a corporation. Usually on private construction projects no tax dollars or public funds are involved in paying for the project. Private construction projects may be announced to contractors, and contracts for these projects may be awarded by any method that the private owner may elect to utilize.

OTHER CLASSIFICATIONS—TYPES OF CONTRACTORS

Construction projects, and the construction contractors who perform these projects, may also be categorized in terms of the type of work that is being performed. While a variety of different categories may be used, and while the differentiation between these categories of types of contractors is not always precise, the following are names for types of construction contractors that are commonly used.

It should be noted that these are names and classifications that construction contractors have chosen to identify themselves. There is neither a taxonomy nor a set of rules that determines the classification for a construction contractor other than the names by which contractors choose to identify themselves.

Residential Contractor

Residential contractors are those who construct residences of different types. Single-family residences, duplexes, apartments, and condominiums may be included in this classification.

Within this category, some residential contractors are speculative builders, while others are custom builders. Some residential builders construct both speculative and custom building projects.

Speculative builders are those who construct a residential building in the role of owner-builder. They acquire real estate, and a design for the residence, and then they build the facility and offer it for sale. Their hope is to sell the property to a new owner during, or soon after the completion of, the construction.

Custom residential builders are those who construct a residence for an owner by the terms of a contract between the owner and the builder. The owner will typically own the real estate on which the residence will be built, and the owner will have obtained a design for the facility to be constructed, either from the builder, from an architect, or from another source.

Commercial Construction Contractor

Commercial contractors are those who build facilities, generally buildings, for businesses and also for other entities and owners. For example, contractors who refer to themselves as commercial contractors may build offices and other buildings for commerce, and may also build schools, government buildings, and churches. Some utilize the category "institutional construction" to refer to the construction of churches and school buildings. Commercial construction contractors also frequently refer to themselves as "general contractors."

Industrial Contractor

Industrial contractors are those who build facilities for industry. Often, these facilities are highly specialized, and frequently the construction is decidedly different from other types of construction. Examples of industrial construction include petrochemical refineries, paper mills, manufacturing plants, chemical process facilities, and so on.

Heavy/Civil/Highway Construction Contractor

Some construction contractors refer to themselves as heavy construction contractors or as civil construction contractors. These contractors build facilities such as power plants, fresh water treatment plants, and wastewater treatment plants, as well as dams, aqueducts, storm drains, and the like. These facilities are typically designed by an engineer, rather than by an architect. Consequently, projects of this type are frequently referred to as "engineered construction."

Highway contractors are builders of roads of all kinds. Usually these contractors also construct the bridges, culverts, storm drains, inlets, and so forth, which are accompaniments to highway construction.

However, as noted earlier in the discussion regarding contractors developing specialties, some contractors have developed a specialty of constructing bridges. These contractors do not perform the other elements of highway construction, only bridges.

Construction contractors make business decisions with regard to the type of construction projects they choose to perform. They acquire the skilled people, and the tools and equipment that are needed to become proficient and competitive and profitable in the segment of the business where they have chosen to operate.

PROJECT DELIVERY METHODS

Design-Bid-Build

The design-bid-build project delivery method has, for many years, been the most commonplace method of project delivery for construction projects of all kinds. The process derives its name from the sequence in which the design and construction functions are performed. This project delivery method is also referred to as linear construction.

In the design-bid-build project delivery method, the process begins with the owner's perceiving a need for the construction of a facility. Typically, the owner considers financing and budget for satisfying the need he has recognized. The owner subsequently enters a contract with an architect or engineer who will provide complete design services. The designer's basic responsibility, as typically defined by the owner, is the creation of an original design for a construction project that will satisfy the needs of the owner, within the owner's budget.

The owner's contract with the designer usually includes responsibility for production of a complete set of drawings and specifications, which will communicate the design to the owner as well as to the contractor and which will define the deliverables in the construction contract. Additionally, complete design services usually include the designer's authorship of all of the bid documents and contract documents for the project. The typical content of the set of bid documents and contract documents for a project is depicted in Figure 16.1.

In addition to providing the services noted above, the designer will typically provide contract administration services during construction and representation of the owner's interests through observation of the work during construction, as well as closeout of the project, and administration of any warranty issues after construction is complete. As noted above, the exact nature of the services to be provided is set forth in the contract between the owner and the architect.

When all of the bid documents and contract documents have been produced and have been approved by the owner, the designer assists in announcing the project to construction contractors by means of an advertisement for bids, an invitation to bid, or a notice to bidders. Through these documents, contractors are made aware of the existence of the project, and they are provided a brief description of the project and the form of contract to be employed. Additionally, the contractors are provided information with regard to how to obtain bid documents and contract documents, and they are informed with regard to the date, time, and place where contractors' proposals are to be submitted.

The designer administers the process of making bid documents and contract documents available to the contractors. In addition, the designer will answer contractors' questions, and will provide interpretations and clarifications regarding the information in the contract and bid documents as requested, during the time when contractors are preparing their estimates, a time known as the bid period. Additionally the architect will respond to written requests for information (RFIs) from the contractors during this period.

In addition, the designer will issue addenda as he or she deems necessary during the bidding period. An addendum (the plural form of the word is addenda) is any modification to any provision of the contract documents or the bid documents for the project, issued by the designer during the bidding period. Addenda are issued in consecutively numbered sequence, and the designer will ensure that he or she sends each new addendum to all of the contractors who have received bid documents and contract documents. Usually when addenda are issued during the

---

**BID DOCUMENTS AND CONTRACT DOCUMENTS
FOR A TYPICAL
BUILDING CONSTRUCTION PROJECT**

BID DOCUMENTS
1. ADVERTISEMENT FOR BIDS, NOTICE TO BIDDERS, INVITATION TO BID
   INSTRUCTIONS TO BIDDERS
   PROPOSAL FORM

CONTRACT DOCUMENTS

1. CONDITIONS OF CONTRACT
       GENERAL CONDITIONS
       MODIFICATIONS TO THE GENERAL CONDITIONS
       SUPPLEMENTARY GENERAL CONDITIONS
       SPECIAL CONDITIONS
2. DRAWINGS
3. SPECIFICATIONS
4. ADDENDA
5. ALTERNATES
6. AGREEMENT
7. MODIFICATIONS
       AMENDMENT
       CHANGE ORDER
       CONSTRUCTION CHANGE DIRECTIVE
       WRITTEN ORDER FOR A MINOR CHANGE

Figure 16.1 Bid Documents and Contract Documents for a Construction Project

bidding period, the architect will require written and signed acknowledgment from each contractor on their proposal submitted on bid day, that they have received and considered all of the addenda, identified by number, which were issued.

The designer will usually administer the process of receiving the contractors' proposals on bid day and will conduct the bid opening process. The architect or engineer will then assist the owner with selection of the contract recipient, and with execution of the agreement between the owner and the contractor.

After the contract has been signed between the owner and the contractor, the contractor will proceed to performance of the contract requirements. Throughout this time, the architect or engineer will provide contract administration services for the owner as defined in the owner-architect contract.

Design-Build

The design-build method of project delivery has been rapidly gaining in popularity as well as in commonness of use in recent years. Most professional practitioners believe that this project delivery method will soon become the most prevalent method for delivery of design and construction services. In fact, there are many who believe that the use of the design-build method has already surpassed the use of the design-bid-build method of project delivery.

The design-build method may also be referred to as the turnkey process or as the turnkey project delivery method. In the design-build system, the owner enters one contract with a single professional entity, and that firm has the responsibility for providing both design and construction services for the owner.

The design-build method of project delivery hearkens back to the time many years ago when there was a master builder who provided both design and construction services to the owner. Today, the design-build firm may be an architecture or engineering firm that is collaborating with, or that has entered a partnership arrangement or a joint venture arrangement with, a contracting firm to provide design-build services. Or, it may be a construction firm that has a collaboration agreement with a design firm, or it may be a construction company that has in-house design capability. Additionally, construction management firms (to be discussed in the next section) sometimes offer both design and construction capability.

The frequency of use of the design-build method is explained by the fact that there are numerous benefits to be derived by the owner through the use of this method of project delivery. Perhaps the most important of these benefits is the inherent communication and collaboration that are central to this process, between those who will produce the design for the project and those who will perform the construction, from the inception of the project to final completion. The contractor's understanding of all aspects of the construction process and knowledge of construction materials, methods, connections, and details, as well as the contractor's understanding of costs, estimating, and scheduling, are important assets that the contractor can bring to the project and that can be provided as input to the design process, from the time the project is first conceived.

Additionally, when owners utilize the design-build method of project delivery, they derive the benefit of one firm having singular responsibility to the owner, for all aspects of both the design and the construction. By contrast, in the design-bid-build system, designers usually produce the design, as well as authoring the bid documents and contract documents for the project, without any assistance or input from a contractor. Sometimes, after the construction contract is formed, there are miscommunications and misunderstandings, and at times there are disputes between the designer and the constructor regarding various elements of the project design, or with regard to a variety different contract issues. This may lead to delays, and additional costs, and perhaps claims. Additionally, these occurrences leave the owner, who simultaneously has a contract with the designer and the constructor, in the uncomfortable position of not knowing which of the professionals whose services it has engaged is correct. Owners who have endured such an experience are quick to say that the design-build method, with its single-point responsibility for all design and construction issues residing in one firm, can provide a significant benefit to the owner.

Many owners are electing to utilize the design-build method of project delivery today, because of the numerous benefits that this method provides for the owner. The use of design-build is now well established and is predicted to continue to grow.

Construction Manager

The construction manager is a professional who enters a contract with the owner to provide a variety of different services to the owner, as defined by the contract between the two parties. The concept of construction management became part of the design and construction process some years ago when design-bid-build was the predominant project delivery method, as owners sought the services of a third party to represent their interests in the owner's contract with the architect and in the contract with the construction contractor. Construction management contracts were utilized when the concept originated, and they continue to be used today, in both the single contract system and in the separate contracts system.

The single contract and separate contracts systems are discussed in a subsequent section of this chapter. Figure 16.2 and Figure 16.3 provide a schematic indication of where the construction manager functions in each of these systems.

The construction manager as agent is graphically depicted below in Figure 16.2 and is further discussed in the paragraphs that follow.

The construction manager at risk is graphically depicted below in Figure 16.3 and is further discussed in the paragraphs that follow.

As the use of construction management contracts has continued and evolved, a variety of different services have come to be included in these contracts. The exact nature of the services provided by construction managers varies considerably; these services are, however, fully defined in the owner-construction manager contract. The fact that the concept has endured and has evolved into several variations is indicative of the fact that owners have recognized, and are willing to pay for, a series of services beyond those defined in traditional historical design contracts and construction contracts.

As the concept has evolved, construction management has come to be defined in two basic variations: construction management agency, which is often referred to as CMA, and construction management at risk, frequently referred to by the acronym CMAR. In the construction manager as agent arrangement, the construction manager enters a contract with the owner and by the terms of that contract, he represents the owner's interests in the owner's contracts with the architect and with the prime contractor(s). The construction manager has no contractual relationship with the architect, nor with the general contractor(s); he provides his counsel and assistance to the owner, and the owner then decides whether to take action or not.

In the construction management at risk (CMAR) contract form, the construction manager enters a contract with the owner whereby he has the responsibility for completing the construction project on time, at or under the stipulated price or the

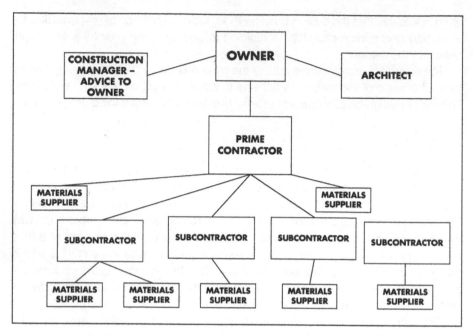

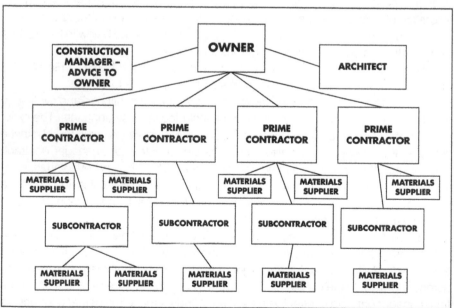

Figure 16.2   Construction Manager as Agent

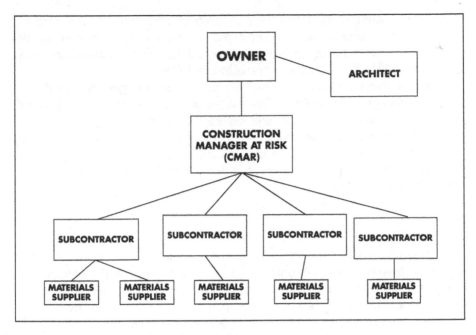

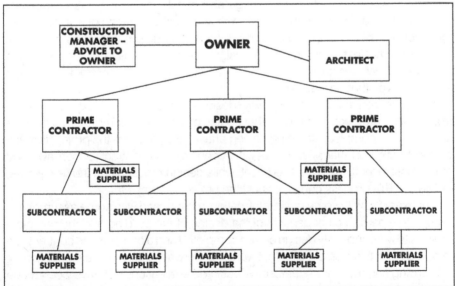

Figure 16.3 Construction Manager at Risk

guaranteed maximum price. Often, this form of construction management contract is written by the owner in such a way that the construction manager can provide consulting services to the owner during design, and the construction manager may actually assist the owner in the selection of the design firm.

In construction management at risk, the construction manager acts as the general contractor during construction. The construction manager at risk contract form may be utilized in both the single and separate contracts systems, as noted in Figure 16.3, and may also be used in the design-build method of project delivery. Thus, the construction management firm frequently has responsibility for delivery of the project to the owner, from the onset of design through final completion of the construction of the project.

Fast Track

Fast track is a method of project delivery that is sometimes employed when the objective is to reduce to a minimum the time required for design and construction of a project. This method is also referred to as phased construction. Fast track involves the assumption of considerable risk on the part of the owner, with the objective of reaping a return in dollars and/or time sufficient to justify the exposure or risk.

In fast track construction, the design and construction functions for parts or phases of the project are "leapfrogged" with one another. Instead of waiting to start construction until a completed design for the entire building has been produced, those using the fast track approach design one part or phase of the work, and as soon as that design is complete, award a construction contract and begin construction on that phase.

While that construction is underway, design proceeds on the next phase of the project, with the objective of having that portion of the design completed by the time construction is complete on the preceding phase, or by the time the project is ready for construction on that phase to begin. A construction contract is then awarded for the most recently designed segment, while design proceeds on the next phase, and so on. This process continues until the project is complete.

For example, with some basic information in hand, and with some assumptions made with regard to other aspects of the design of the building and the complete facility, site work and utility design can be completed and a construction contract can be awarded for this phase. While that work is underway, design of the foundation for the building commences. Again, some basic determinations will have been made regarding building's size and footprint, structural loads, and so forth, sufficient to allow a proper foundation design. As soon as that design is complete, and as soon as there will be no interference with site work and utility operations that may still be underway, a foundation construction contract is awarded. While foundation construction is underway, design work continues for the building's structural system. In this fashion, the design and construction functions are "leapfrogged" with one another throughout the project.

The advantage of the fast track method is that construction of the project can begin at the earliest possible time, and the construction of each phase of the building can begin without the need to wait for a complete design of the entire facility. This can significantly reduce the overall time required for design and construction.

The disadvantage, of course, is that when there is not a complete and integrated design for the building before construction commences, some retrofitting may be necessary, and some work may have to be removed and replaced or some parts of the project may be overdesigned because of the assumptions that needed to be made early, in order to allow construction to commence at the earliest time. The objective of fast track is to have the overall time and dollar savings that result from having the design and construction completed at the earliest possible time, be sufficient to compensate for the errors and the retrofitting and the "tear out and redo" that may result from not having a complete and integrated set of design documents in hand prior to the onset of construction, so that in the end there is a net gain for the owner.

Fast track is most commonly employed with the separate contracts system in use. Construction management services are also frequently used by the owner in the fast track method.

Value Engineering

Value engineering is not as much a project delivery system as it is an accompaniment to many of the project delivery systems in use today. Value engineering may or may not be utilized on a project at the discretion of the owner.

Conceptually, value engineering involves the owner and the architect seeking the input of the contractor with regard to his recommendations for alternative materials or systems that could be used on the project. The objective may be to reduce construction cost or to seek better value for the owner.

The circumstances in which value engineering is employed may vary considerably. Sometimes in competitive bid contracting, especially when the bids that are received from contractors exceed the owner's budget, the contractor who submitted the lowest proposal price will be asked to "value engineer" the project. This endeavor will be focused upon the contractor putting forth alternative materials or systems that he may recommend, which could be delivered at a lower cost than what was designed and specified in the contract documents prepared by the architect that the contractor used to prepare his original proposal. The architect and the owner will assess the contractor's recommendations and will decide whether to accept or to reject any or all. For those accepted, the contract documents and the contract price will be modified accordingly.

At other times in competitive bid contracting, the contractor may be asked to submit "value engineering" proposals as an accompaniment to the bid. These take the form of proposed alternates to the original contract documents and, again, usually include the contractor's recommendations as to alternative materials or systems being proposed in lieu of what the original contract documents set forth, along with the change in price that would result from acceptance by the owner and architect.

In like fashion, sometimes in negotiated contracting and in the competitive sealed proposals method of contract formation (to be discussed later in this chapter) the contractor will submit value engineering proposals for consideration. In each case, the contractor recommends alternative materials or systems that could reduce the contract price or that will provide better value for the owner. If the contractor's recommendations are accepted by the architect and the owner, they are written into the contract agreement and are reflected in the contract price.

TYPES OF CONSTRUCTION CONTRACTS

Construction projects may be built by means of a single contract system or a separate contracts system. Various specific forms of contract may be utilized within the single contract system and the separate contracts system.

Single Contract System

In the single contract system, there is one prime contractor who has a contract with the owner. This contractor is responsible to the owner for the construction of the entire project, and for fulfilling all of the requirements set forth in the contract documents. The single contract system, illustrated in Figure 16.4, is the most common type of contract system in use today.

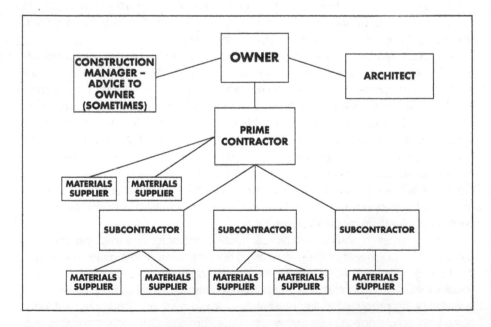

Figure 16.4 Single Contract System

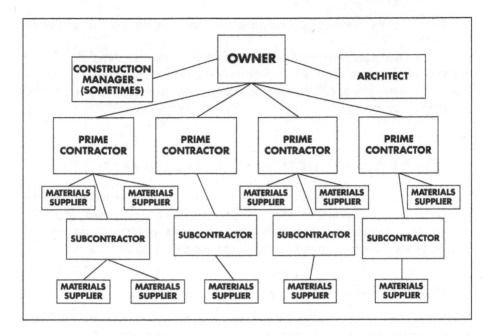

Figure 16.5   Separate Contracts System

Separate Contracts System

In the separate contracts system, depicted in Figure 16.5, there is more than one prime contractor who has a contract with the owner. Each separate prime contractor is responsible for the performance of a scope of work as defined in his or her contract, but that scope of work includes only a specific defined portion of the overall project.

When the separate contracts system is employed, the owner will subdivide the project into well-defined segments or phases, and will define a work package for each of the various parts. This will in turn define the scope of work for each of the separate prime contracts. A separate prime contract will then be awarded for each of the work packages. Sometimes the owner performs the function of subdividing the overall work to be done and defining the various work packages and the contents of the several separate prime contracts, if the owner has the experience and the expertise that are necessary to do so. At other times, the owner will utilize the services of a construction manager, in either a CM agency or a CM at-risk capacity, to perform this function.

When the separate contracts system is employed, it is imperative that there be some method in use for providing coordination among the various separate prime contractors. Again, the owner may sometimes perform this function, provided the owner has the experience and the expertise necessary to do so. Alternately, the owner may utilize the services of a construction manager to perform this function in its behalf. Additionally, the owner may structure the separate contracts in such a way that one of the separate contractors is designated as the "coordinating prime contractor."

All of the separate prime contractors will be responsible the completion of their scope of work in accord with the provisions of their contracts. All of the separate contractors may elect to self-perform the work or may award subcontracts for the performance of some of the work in their defined work packages.

FORMS OF CONSTRUCTION CONTRACT AWARD

There are three basic methods by which the contract for construction may be awarded to a construction contractor: competitive bidding, negotiation, and competitive sealed proposals. Within each of these contract award methods, different specific forms of contracts may be utilized, as indicated in Figure 16.6.

Competitive Bid Contracting

Competitive bid contracting has been used historically and continues to be very widely utilized today for contracts for engineered construction, as well as for building

COMPETITIVE BID	NEGOTIATED	COMPETITIVE SEALED PROPOSALS
LUMP SUM	LUMP SUM	LUMP SUM
UNIT PRICE	UNIT PRICE	UNIT PRICE
	COST PLUS A FIXED FEE	COST PLUS A FIXED FEE
	COST PLUS A PERCENTAGE OF COST	COST PLUS A PERCENTAGE OF COST
	COST PLUS A FIXED FEE OR PERCENTAGE OF COST WITH A GUARANTEED MAXIMUM	COST PLUS A FIXED FEE OR PERCENTAGE OF COST WITH A GUARANTEED MAXIMUM
	COST PLUS A FIXED FEE OR PERCENTAGE OF COST WITH A GUARANTEED MAXIMUM AND A SAVINGS OR INCENTIVE CLAUSE	COST PLUS A FIXED FEE OR PERCENTAGE OF COST WITH A GUARANTEED MAXIMUM AND A SAVINGS OR INCENTIVE CLAUSE

Figure 16.6 Project Delivery Methods and Forms of Contracts

construction. Competitive bid contracting has long been the most prevalent form of contract award.

In this method of contracting, the owner and the designer will prepare complete drawings, specifications, and contract documents, which will describe the project in detail and which will fully define the expectations of the owner in terms of what the contractor is to deliver in performing the project. Bidding documents will also be prepared, which will set forth the procedure by which contractors are made aware of the project and which will define all of the details and stipulations regarding the bidding and contract award procedures to be used.

Contractors, after they have learned of the existence of the contract documents and bid documents, will, if they are interested, obtain a complete set of these documents from the architect or engineer in accord with the procedures defined in the Advertisement for Bids or in the Invitation to Bid. They will examine these documents carefully and will then make a decision as to whether they are interested in proceeding, or not. If they are not interested, the contractors will return the bid and contract documents to the architect or engineer.

If a contractor is interested in proceeding with the effort to secure a contract award for the project, the contractor will embark upon the estimating process. As described in Chapter 12, the contractor will prepare a detailed estimate for the project and will then prepare a proposal for submittal to the architect or engineer on the designated day, which is called bid day, at the designated time and place.

The contractor will independently prepare a proposal and will submit it to the architect and owner on bid day, in competition with the other contractors who have made a decision to submit a proposal for the project. The architect or engineer will receive all of the proposals from all of the bidding contractors and will analyze each proposal as well as the credentials of the contractor who submitted it, and will, within the number of days set forth in the bid documents, decide who the contract recipient will be.

The contractor whose proposal has been accepted will be notified of this fact and will be notified to meet with the architect or the engineer and the owner in order to formalize the contract by signing the agreement. The contractors who were unsuccessful bidders will likewise be notified that their proposals were not accepted and that they will not be receiving the award of the contract.

After the successful contractor has signed the agreement, a letter of intent or notice to proceed will be issued by the designer and the owner to the contractor. This document will authorize the contractor to commence work on the project, and will denote the beginning of contract time. The contractor will then occupy the construction site, and will commence construction operations.

The competitive bid contract award procedure has, as noted above, been in use for many years. It offers the advantage for the owner that, if the contract documents fully describe the project and all of the contractor's obligations (and all that the owner will receive) under the contract, the owner will then be the beneficiary of competition among the contractors on the basis of which contractor can deliver the work as defined in the contract documents for the lowest price.

Lump Sum Competitive Bid Contracting

The lump sum competitive bid method of contract award has, for many years, been the most prevalent method of contract award for building construction work. This system is also referred to as "hard money contracting."

In this system, the contractor's estimating procedure results in a single lump sum figure, which represents the exact amount of money for which the contractor is willing to enter a contract to fulfill all of the requirements set forth in the contract documents. Each contractor will independently prepare an estimate for the project and will submit his lump sum figure on his proposal form on bid day, in competition with all of the other contractors who are bidding the project. The expectation is that the dollar amount that the contractor has entered on the proposal will become the contract amount if the contractor is selected by the owner to be the contract recipient.

Lump sum competitive bid contracting is possible in building construction projects because when the drawings and specifications and other elements of the contract documents completely and accurately describe the expectations of the designer and the owner, the contractors can accurately determine the quantities and prices for all of the materials and equipment to be provided and for all of the work to be done. Contractors are able to use their estimating process to arrive at a lump sum proposal amount for the project. This method of contract award has served owners well for many years and, thus, remains in common use today.

Unit Price Competitive Bid Contracting

Unit price competitive bid contracts are most commonly used on highway construction projects, as well as on earthwork cut-and-fill projects such as dams and levees. These types of projects are almost always designed by an engineer rather than by an architect and, as noted earlier, are often referred to as "engineered projects."

This contracting method is often used by owners because it offers the owner the same advantages as competitive bid contracting, as discussed previously. Additionally, because the types of projects where these contracts are employed are often public projects, laws typically require the use of competitive bid contracts.

Unit prices, dollars per unit of quantity, are used as the basis for contractor selection, and after the contract is formalized, these same unit prices are used as the basis for payments made to the contractor by the owner. The reason for the use of the unit price method is that on the types of projects where the use of these contracts is commonplace, actual quantities of materials and actual quantities of the work to be performed cannot be determined in advance with sufficient accuracy to permit lump sum estimating and pricing.

On projects of this kind, the design engineers will derive approximate quantities and will provide them to the bidding contractors for use in preparing their proposals. When the contract recipient has been selected and the work is performed, the actual quantities that are necessary to meet project design requirements are measured and verified, and then paid for by the owner on a unit price basis.

In the unit price system of contract award, the owner and design engineers will typically provide a listing of the major activities, steps, or operations to be performed in the work. With each activity, the engineer provides an estimated approximate quantity of that material or work activity. The owner will require the contractor to enter a unit price, that is, dollars per unit of quantity, for which the contractor proposes to complete each activity or work item named on the proposal form in such a way as to satisfy the requirements of the contract documents. An example of a typical unit price proposal form is provided in Figure 16.7.

In preparing the estimate, the contractor will analyze each activity that the owner has named and will consult the drawings, specifications, and other contract documents for all of the requirements that pertain, and then will compute a price for performing that activity or work item in such a way as to comply with the requirements of the contract documents. This price will be the sum of the contractor's anticipated costs for materials, labor, equipment, project overhead, and general overhead, as well as his markup. The contractor will then divide the total estimated cost for each activity by the owner's estimated quantity (or perhaps by his own determination of estimated quantity) in order to calculate the unit price (dollars per unit of quantity) that will be entered on the proposal form. The contractor will use the same procedure for determining the unit price for each of the activities and work items listed on the proposal form.

The contractor will complete the estimate, prepare the proposal, and will submit the proposal, usually in a sealed envelope, on the designated date and at the designated time and place, in accord with the instructions to bidders in the bid documents that are received from the engineer. The engineer will receive all proposals from all of the contractors who are bidding the project at that time.

After all proposals have been received, each will be analyzed by the engineer and the owner to determine which of the contractors has submitted a valid proposal that will result in the lowest total cost to the owner. That contractor will usually be named the contract recipient.

When the agreement is signed, the unit prices on the contractor's proposal will become the contract prices for the performance of the work in each named activity or work item. As the work on the project is performed, both the contractor and the engineer will tabulate quantities of each named activity or item of work that is actually performed in fulfilling the requirements of the contract documents.

Prior to each monthly application for payment by the contractor, the engineer and the contractor will review their records regarding actual quantities satisfactorily installed and will reconcile any differences. The engineer will then authorize the owner to make payment to the contractor for the actual quantity of each activity satisfactorily completed, multiplied by the unit price for that activity, less retainage.

This process is continued until the project is complete. It is important to note that the owner will provide, and the contractor will receive, payment for the actual quantity of each activity or work item installed or performed at the contracted-for unit price, without regard to whether this actual quantity is greater than, or is less than, the quantity originally estimated by the engineers during design.

SAMPLE UNIT PRICE PROPOSAL FORM

COURTESY BIG CREEK CONSTRUCTION

PROPOSAL FORM TO BE SUBMITTED ON BID DAY

COUNTY: BRAZOS
PROJECT: HWY FM 60
ENGINEER'S ESTIMATE: $ 1,274,096.98

CONTRACTOR'S ESTIMATING WORKSHEET AREA
NOT SHOWN ON PROPOSAL FORM TO BE SUBMITTED

ITEM NUMBER	BID ITEM DESCRIPTION	UNITS	QTYS.	UNIT BID	EXTENSIONS	Current Extension	Labor Cost Per Unit	Total Labor	Material Cost/Unit	Total Material	Misc Trk Per Unit	Total Misc Trk	Sub Cost Per Unit	Total Sub
														0.00
316	AGGR [TYPB GR.4 OR TYPL GR.4 SAC-E]	CY	894			0.00	0.00	0.00	0.00	0.00	0.00	0.00	0.00	0.00
316	ASPH[AC-15P OR20-5TR OR20XP OR1C-2TR]	GAL	43355			0.00	0.00	0.00	0.00	0.00	0.00	0.00	0.00	0.00
341	D-GR HMA[QCQA] TYC PG70-22	TON	282			0.00	0.00	0.00	0.00	0.00	0.00	0.00	0.00	0.00
354	PLANE ASPH CONC PAV[0" TO 1"]	SY	14177			0.00	0.00	0.00	0.00	0.00	0.00	0.00	0.00	0.00
354	PLANE ASPH CONC PAV[0" TO 2"]	SY	11378			0.00	0.00	0.00	0.00	0.00	0.00	0.00	0.00	0.00
354	PLANE ASPH CONC PAV [1"]	SY	3609			0.00	0.00	0.00	0.00	0.00	0.00	0.00	0.00	0.00
354	PLANE ASPH CONC PAV [2"]	SY	2560			0.00	0.00	0.00	0.00	0.00	0.00	0.00	0.00	0.00
500	MOBILIZATION	LS	1			0.00	0.00	0.00	0.00	0.00	0.00	0.00	0.00	0.00
502	BARRICADES, SIGNS AND TRAFFIC HANDLING	MO	3			0.00	0.00	0.00	0.00	0.00	0.00	0.00	0.00	0.00
662	WK ZN PAV MRK SHT TERM [TAB] TY W	EA	5076			0.00	0.00	0.00	0.00	0.00	0.00	0.00	0.00	0.00
662	WK ZN PAV MRK SHT TERM [TAB] TY Y2	LF	3978			0.00	0.00	0.00	0.00	0.00	0.00	0.00	0.00	0.00
666	REFL PAV MRK TY I [W] 4" [BRK][100MIL]	LF	7435			0.00	0.00	0.00	0.00	0.00	0.00	0.00	0.00	0.00
666	REFL PAV MRK TY I [W] 8" [DOT][100MIL]	LF	236			0.00	0.00	0.00	0.00	0.00	0.00	0.00	0.00	0.00
666	REFL PAV MRK TY I [W] 12"[SLD][100MIL]	LF	13937			0.00	0.00	0.00	0.00	0.00	0.00	0.00	0.00	0.00
666	REFL PAV MRK TY I [W] 24"[SLD][100MIL]	LF	2691			0.00	0.00	0.00	0.00	0.00	0.00	0.00	0.00	0.00
666	REFL PAV MRK TY I [W] 4" [BRK][100MIL]	LF	1264			0.00	0.00	0.00	0.00	0.00	0.00	0.00	0.00	0.00
666	REFL PAV MRK TY I [W] 4" [SLD][100MIL]	LF	1755			0.00	0.00	0.00	0.00	0.00	0.00	0.00	0.00	0.00
666	REFL PAV MRK TY I [W] 24"[SLD][100MIL]	LF	29280			0.00	0.00	0.00	0.00	0.00	0.00	0.00	0.00	0.00
666	REFL PAV MRK TY I [W] 4" [SLD][100MIL]	LF	676			0.00	0.00	0.00	0.00	0.00	0.00	0.00	0.00	0.00
668	PREFAB PAV MRK TY C [W] [ARROW]	EA	44			0.00	0.00	0.00	0.00	0.00	0.00	0.00	0.00	0.00
668	PREFAB PAV MRK TY C [W] [DBL ARROW]	EA	1			0.00	0.00	0.00	0.00	0.00	0.00	0.00	0.00	0.00
668	PREFAB PAV MRK TY C [W] [RR XING]	EA	2			0.00	0.00	0.00	0.00	0.00	0.00	0.00	0.00	0.00
668	PREFAB PAV MRK TY C [W] [WORD]	EA	27			0.00	0.00	0.00	0.00	0.00	0.00	0.00	0.00	0.00
668	PREFAB PAV MRK TY C [W] [BIKE ARROW]	EA	36			0.00	0.00	0.00	0.00	0.00	0.00	0.00	0.00	0.00
668	PREFAB PAV MRK TY C [W] [BIKE SYMBOL]	EA	606			0.00	0.00	0.00	0.00	0.00	0.00	0.00	0.00	0.00
672	REFL PAV MRKR TY I-C	EA	672			0.00	0.00	0.00	0.00	0.00	0.00	0.00	0.00	0.00
672	REFL PAV MRKR TY II-A-A	LF	1512			0.00	0.00	0.00	0.00	0.00	0.00	0.00	0.00	0.00
688	VEH LP DETECT [SAWCUT]	TON	459											
3131	CAM ASPHALT [PG 70-22]	TON	5678											
3131	CAM AGGREGATE	SY	31724											
3173	SALVAGED RAP [CREDIT ITEM][0"-3"]													

Figure 16.7 Unit Price Proposal Form

Negotiated Contracting

Construction contracts can also be, and frequently are, formed through negotiation rather than through competitive bidding. While negotiations can take many forms and can be performed in numerous different ways, the underlying concept is that the owner and the prime contractor will negotiate, or bargain for, the terms and provisions of their contract. Whatever the two parties can bargain for and decide, absent fraud or any other kind of criminal activity, becomes the contract.

Negotiated contracts are almost never used for public construction projects because the laws that regulate contract formation on these projects require the use of open competitive bidding. On private construction projects of all kinds, however, negotiated contracts are commonly employed.

Forms of Negotiated Contracts

Lump Sum

Lump sum contracts for construction projects can be, and sometimes are, negotiated. The contractor and the owner negotiate for the conditions of the contract, and oftentimes may include discussions regarding alternate materials for use on the project on the basis of value engineering. The negotiations culminate in the contractor providing a lump sum dollar amount for which the contractor is willing to complete all requirements of the contract as negotiated and agreed to.

It is important to note that if a lump sum negotiated contract is to be agreed to, the owner must be able to provide the exact scope of work, and the exact specifications for the materials and building systems to be employed, as well as definitive standards for the quality of work to be performed. Only when these conditions are in place can the contractor determine a lump sum amount for performing the requirements of the contract.

Unit Price

The terms of unit price contracts for construction projects can also be negotiated between the owner and the prime contractor. The parties can negotiate for the conditions of the contract, and they can also negotiate the activities or work items to be defined in the work, as well as the unit prices for the performance of each of these work items. This can be especially useful on engineered projects, where exact quantities of the materials required and/or of the work to be performed cannot be determined in advance of construction. Additionally, this method can be useful on restoration, renovation, and remodeling projects, where exact quantities cannot be determined in advance of performing the work.

Cost Plus or Cost Reimbursable

Cost plus contracts, also known as cost reimbursable contracts, are very frequently negotiated and are utilized for a variety of construction projects. While they may contain many different provisions, and may take a number of different forms, the basic provisions of cost plus contracts are that the owner will pay or will reimburse the contractor's costs of construction, and in addition will pay the contractor an agreed-upon fee for his services.

Cost plus contracts are commonly employed for any of the following reasons:

1. When the work to be done does not lend itself well to the preparation of complete drawings and specifications in advance of construction
2. When the exact scope of work is unknown at the time construction commences
3. When the nature of the work does not lend itself to exact quantity determinations and/or price determinations before construction is to get underway
4. When speed in commencing construction is an objective
5. When one of the objectives is to remove or minimize risk in the project for the contractor, thereby making the project more attractive and/or resulting in a better price

When cost-plus contracts are negotiated, there are several considerations that should always be included as part of the negotiation and contract formation. Among these are the following:

1. Mutual understanding and firm definition of costs that are reimbursable to the contractor and costs that are nonreimbursable.
2. Whose accounting person or department will be used for project accounting.
3. Generally an "open books" accounting method is employed, whereby all accounting documents for the project are available and transparent to both the owner and the contractor.
4. Regular audits of the contractor's payroll records and/or materials invoices on the part of the owner.
5. Clear understanding must be derived regarding contractor's general overhead and project overhead costs that are to be allocated to the particular project.
6. The subcontract award and subcontract payment procedures to be employed.
7. Responsibility for errors in the work, and for rework.
8. Provisions regarding the requisite quality in the workmanship and for rejection and replacement of nonconforming work.
9. When the contractor's fees are payable.
10. Termination of the contract.
11. Warranty provisions.

Basic Forms of Cost Plus Contracts
Cost Plus a Fixed Fee

The owner will reimburse the contractor's costs of constructing the project as the costs have been defined and agreed upon, as discussed previously.

Over and above those costs, the owner will pay the contractor an agreed-upon lump sum fee.

In order for this variation of cost plus to be workable for the contractor, a relatively firm definition of scope of work is required. A significant drawback of this form of contract, which is often troublesome for owners and for the financial institutions that provide construction financing and permanent financing for the project, is the fact that the contract form places no upper limit on what the total cost of construction will be. This form of contract is referred to as an open-ended contract.

Cost Plus a Percentage of Cost

The owner will reimburse the contractor's costs of construction and in addition will pay a negotiated agreed-upon percentage of all of the project costs to the contractor, as a fee for performing the work on the project.

This form of contract is very useful when the scope of work cannot be accurately defined in advance or when the intent is to get the work underway at the soonest possible time without awaiting complete definition and determination of scope of work. This contract form is also very widely used in restoration, renovation, remodeling, and adaptive reuse projects, where the variables that will be encountered in the performance of the work cannot be defined and predicted in advance. This is also a form of open-ended contract.

Cost Plus a Fixed Fee or Percentage of Cost with a Guaranteed Maximum

The owner will reimburse the contractor's costs of construction, and over and above those costs will pay either an agreed-upon lump sum fee or an agreed-upon percentage of those costs to the contractor as a fee for performing the work on the project. The contract also contains an additional provision for a guaranteed maximum on the part of the contractor. This means that the contractor guarantees to the owner that the total cost of the construction project will not exceed the guaranteed maximum amount.

If the contractor's costs for performing the agreed-upon scope of work to the agreed-upon level of quality exceed the guaranteed maximum amount, the contractor still must complete the project and must fulfill all of the requirements set forth in the contract documents. However, the contractor will pay for all additional costs beyond the guaranteed maximum amount. This places the contractor at risk for completing the project for a price at or less than the guaranteed amount. This contract form also avoids the difficulties associated with open-ended contracts. Complete definitions of the scope of work, and of the quality of materials and workmanship to be provided, are required in order to make this contract form workable.

In this form of contract the contractor is in a risk-taking position, and the owner has a guarantee of the upper limit of the cost of construction. Sometimes the point has been offered for applying, however, that there is little incentive for the contractor to complete the work at a price less than the guaranteed maximum, and that there is no reward for the contractor for applying skill and good management in such a way as to complete all project requirements for an amount less than the guaranteed maximum.

Cost Plus a Fixed Fee or Percentage of Cost with a Guaranteed Maximum and a Savings or Incentive Clause

The owner will reimburse the contractor's costs of construction, and over and above those costs will pay either an agreed-upon lump sum fee or an agreed-upon percentage of those costs to the contractor as his fee for performing the work on the project. Additionally, a guaranteed maximum amount is negotiated, so that the contractor must complete the work to satisfy contract requirements for an amount not greater than the guaranteed maximum amount. If the contractor cannot complete the project for the guaranteed maximum amount or less, the contractor still must complete the defined work and must fulfill all contract requirements; however, all additional costs to complete, beyond the guaranteed maximum amount, must be borne by the contractor.

In addition, in this form of contract, the contractor and the owner will negotiate a number, called a target figure, also referred to as an upset figure, which is equal to or less than the agreed-upon guaranteed maximum amount. The owner and the contractor agree (and write into the contract) that for each dollar less than the target figure for which the contractor can satisfactorily complete all of the requirements of the contract, there will be some split of those dollars between the owner and the contractor. The amount of the split payable to the owner and to the contractor is negotiated and agreed-upon, and is written into the contract.

There are advantages as well as disadvantages for the parties to the contract associated with each of these variations of cost plus contracts. A basic summary of these considerations is provided in Figure 16.8.

Competitive Sealed Proposals

Competitive sealed proposals is a relatively new method of contract award that is, however, widely used today and that is growing in popularity. When this method is used, the owner invites contractors to prepare proposals for a project for the owner. Each contractor will independently prepare a proposal in accord with instructions and guidelines that the owner has provided, and will submit that proposal to the owner at the time specified by the owner.

These proposals may include construction services, construction management services, or design-build services. They may include value engineering of a design that the owner has furnished. These competitive sealed proposals also will commonly include the contractor's proposed time schedule for the performance of the work. The price that the contractor submits may be a lump sum, a series of unit prices, or one of the variations of cost plus.

In addition to describing the professional services that the contractor will provide, the competitive sealed proposal will contain a great deal of additional

VARIATIONS IN FORMS OF COST PLUS CONTRACTS, AND ADVANTAGES AND DISADVANTAGES INHERENT IN EACH FORM		
	ADVANTAGES	DISADVANTAGES
COST PLUS A FIXED FEE	Some flexibility for the owner to make changes.	Open-ended contract for owner. Owner does not know what the final cost of the construction will be, nor what the maximum cost of the construction will be.
	Some flexibility for the contractor to accommodate changes by the owner.	Open-ended contract form is very troublesome for lenders who provide construction and permanent financing for construction projects.
		Open-ended contracts are also very troublesome for those who provide insurance and bonds for construction projects.
		Since his fee is predetermined and is written into the contract, the contractor will require relatively firm definition regarding the scope and duration of the work, as well as the quality of materials and workmanship to be provided.
COST PLUS A PERCENTAGE OF COST	Maximum flexibility for the owner to make changes in the work, including making changes in the scope of the work.	Open-ended contract for the owner. No definition of what the final cost of the project will be.
	Maximum flexibility for the contractor in accommodating changes made by the owner.	Open-ended contract form is very troublesome for lenders who provide construction and permanent financing for construction projects.
	Minimum risk for the contractor.	Open-ended contracts are also very troublesome for those who provide insurance and bonds for construction projects.
		No incentive, at least by the terms of the contract, for the contractor to be efficient, or effective, or cost-conscious.
COST PLUS A FIXED FEE OR PERCENTAGE, WITH A GUARANTEED MAXIMUM	Provides definition for the owner, and for those financing the project, for what the maximum cost of the project will be.	Requires the owner to provide firm definition of project scope and duration. Owner also must provide complete definition of materials quality and standards of workmanship in the form of specifications.
	Provides definition for the owner, and for those financing and insuring and bonding the project, for what the maximum cost of the project will be.	Places the contractor at risk for completing the project requirements for a sum not to exceed the guaranteed maximum amount.
		No reward for the contractor, if he exercises his skill and good judgment, and thereby completes the project for significantly less than the guaranteed maximum amount.
		Requires the owner to provide firm definition of project scope and duration, as well as for quality levels of materials and workmanship to be provided.
COST PLUS A FIXED FEE OR PERCENTAGE, WITH A GUARANTEED MAXIMUM AND A SAVINGS OR INCENTIVE CLAUSE	Provides definition for the owner, and for those financing and bonding the project, for what the maximum cost of the project will be.	Places the contractor at risk for completing the project requirements for a sum not to exceed the guaranteed maximum amount.
	Provides an incentive for the contractor to complete the project with maximum efficiency and cost consciousness, while satisfying scope and quality requirements as established by the owner.	Requires the owner to provide firm definition of project scope and duration, as well as for quality levels of materials and workmanship to be provided.
	Provides the owner an opportunity to share in the savings resulting from the contractor's efficiencies.	

Figure 16.8 Advantages and Disadvantages of Various Cost Plus Contracts

information about the contractor's background and experience, portfolio of projects completed, and the history of owners the contractor has contracted with and performed work for, along with those owners' contact information. The credentials of the contractor's estimating and scheduling and management personnel will be provided, often including the resumes of the project manager as well as the superintendent or foremen who will manage the owner's project.

Contractors will typically be required to submit information regarding their safety policies, as well as their accident history, and their Experience Modifier Rating. Additionally, the owner will usually request a copy of a contractor's quality assurance policy or total quality management (TQM) program.

In addition, owners commonly require the prime contractors to include in their submittal, the names and credentials of the major subcontractors whom they will plan to use on the project. Lists of projects completed, including projects of the size and type the owner envisions, are commonly required for the key subcontractors. Additionally, lists of references from owners, general contractors, construction managers, and others must commonly be furnished, as well as copies of the subcontractors' written quality management and quality assurance programs. In addition, usually the subcontractors' key management and craft labor personnel and their credentials must be provided, including the names and qualifications of the subcontractors' project manager and supervisor who will be assigned to this project. This allows the owner to evaluate all subcontractors before the project is awarded, including their financial capability to handle a project of the size and type the owner is planning to construct.

The owner will receive this comprehensive package of information in the form of competitive sealed proposals from each of the contractors that it has invited to submit on the project. The owner, frequently with the assistance of an architect, engineer, or construction manager, will review each competitive sealed proposal in detail, in order to determine who the contract recipient will be, who will be selected to enter a contract to perform the owner's project. Sometimes a point system is established by the owner, in order to provide the owner a means of quantitatively assessing various components of each contractor's submittal.

Sometimes the owner will make a preliminary assessment of all of the submittals, and then will invite some of the contractors who have received good evaluations on their submittals, and who are now finalists to receive the contract award, to prepare verbal presentations for the owner and its design team or management staff. At the conclusion of this process, the owner will make the selection of the contract recipient who will perform his project.

The intent of competitive sealed proposals is to provide the owner with a great deal of information with regard to each of the contractors who is submitting a proposal, to assist it in making the best choice regarding the contract recipient. Owners frequently use the expression "best value" in describing their process of evaluating competitive sealed proposals and determining who the contract recipient will be. Contractors are afforded the opportunity to provide not only their price but also information regarding their experience and their qualifications. Additionally, this method of contract award allows contractors to input their expertise regarding materials and systems for use in the project.

OTHER FORMS OF CONTRACTS

Time and Materials Contracts

Owners and contractors sometimes enter contractual arrangements wherein construction work is to be done on a "time and materials" basis. These arrangements are essentially the same as cost-plus contracts.

The owner agrees to reimburse the contractor for the cost of materials, and beyond those costs, agrees to pay the contractor an agreed-upon number of dollars per hour, per day, per week, per month, and so on for the work. The dollars per unit of time agreed upon will compensate the contractor for labor costs, as well as equipment costs and overhead costs, and will include markup.

Job Order Contracting

This contracting arrangement is often used for building maintenance or for facilities management contracts but is applicable to other forms of construction work as well. In this form of contract, the owner will compile a list or a schedule of operations or items of work that it has need for, or that it anticipates it may have need for, during a defined period of time, such as one year. The description of each item of work will include a complete definition of the scope of work, as well as specifications for the level of quality of materials and workmanship to be provided in performing this work package.

The owner will then enter into an agreement with a contractor, often after receiving proposals or price quotations from several different contractors, on a dollar figure for the performance of each work package, in compliance with the specifications. Response times and durations of work for each of the activities are also typically included in the agreement.

Subsequently, when the owner finds itself in need of one of the services described in the job order contract, it will contact the contractor, who will then respond and will perform the work as described and specified. Following completion of the work, the contractor will invoice the owner for the work at the agreed-upon rate, and the owner will make payment to the contractor.

PROJECT ORGANIZATION AND RELATIONSHIPS

In addition to understanding the project procurement and project delivery methods, and the forms of contract that were discussed in the preceding section, the supervisor should also be familiar with the organizational relationships that exist among the people who are involved on a construction project. Chapter 1 defined the typical roles of the owner, architect, engineer, consultant, construction manager, prime contractor, subcontractor, sub-subcontractor, and vendor. The contractual relationships among these people who are the typical participants on a construction project were also depicted visually. For reference, that representation is again depicted in Figure 16.9.

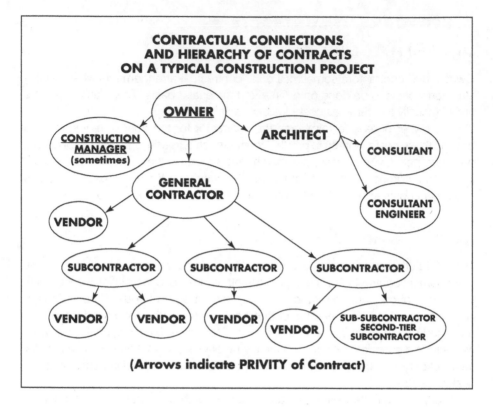

Figure 16.9   Contractual Connections and Hierarchy of Contracts on a Building Construction Project

Additionally, Chapter 2 discussed and depicted the typical functional organization of a construction company home office as well as the typical functions of the people who work there. For reference, those functions and relationships are again shown in Figure 16.10.

Prime Contractor's Project Organization

The prime contractor on a project, who is often referred to as the general contractor, will typically have a project superintendent as his chief management person on the construction site. The superintendent is the person responsible for the day-to-day on-site management of all aspects of the construction project. He has the responsibility of managing all of the other people and firms who are performing work on the site. The superintendent will typically interact with the project manager for his company as he performs his work, and often may interact with the contractor's home office as well.

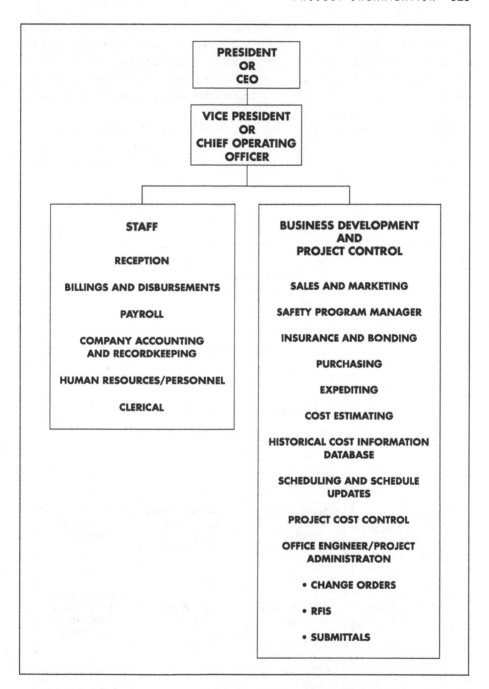

Figure 16.10   Typical Construction Company Home Office Organization

The superintendent for the general contractor will manage, schedule, and coordinate the work of all of the subcontractors who are performing work on the project. To accomplish this function, the superintendent usually works with the supervisor for each subcontractor's workforce at the site. At times, the superintendent may interact with the project manager for the subcontractors' companies in managing the work of the subcontractors.

Additionally, if the general contractor is self-performing some of the work on the project, the superintendent will manage and coordinate the supervisor or foreman of the general contractor's crew(s) of craft workers, as well. These relationships are illustrated in Figure 16.11.

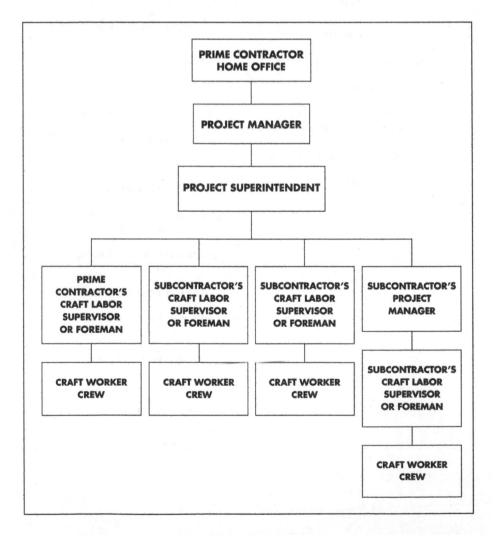

Figure 16.11   Prime Contractor's Project Organization

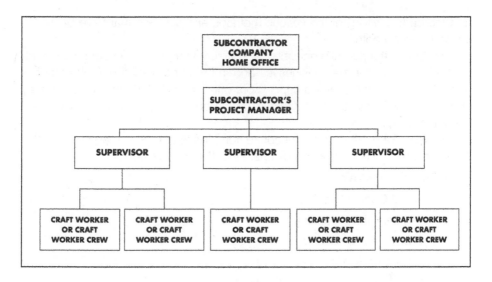

Figure 16.12 Subcontractor's Project Organization

Subcontractor's Project Organization

Each of the subcontractors on a construction project will typically employ a management structure similar to that depicted in Figure 16.12. A project manager will interface between the home office of the subcontracting company and the people who perform the work at the construction site. These people will be the supervisor or foreman and the craft workers who compose the crew(s) who perform the work. Depending upon company policy and the size of the individual construction projects, a subcontractor's project manager may manage and coordinate a single project, or more typically, may be coordinating a number of different projects.

For each of the construction projects that the subcontracting company performs, there will typically be a supervisor who is responsible for all aspects of the subcontractor's work on the construction site. This person may be referred to as a supervisor, a foreman, or perhaps a site superintendent or superintendent. Depending on company policies, this supervisor may be responsible for performing management duties exclusively, or he or she may perform some craft labor work along with performing management functions at the job site. Depending on the size of the project, the supervisor may manage one crew of craft workers, or he or she may manage several crews on the construction site.

SUMMARY

Numerous project procurement and delivery methods and numerous different forms of contracts are employed in the construction industry. The supervisor may be involved with any one of them, or any number of them, during his or her management career. It is, therefore, important that he or she be conversant with the terminology and the basic working of the various methods and systems that have been discussed in this chapter.

In addition, the typical organizational structure that prime contractors and subcontractors utilize for their businesses and in the performance of their contracts were discussed. This provides supervisors with a context for knowing where they and their work fit into the larger picture of the business enterprises that perform construction work.

The types of projects, types of contractors, forms of contract, and methods of project delivery discussed in this chapter include the following:

- Public and Private Projects
- Types of Contractors
  - Residential Contractors
    - Speculative Builders
    - Custom Builders
  - Commercial Construction Contractors
  - Industrial Contractors
  - Heavy/Civil/Highway Construction Contractors
- Project Delivery Methods
  - Design-Bid-Build
  - Design-Build
  - Construction Manager
    - Construction Manager as Agent
    - Construction Manager at Risk
  - Fast Track
  - Value Engineering
- Types of Construction Contracts
  - Single Contract System
  - Separate Contracts System
- Forms of Construction Contract Award
  - Competitive Bidding
    - Lump Sum Competitive Bid
    - Unit Price Competitive Bid
  - Negotiated Contracting
    - Lump Sum
    - Unit Price
    - Cost Plus
      - Cost Plus a Fixed Fee
      - Cost Plus a Percentage of Cost

- ○ Cost Plus a Fixed Fee or Percentage with a Guaranteed Maximum
- ○ Cost Plus a Fixed Fee or Percentage with a Guaranteed Maximum and a Savings Clause
- ■ Other Forms of Contracts
  - ■ Time and Materials Contracts
  - ■ Job Order Contracting

Learning Activities

1. Find out the specific form of contract that is in effect on the project where you are currently supervising and see how it fits into the context of various contract forms discussed in this chapter.

   If your firm is a subcontractor on this project, find out the form of contract used for the prime contract. If your firm is the prime contractor on your project, find out the contract form used for the subcontracts.

2. Understanding more about risk analysis.

   Look again at the various forms of contract methods that are employed in the construction industry. They are recapitulated in the end-of-chapter summary above.

   Think of each of them now in terms risk analysis, a concept that is discussed in several sections of this book.

   Think about where risk resides for the parties involved in the various forms of contracts that are used, and think about how that risk changes as the forms of contracts change.

CHAPTER 17

MOBILIZATION

INTRODUCTION

Mobilization—moving onto the construction site and setting up facilities and operations for the construction of a project—is a very important time for the supervisor. It involves a series of functions and activities with which the supervisor will, quite literally, set the stage for all of the work and all of the management activities on the project. Thus, the supervisor will do well to give his or her attention to this important matter and allow sufficient time for careful planning.

HANDOFF FROM PRECONSTRUCTION TO CONSTRUCTION—HANDOFF MEETINGS

Supervisors should plan to meet with the estimators who prepared the proposal for the project and/or with those who negotiated the terms of the construction contract, prior to starting the project. Further, they should factor what they learn from these people into their own thinking, planning, and decision making. Such meetings are referred to as "handoff meetings," where those who estimated the project cost and who assisted with contract formation hand off the project to those who will construct the project. These people will already have given a great deal of thought to the project, will have carefully examined the contract, and will have gathered large amounts of information regarding all aspects of the project. All of this information will be useful to the supervisor.

Earlier in the book, the work of the estimator was described as "building the project in his mind" and then ascribing costs to the materials and equipment to be used and to the activities to be performed. It is only logical for supervisors to avail themselves of the assumptions the estimator has made, as well as the thinking and planning that the estimator did when preparing the estimate for the project.

Supervisors may ultimately take the same approach to portions of the work that the estimator assumed or may elect to exercise a different plan when actually performing the work. At least, however, they have availed themselves of the thinking of this other person, so that they can then overlay this thinking with their own planning and decision making.

Additionally, it was noted earlier that the costs that the estimator has determined will form the basis for the project budget, and for all of the cost reports and the cost control measures throughout the life of the project. By meeting with the estimator before the project begins, the supervisor can gain an understanding of the manner in which these costs were determined during the estimating process.

In addition, this book has continuously underscored the importance of the supervisor's having access to, and reading and understanding, the contract for the project. Those who estimated the project and those who negotiated the contract or who sold the job will already have made a thorough analysis of all of the contract documents for the project. These people can share with the supervisor key sections or important provisions of the contract that will impact the performance of the work or that the supervisor should know about. In summary, all of the people who have been involved with the project prior to the formation of a contract can share a wealth of information with the supervisor, which will prove to be invaluable as the supervisor manages the work in constructing the project.

PRECONSTRUCTION CONFERENCE

Preconstruction conferences are held prior to the onset of construction on most projects. At these conferences, the general contractor and his management team meet with the owner and the architect or engineer, and with all of the subcontractors and their management teams. Sometimes key suppliers are included in the preconstruction conference as well.

At these meetings, the general contractor typically introduces all of the participants. He will typically provide insight and elaboration regarding the project objectives and will emphasize important aspects of the performance of the project such as teamwork, safety, and quality. He will define certain project procedures, and will typically express his wish to have the project be a successful, and pleasant, and profitable venture for all concerned.

The supervisor should plan to attend the preconstruction conference. Moreover, she should plan and prepare for the meeting. Experience has shown that most often these meetings are what we make of them. When proper planning precedes participation at the preconstruction conference, these meetings can be highly productive and beneficial.

Some principles for maximizing the value of preconstruction conferences are outlined below. The supervisor should use these thoughts as a point of beginning in planning his participation at the preconstruction conference.

1. Bring a schedule, or at least a preliminary schedule, to the preconstruction conference. The schedule has been defined previously as "a graphic representation of the contractor's plan for the work." Whether the supervisor is working for a general contractor or for a subcontractor, the plan for the performance of the work should be prepared as soon as possible following the contract award and should be shared at the preconstruction conference.

   At the preconstruction conference, the plan for the work (schedule) that the supervisor has brought can be shared and coordinated with the schedules of the other contractors on the project. If the other contractors have not yet prepared their schedules and brought them to the meeting (as will often be the case), then the schedule of those who have planned and prepared, and who have brought their schedules with them, are often used as the beginning point in defining the plan for the project and may in fact become the basis for the project schedule.

   It is sometimes argued that, at the time of the preconstruction conference, there are too many variables in the planning of the work of the general contractor and all of the subcontractors on the project to warrant preparing and bringing even a preliminary schedule to a preconstruction conference. Actually, it has been shown that the opposite is true.

   If the supervisor is working for the general contractor, the sooner the general contractor and his or her management team prepare their schedule and begin to coordinate the plan for their work and that of all of the subcontractors, the better. If the supervisor is working for a subcontractor, the sooner the subcontracting firm's plans for the work (their own schedule for their portion of the project) can be shared with and integrated into the plan of the general contractor and the other subcontractors, the better for everyone concerned.

2. Bring scope of work issues to the preconstruction conference for discussion and resolution. Given the complexity of construction projects today, and given the fact that a large number of subcontractors are typically involved in the performance of the work on the project, issues and conflicts regarding scope of work and coordination of the work are commonplace occurrences. There is no better time or place to bring these issues to the fore, and then to have them discussed and resolved, than at the preconstruction conference.

   As solutions and agreements are worked out, they should be written down and should be included in the supervisor's project documentation files.

3. The supervisor should never attend a preconstruction conference without saying something. If the supervisor has done the planning and preparation as recommended in the steps above, it is almost inevitable that he will have something to say at the preconstruction conference. But if not, the supervisor should watch for the opportunity to offer a comment or suggestion, or to

provide input regarding a matter that is under discussion. He should let his voice be heard.

Not only does this provide opportunity for the supervisor to provide input, but it also gains the attention of the other participants at the meeting and establishes the fact that the supervisor and his or her company plan to be active participants in the management and coordination of the project.

4. The supervisor should never leave the preconstruction conference without asking something. As noted above, if the supervisor has planned and prepared for the preconstruction conference, it is almost certain that she will have a list of questions on her agenda for the meeting. Additionally, it is certainly a commonplace occurrence that discussion at the meeting will lead to questions on the supervisor's part.

The supervisor should ensure that she does not leave the meeting with any question unanswered. As questions are answered, the supervisor should make note, should write down the answers provided, and should include this documentation in her project files.

As noted previously, the supervisor's asking questions at the preconstruction conference will not only provide her the information or resolution she needs, but will also establish among the other project participants that the supervisor and her company plan to play an active role in the planning and coordination of the project.

SETTING UP THE SITE

Planning is the key to success for supervisors in setting up the site. Knowing that the decisions they make and the things they do will provide the framework for all of their operations on the site for the duration of the project, should spur supervisors to give care and attention, and proper planning, to the matter of site setup.

As was noted in Chapter 11, there are numerous issues and considerations inherent in proper site planning and layout. Included is the allocation of space for the field office and for the lay-down area, for materials storage outside, as well as in a secure and protected environment, for tools and equipment, for gang boxes, for employee parking, and so forth. Planning is the key to success for all of these considerations.

It is recommended that a site plan be drawn on grid paper and that space be planned for all of the needs that the supervisor can envision during the planning for site setup. Additionally, consideration must be given to matters such as job gates for vehicles and personnel, temporary roadways, parking areas, unloading areas and turnaround areas, locks and other security features, temporary utilities, security lighting, sanitary facilities, employee break areas, employee lunch areas, and employee dressing rooms.

It is recommended that supervisors prepare a Site Planning Checklist. Such a checklist will assist supervisors in organizing their thoughts and in preparing their plan for the site setup for the project. Additionally, the checklist can be used as the basis for planning site layout and setup on subsequent projects. In this way, over a period of time, supervisors can produce a Master Site Planning Checklist. Such a checklist will not only assist in ensuring that nothing is overlooked but will also provide the added benefit of freeing supervisors from having to remember all of the elements and aspects of site planning. This will, in turn, allow supervisors to devote more of their time to the planning, analysis, and decision making regarding the project and their management of the work on it, which will be the hallmarks of their success.

SETTING UP THE FIELD OFFICE

The field office, often referred to as the "job shack," should be envisioned by supervisors as their management headquarters for the duration of the project. Accordingly, the setup and organization of the field office merit planning and careful attention.

Careful consideration should be given to communications in the field office. Any or all of the following may be employed: wired telephones, wireless phones, facsimile machines, intercom systems, radios, and personal digital assistants (PDAs).

Additionally, computer facilities for the field office need to be planned for, and obtained, and installed. Desktop computers, laptop computers, notebook computers, and tablet computers, as well as peripheral equipment such as printers, scanners, plotters, and digital routers may play an important role in the supervisor's management plans. Other electronic devices, such as digital still cameras, digital video cameras, liquid crystal diode (LCD) projectors, and so forth, may be included in the supervisor's planning.

Along with planning for computer hardware, the supervisor should give consideration to the software that will be needed on the computing equipment at the job site. Training may be necessary for proper use of the software.

Provision for the storage, handling, and layout of drawings, specifications, and other contract documents will need to be considered. A list of contacts, organized by company, by service or product, and by person, will need to be prepared. A list of telephone numbers for emergency services (police, fire, ambulance, and medical) likewise must be prepared and posted. A bulletin board, announcement center, and message center will need to be provided.

One of the most important aspects of field office setup is the filing system that will be employed for organizing and storing project documentation and information of all kinds. The supervisor's company may have a standardized set of files and file names that the company uses for job site files. If so, the supervisor should, of course, use the company system. Additionally, the supervisor should add to it and customize it, so as to make it efficient and workable so that it meets his or her needs.

If the company does not have a set of standard files, supervisors should build a filing system of their own to meet their needs. The project manager may be able to provide insight and assistance in this regard.

The examples in Figure 17.1 are provided to assist supervisors in determining what types of files they might find useful. While not intended to be comprehensive or

PROJECT CONTRACT DOCUMENTS	SAFETY INFORMATION
REQUESTS FOR INFORMATION (RFI'S) – FILED BY DATE	ATTENDANCE ROSTERS – SAFETY MEETINGS
CORRESPONDENCE – INCOMING	MINUTES OF SAFETY MEETINGS
CORRESPONDENCE – SENT/OUTGOING	MSDS SHEETS
PURCHASE ORDERS	MATERIALS FILES – BY MATERIAL
PROJECT SCHEDULE AND SCHEDULE UPDATES	MATERIALS FILES – BY SUPPLIER
COST REPORTS	PROJECT COORDINATION MEETING MINUTES
SUBMITTALS LISTING	SCHEDULE OF VALUES
SUBMITTAL TRACKING CHART	PAYMENT REQUESTS
PROJECT BUDGET	DELAYS DOCUMENTATION
CHANGE ORDERS – BY DATE	CLAIMS AND POTENTIAL CLAIMS
CHANGE ORDERS – BY NUMBER	

Figure 17.1   Typical Project Files and File Names

complete, this list will serve to provide illustrations of, and perhaps a starting point for, files and file folders that will prove valuable to the supervisor for organizing and storing the papers and elements of documentation which will be generated during the course of the performance of the project.

In a related note, it is recommended that supervisors also develop a filing system for use in their vehicle. File boxes or hopper boxes are available at office supply stores for this purpose. Such boxes accommodate a number of standard manila file folders and their contents. The supervisor's use of such a hopper box, along with the use of manila file folders labeled as appropriate and kept inside the hopper box, is highly recommended. Their use greatly facilitates organization on the part of the supervisor. Additionally it avoids the all-too-common occurrence of important papers or notes being lost or misplaced.

ESTABLISHING FIELD PROCEDURES

The supervisor will need to plan for and to operationalize the procedures to be utilized in managing and controlling the project. Some of these procedures will be mandated by company policy or custom, and others may be established by the project manager. It is emphasized that beyond what the company policies require, the set of procedures for managing the project should be planned, determined, and implemented by the supervisor.

Some of the typical project procedures that should be included in the supervisor's planning and thinking are listed in Figure 17.2. These are examples of matters for which the supervisor should plan operational procedures.

The supervisor undoubtedly will encounter others that need to be handled in the course of the management of the project. When that happens, the supervisor should enter them on the Planned Procedures List so that, in time, this list can become a master checklist. It should also be pointed out that, as the supervisor makes plans for the implementation of these procedures on the project, he or she could also be thinking in terms of delegating some of these tasks.

PROJECT OBJECTIVES

All of the field procedures, and all of the management actions taken by the supervisor, should remain directed toward accomplishment of the project objectives. There are five objectives that are common to every project that the supervisor manages. These project objectives are:

1. Completing the project on time
2. Completing the project at or under budget
3. Completing the project safely
4. Providing quality in all aspects of the construction of the project
5. Making the customer a satisfied client

Attendance and Representation at Project Coordination Meetings	Quality Control -- All Aspects
Verbal and Written Communications with Owner, Architect, General Contractor, Subcontractors	Coding Labor on Time Cards
Analyzing Periodic Cost Reports and Determining Appropriate Actions to be Taken	Verifying and Submitting Time Cards
Analyzing Schedule Updates, and Determining Plans of Action	Payroll Checks
Preparing Short Interval Schedules	Purchase Orders
Preparing, Following Up On, and Filing Incident Reports	Verifying Supplier Billings
Handling Submittals and Maintaining a Submittal Tracking Log	Safety Meetings
Ordering Materials	Safety Training
Deliveries of Materials and Equipment, Unloading, Checking for Correctness of the Order, Checking for Damage, Placing Delivered Items into Temporary Storage on the Jobsite	Accident Reports
Invoices	Documenting Near Misses
Preparing and Documenting Requests for Payment	Employee Hiring and Dismissal
Handling Change Orders	New Employee Orientation
Tool Inventories and Tool Security	Employee Training and Development
Tool Maintenance, Repair, and Replacement	Handling Employee Complaints
Obtaining, Storing, and Maintaining Personal Protective Equipment	
Planning and Communicating Employee Work Assignments	Equipment Repair and Maintenance
Monitoring the Work in Progress	Maintaining the Jobsite Filing System
Jobsite Communications Systems – Wired Telephones, Wireless Telephones, Intercoms, Radios, etc.	Jobsite Computer Equipment – Desktops, Laptops, Tablet Computers, PDA's

Figure 17.2   Examples of Typical Project Procedures to Be Planned, Determined, and Implemented by the Supervisor

Supervisors should recognize, and should remain mindful at all times, that the success of the projects they manage will be measured by whether or not these objectives are fulfilled. Additionally, supervisors should know that their personal and professional success will be determined by how effective they are in consistently accomplishing these five unchanging objectives.

SUMMARY

Some of the key points of learning that have been covered in this chapter are:

- Handoff meetings, and their value to the supervisor;
- Preconstruction conferences, and their functions, and their relevance;
- Elements of guidance for the supervisor, to help her make effective use of preconstruction conferences;
- Guidelines for setting up the construction site;
- Guidelines for setting up the construction field office;
- The importance of documentation, and of making and maintaining filing systems to facilitate effective documentation;
- Importance of properly managing field procedures, guidance to make a checklist;
- List of project objectives which pertain to every project, and emphasis on the fact that the supervisor's success will be directly related to his or her effectiveness in consistently accomplishing the project objectives.

Learning Activities

1. Start your own "mobilization checklist" as described in this chapter.

   Think about when you started the project you are working on at the present time, and the things you did in order to mobilize, and to prepare for beginning the actual construction work.

   List the steps you took and the activities you conducted.

   Write these down in list form, and keep them in a file.

   Save the file as "Mobilization Checklist for Job XXX (job name and/or job number)."

   The next time, you mobilize to start a project, consult your list, and see for yourself how effective it is, to have something to go by instead of having to remember everything.

   (continued)

If there are things to be done or elements of mobilization for this project that are not on your list, then add them to your list, and again file the list for your use on the next project. Save this file as "Mobilization Checklist for Job YYY (job name and/or job number)."

Combine the two files (and others for subsequent jobs) into your "Mobilization Checklist Master File."

2. Look again at Figure 17.2, "Examples of Typical Procedures to Be Planned, Determined, and Implemented by the Supervisor."

Make a copy of this list, or write down the procedures which appear in the list.

Think of the project you are supervising at the present time. Add to the list in the example, other procedures you are handling in managing this job. Begin to make a master checklist of procedures to be handled.

If you have yet to supervise your first project, do some brainstorming, and/or talk to a supervisor or project manager you know, or to one of your mentors, and begin to think of additional procedures you will need to handle in managing a project.

Begin to make the master checklist referred to above.

CHAPTER **18**

ONGOING OPERATIONS

INTRODUCTION

During the performance of a construction project, four of the supervisor's most important functions will be documenting, reporting, and assuring safety in all operations, as well as assuring quality in the work. How well supervisors manage in performing these four functions determines how successful they will be in achieving four of the five objectives which are consistently the same for all construction projects. As noted in previous discussions, these first four project objectives are: completing the project on time, completing the project at or under the project budget, completing the project safely, and providing quality in all aspects of the project. The supervisor's success in managing so as to fulfill these four project objectives will lead directly to achieving the fifth project objective: making the customer a satisfied client. The supervisor's accomplishment of these five objectives will result in a successful project, and will in turn, become the hallmarks of the effectiveness and success of the supervisor.

Documenting consists in creating a timely and accurate record of all of the important events that occur during the performance of the project, and recording and filing this critically important information in such a manner that it can be retrieved if and when it is needed. Documenting is one of the most important functions that a supervisor performs. Having accurate and reliable information is the key to avoiding or resolving questions and disputes of all kinds. The importance of proper documentation, which produces and archives this important, accurate, and reliable information, simply cannot be overstressed.

Reporting involves compiling information concerning the project's status and sending that information to the project manager and/or to the supervisor's company office for analysis and review. Reporting, management analysis, management decision making, and the management actions that follow are among the most important controls to ascertain that the construction project is within the project budget, is on schedule for timely completion, is being built safely, and is being constructed with high quality.

DOCUMENTING

The Job Log

The Job Log, which may also be referred to as the Job Diary or the Daily Log, is one of the single most important documents that the supervisor prepares during the course of every construction project. The Job Log is considered by all levels of company management, and by the legal system as well, to be the primary, complete, comprehensive, and accurate record of everything of significance that occurred every day throughout the duration of a construction project. It is *absolutely imperative* that this Job Log be completed with regularity every single day by the supervisor and that it contain a complete and accurate account of all of the important events that occurred on the project on every day.

If, at any point in the future, there is any sort of question or controversy regarding any occurrence on the construction project, the Job Log is the first and primary source of information that will be consulted. In any kind of claim, mediation, arbitration, or civil or criminal legal proceeding, the Job Log will be among the first elements of information or evidence to be examined.

Additionally, the Job Log is an invaluable asset to the supervisor and to construction company management, in assessing the effectiveness of the management of the project and in benchmarking the success of management actions taken during the performance of the work on the project. As a component of the project review, discussed in another section of this book, the Job Log contains an enormous amount of useful information.

Because of its value and because of its importance, the Job Log must be as complete and as comprehensive as possible. In addition, in the event it may become a part of the information or evidence used in a legal proceeding, it must be as defensible as possible. To assist the supervisor in ensuring that his Job Log is as useful and complete as possible, some guidelines for the supervisor's Job Log preparation are listed in Figures 18.1 and 18.2.

Figure 18.1 provides some important guidelines for the supervisor with regard to daily Job Log completion.

A basic checklist of elements that should be included with each day's entries in the Job Log includes the items listed in Figure 18.2. This guidance pertains to the basic content of each day's Job Log entries.

The supervisor should understand that the list of Job Log guidelines provided above is intended to provide an indication of the kinds of events that should be included in the log. These guidelines, and any others the supervisor may encounter, cannot possibly include everything that may occur on the job site and that should be included in the log. Therefore, the supervisor will be best served by the following

1. The Job Log should be completed by the supervisor not later than the end of each day of work on the project.
    Ideally, as soon as possible after the occurrence of any event or activity of importance on the project it should be entered into the Job Log.
    At the end of the day, and without fail, and before leaving the job site, the supervisor should review all of the day's entries in the log, and should add any additional notations regarding anything of import that occurred regarding this project on this day.

2. Any matter of significance should be entered into the log.
    Supervisors and other managers are sometimes heard to inquire, "How do I know if an event is important enough to enter into the Job Log?" As a matter of general practice, good guidance to follow is, "if you have asked that question regarding an occurrence, it likely merits inclusion in the log."
    Or, stated another way, if an event seems as if it might have significance, it should be entered into the Job Log.
    It is generally far better to err on the side of inclusion than to omit an occurrence that is later determined to be of importance.

3. Ideally, the Job Log should be kept in a bound book having consecutively numbered pages. This makes it much easier to defend the Job Log as being a complete, and chronological, and comprehensive record of the project.
    Should it happen however, that the bound log is not available on a certain day (e.g., If the log was left in another vehicle, etc.) the supervisor should by all means chronicle that day's activities on the project, in writing. This can be done in a notebook, or on sheets of paper that are stapled together at the end of the day. These pages can then be stapled or taped into the Job Log as soon as the log is again available, accompanied by an explanatory entry stating the reason for the change in recording method.
    Some prefer to keep the Jog Log in computer files, and this practice is becoming increasingly commonplace. Frequently, companies that have adopted this policy, require the supervisor to e-mail the daily Job Log entry to the project manager and/or to the company home office on a regular basis.

4. Ideally, the Job Log should contain a minimum of strikethroughs or erasures. However, if the supervisor realizes after making an entry in the log that an error was made in recording an entry, or if an incomplete entry was made, by all means the Job Log should be corrected so that it reflects the most complete and accurate record of all that actually happened.
    Where an erasure, strikethrough, or correcting entry is made, the supervisor should add an explanatory note, stating that he is making a correction or addition specifically so as to provide an accurate and complete chronicle of what actually took place.

5. All entries should be legible, and all entries should be written with attention to proper word use, as well as to proper grammar, punctuation, and spelling.

6. The supervisor should give careful thought to each entry to be included in the Job Log, so that all entries are coherent and logical.

7. Care should be exercised by the supervisor to record the facts and to avoid expressing opinions and including emotional entries.

Figure 18.1 Guidelines for Job Log Completion

Day and date.	Nonconforming work encountered, and instructions given and actions taken.
Weather conditions at the start of the work and significant weather changes that occur during the day	Safety issues encountered and actions taken
List of workers present at the start of the day and notation of any changes during the day	Accidents or near misses with relevant details
Notation of any observed or stated concerns regarding the health or well-being of workers at the start of the work day	Decisions made
Subcontractors who worked on the project during the day and notations regarding their progress, as well as any difficulties they encountered, etc	Commitments made
Equipment in use on the project	Conferences or meetings conducted or attended
Visitors to the job	Inspections conducted and outcome
Work assignments made to individuals and to crews	Important telephone conversations, e-mails, mailings, or facsimiles received or sent
Narrative describing the day's work and activities on the jobsite, and a summation of progress made on those work items and activities	Training conducted
Verbal instructions given	Change orders issued or discussed
Warnings or reprimands issued or disciplinary actions taken	Any conversations with the owner or the architect or engineer

Figure 18.2 Checklist of Elements to be Included in Each Day's Job Log Entries

guidance: If an event is or seems to be important, it should be included in the Job Log.

It is also very important for the supervisor to understand the critical importance of, and thus the absolute necessity of, completing the Job Log daily and of ensuring that its entries reflect an accurate and complete history of all that occurred on the job site that day. Some supervisors, especially those who have come to supervision from having previously worked in the trades, express a dislike for "having to do paperwork" like the Job Log. The supervisor should understand clearly that documentation of all kinds, including proper preparation of the all-important Job Log, is a fundamental supervisory management responsibility.

Photographs for Documentation

Another very valuable tool that the supervisor can employ for documentation on a construction project is photographs. Photographs can record, and therefore can provide documentation for, all manner of events, situations, details, and occurrences on a construction project.

Especially given that high-resolution digital cameras are readily available and inexpensive, the supervisor should regularly make use of photographs for documentation. Digital photographs can be filed and stored on a variety of media, including computer hard drives, flash drives, and CD and DVD discs.

Additionally, digital photographs can be embedded in or attached to electronic communications and reports of all kinds. In addition, electronic images can be printed in hard copy if desired, and digital images can also be projected onto a screen by an LCD projector for use in group meetings and group settings. Digital photographs provide an easy-to-use, yet powerful and effective, tool that can further assist the supervisor in the all-important function of documenting all aspects of the projects that he or she manages.

REPORTING

The supervisor should expect to be involved in submitting and receiving a variety of reports relating to events at the job site and the status of the job. These reports provide critical information for the management of the project and for fulfilling the project's objectives. Some of the primary reports that are used on a consistent basis in the construction industry are summarized here.

Cost Reports

Cost reports are one of the primary project control mechanisms utilized by construction companies. Cost reports provide the means by which the company ensures that the project will be completed at or under the project. Cost reporting is also discussed in Chapter13.

While various components of construction cost on a project may be monitored and reported for analysis in cost reports, the element of cost that is almost invariably tracked, reported, and analyzed is labor cost. Labor is typically one of the largest components of cost on a construction project. Additionally, labor costs and labor productivity are subject to many influences and, therefore, are highly variable. Labor cost is, therefore, a significant risk factor for the contractor on nearly every project that he performs. For these reasons, management typically includes labor cost reporting as a vitally important management control on construction projects. Labor cost reports are typically prepared on a weekly basis.

The supervisor is a direct participant in this important process. Therefore, it is important that the supervisor understand the workings of the process and adhere to some fundamental guidelines to ensure the accuracy and, therefore, the usefulness of the labor cost reports. Chapter 13 discussed the workings of the labor cost reporting system that most contractors employ. For reference, the cycle of historical information, estimating, labor coding, cost reporting, and so forth is again depicted in Figure 18.3. In this chapter, we will emphasize a few additional aspects of the supervisor's role in this important reporting process.

As discussed in Chapter 13, supervisors will be responsible for preparing coded time cards for each craft worker in the company who performs work on the project

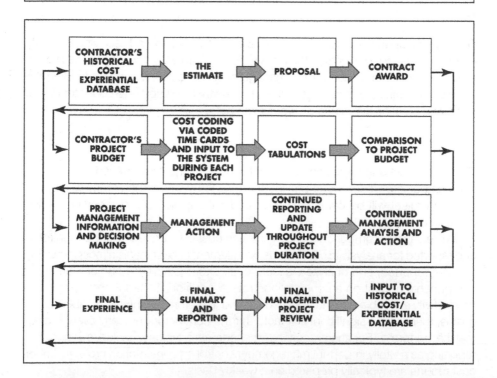

Figure 18.3 Project Cost Accounting, Cost Reporting, and Cost Control System

each day. Therefore, the supervisor must be closely familiar with the system of labor cost codes that has been developed by the company so that they can accurately apply the correct cost code number to each of the activities on which craft labor is performed on the project.

Moreover, the supervisor must prepare the coded time cards daily on a consistent basis, while the memory of which craft worker performed work on which activity or activities is fresh in his or her mind and can be accurately reported. The supervisor must not fall into the practice of planning to code labor time cards at some later time or "tomorrow, for sure" or on the day when time cards are to be submitted to the home office for payroll purposes. Accurate labor coding is absolutely crucial to the reliability and success of the estimating, cost reporting, cost accounting, and cost control system described in Chapter 13 and should, without fail, be performed on a daily basis.

Another important facet of the supervisor's reporting of labor costs is honesty in reporting. As labor hours and labor costs are accrued and reported, and as cost reports are produced, it may become very tempting for the supervisor to examine the cost report and to determine that, for the activities where the labor is over budget, he will not code any more labor that is performed on that activity to that cost code number. Rather than further increasing the overbudget amount for that activity by reporting labor that was performed that day on that activity to its actual assigned code number, it is a simple matter for the supervisor to simply code that labor to the cost code of another activity where work is ahead of budget, or to an activity that has zero labor hours reported to date. Some supervisors reason "What is the difference? It will all come out of the same project budget." This is dangerous and fallacious logic, which can have dire consequences.

If one refers again to the link between the labor coding performed by the supervisor during the performance of a project and the historical information database and the contractor's estimating process as discussed in Chapters 12 and 13, one can see the danger in this practice. If supervisors code labor incorrectly, no matter whether they do so intentionally or unintentionally, the result is that the cost reports do not reflect the actual status of the labor cost on the project. Additionally, management decisions that are based on these labor cost reports are then founded on incorrect information.

Moreover, this errant information regarding labor cost comes to be stored in the contractor's historical information database and, in turn, becomes the basis for future estimates that the contractor prepares. So the contractor could fail to win a contract award because of an inaccurate estimate, caused by inaccurate labor costs stored in the system. Or, the contractor could win a contract award, and then the supervisor on that project could find that it is virtually impossible to meet the project's labor budget, because of the fact that the labor costs in the budget are not reflective of the true cost of the labor for performing certain activities. These occurrences are the direct result of the fact that the supervisor on the present project did not accurately and honestly code and report the craft labor for the project activities.

It should be clear to the supervisor who understands costs and their derivation, and who understands the cycle depicted in Figure 18.3, that the system of historical information, estimating, project budgeting, cost reporting, and so forth is literally poisoned by the input of erroneous labor cost data to the system. The importance of the supervisor's accurately and honestly coding labor and reporting costs should now be apparent to the supervisor.

Schedules and Schedule Updates

As was noted in Chapter 14, the project's network schedule is the primary scheduling tool for managing time throughout the life of the construction project. The contractor, and the supervisor, simply *must* know at all times during the project, what the status is relative to completing the project on time. Additionally, they must know which activities are critical and should be given priority for completion. And finally, they must know which activities are upcoming, so as to be able to properly plan their work. The network schedule provides all of this important information.

As was noted in Chapter 14, in actuality multiple versions of the network schedule typically are, or certainly should be, produced during the course of construction of the project. An initial schedule is prepared, and the work begins. As activities are completed, and as circumstances change on the project, schedule updates are produced, in order to reflect the progress as well as changes in the work and to produce the most current plan for the completion of the work.

As was recommended previously, supervisors should always make sure to check the date of last revision on the network schedule that they are using in planning the work or any time they are communicating with others regarding the schedule, in order to ensure that they have, and are working from, the most current revision of the schedule. A great deal of confusion results, and effective outcomes are impossible to achieve, unless everyone in management is working from the same schedule update.

It should also be noted that changes in the network schedule will also necessitate the corresponding changes to the bar chart schedule and to the short-interval schedule, which are typically used by the supervisor in planning the work. Schedules of all kinds represent the plan for the performance of the work. These plans must be continually updated in order to reflect current events as well as changes that have taken place on the project.

Incident Reports

Many construction companies have a policy whereby the supervisor has the responsibility for preparing and submitting to the project manager and/or to the company's office for review, special reports that are referred to as "incident reports" when events occur or when special conditions warrant on a project. While incident reports may be required for a variety of occurrences, one type of event for which it is recommended that the supervisor always prepare an incident reports is "near misses."

Near misses can be defined broadly and in a variety of different ways. Generally the terms are used to describe an event wherein something unexpected has occurred on the job site. Usually the implication is that no reportable injury or serious damage occurred but easily could have.

Near misses almost always serve as a signal that something was amiss in the planning or in the execution of an activity on the project. Therefore, such occurrences also signal the need for a root cause analysis (as discussed in other sections of this book), in order to determine the cause or the basis for the event. Additionally, near misses usually indicate the need for corrective action, in the form of training or other management action, in order to ensure that such a mishap does not recur, with perhaps more serious consequences the next time.

Physical Progress

During the course of the construction of the project, another of the responsibilities supervisors have is monitoring the progress of the work. They will do so for two primary purposes: to keep the schedule updated, as discussed in the previous section, and to submit periodic payment requests. These payment requests are referred to as "Applications for Payment," and they are the basis for the contractor's receiving revenue during the course of the work on the project.

All of the elements regarding Application for Payment by the contractor and the subsequent payment by the owner are spelled out clearly in the "Payment" section of the General Conditions and Special Conditions of the contract documents. Because of the importance of this matter, and because the supervisor must have a clear understanding of the process by which the contract amount for a project becomes revenue for the contractor, the basic elements of the payment process will be set forth here. While the details of the process will vary somewhat, depending upon the structure provided by the architect and the owner in the contract documents, and while they may also vary with the project delivery method being employed, the lump sum competitive bid contract system that, as has been noted, is very commonly used for building construction projects, will be the basis for the process as described here.

Soon after the award of the contract, the general contractor will be required submit to the designer and the owner a document called the Schedule of Values. The Schedule of Values will form the basis for all payments made by the owner to the contractor during the course of the project. It structures the payment of the contract amount to the general contractor, and by the general contractor to the subcontractors. It forms the basis for the cash flow of the general contractor and subcontractors throughout the construction of the project. This document is, therefore, prepared with the greatest of care.

After the general contractor has prepared the Schedule of Values as he sees fit, this document must be submitted to the architect or engineer and the owner for review and approval. Once approved, it is not changed, and it will become the basis for all of the contractor's payments during the life of the project.

The general contractor will include in the Schedule of Values a listing of all of the activities or items of work that he will be performing during the construction of the project. Along with each listed activity, the contractor will provide a dollar amount. This amount represents the amount of money that the contractor expects to be paid upon the completion of that activity. The Conditions of the Contract may also provide that the contractor is authorized to include in his "Applications for

Payment," a request to be paid for materials delivered and properly stored on the site. If this provision is included in the "Payment" clause of the contract documents, the contractor will typically include a line item labeled "Materials Delivered and Properly Stored" in the Schedule of Values.

The total of all of the Schedule of Values amounts for all of the activities in the project will equal the contract amount. It should be noted that there will be no line item in the Schedule of Values for project overhead, for general overhead, or for markup. The dollar amounts for these elements are embedded and are distributed in the dollar amounts of the activities throughout the project.

For items of work that the general contractor will subcontract, he will require each subcontractor to prepare a Schedule of Values for their work activities and for their subcontract amount in an amount of detail suitable to the general contractor and also as may be required by the architect. The general contractor will then include each subcontractor's Schedule of Values activities and dollar amounts in his own Schedule of Values for submittal to the owner.

The architect and owner will review the general contractor's Schedule of Values submittal, will analyze it, and will determine whether to approve it or not. Sometimes approval is readily forthcoming; at other times, the architect or engineer will require revision and resubmittal, sometimes in several iterations, until approval is forthcoming.

If there is a requirement for resubmittal, it will usually be based on the amount of clarity or the amount of detail provided in the activity descriptions and subcontract items, and/or in the amounts of money associated with the named activities in the project. Just as the Schedule of Values structures the contractor's cash flow during the project, so too does it structure the owner's revenue disbursement and, thus, the owner's cash flow. This document is, therefore, of great importance to the owner, as well as to the general contractor, and all of the subcontractors.

When the Schedule of Values has been approved, it will stand unchanged throughout the project, except as it may be affected by change orders. It will form the basis for all payments made by the owner and for all revenue received by the contractor for the performance of the project. Figure 18.4 shows a typical Schedule of Values which accompanies a Periodic Payment Request.

The payment provisions of the Conditions of the Contract typically provide that payment will be made by the owner to the general contractor on a monthly basis for the duration of the project, and that payment will be based upon "earned value" in the contractor's work for the previous 30 days. A specific time of the month will be designated for the contractor's submittal of the Application for Payment to the designer. The Application for Payment will be a duplication of the Schedule of Values, with some additional columns added to the right of the "Amounts" column in the Schedule of Values. There will be a column labeled Percentage Complete. The next column will be called Gross Earned Value, and next to that will be a column labeled Less Amounts Previously Paid. The next column will be Amount Currently Due, or Earned Value This Period. Next will be a column labeled Less Retainage, and finally a column labeled Net Amount Currently Due. Figure 18.4 provides an illustration of the typical structure of a Periodic Payment Request, which is also known as an Application for Payment.

APPLICATION AND CERTIFICATION FOR PAYMENT

AIA DOCUMENT G702   PAGE ONE OF **Three** PAGES

To Owner:

Valley Realty, LLC
678 Bryan Avenue
Bryan, Texas 77803

Project: TAMU HSC
Clinical Building 1

Application No.: 3

PERIOD TO: July 31, 2010

Distribution to:

X	OWNER
X	ARCHITECT
X	CONTRACTOR

From Contractor:

USA Building Inc.
7776 York Steet, Suite 678
Houston, Texas 77080

Via Architect

ABC Architects, LLC

Dallas, Texas 75231

Project No: 349133
Project No: 09048.400
Contract No. 23-3065

CONTRACT FOR:

CONTRACTOR'S APPLICATION FOR PAYMENT

Application is made for payment, as shown below, in connection with the Contract.
Continuation Sheet, AIA Document G703, is attached.

1. ORIGINAL CONTRACT SUM	$	15,409,933.00
2. Net change by Change Orders		0.00
3. CONTRACT SUM TO DATE (Line 1 ± 2)	$	15,409,933.00
4. TOTAL COMPLETED & STORED TO DATE (Column G on G703)	$	3,026,364.04
5. RETAINAGE:		
Total in Column I of G703)	$	151,318.20
6. TOTAL EARNED LESS RETAINAGE (Line 4 Less Line 5 Total)	$	2,875,045.84
7. LESS PREVIOUS CERTIFICATES FOR PAYMENT (Line 6 from prior Certificate)	$	1,707,7360
8. CURRENT PAYMENT DUE	$	**1,167,685.58**
9. BALANCE TO FINISH, INCLUDING RETAINAGE (Line 3 less Line 6)		12,534,887.16

CHANGE ORDER SUMMARY	ADDITIONS	DEDUCTIONS
Total changes approved in previous months by Owner		
Total approved this Month	$0.00	$0.00
TOTALS	$0.00	$0.00
NET CHANGES by Change Order	$0.00	

AIA DOCUMENT G702: APPLICATION AND CERTIFICATION FOR PAYMENT · 1992 EDITION · AIA · ©1992
Users may obtain validation of this document by requesting a completed AIA Document D401 - Certification of Document's Authenticity from the Licensee.

The undersigned Contractor certifies that to the best of the Contractor's knowledge, information and belief the Work covered by thi s Application for Payment has been completed in accordance with the Contract Documents, that all amounts have been paid by the Contractor for Work for which previous Certificates for Payment were issued and payments received from the Owner, and that current payment shown herein is now due.

CONTRACTOR :

By: _____ Date: _____

State of: _____ County of: _____
Subscribed and sworn to before me this   day of
Notary Public:
My Commission expires:

ARCHITECT'S CERTIFICATE FOR PAYMENT

In accordance with the Contract Documents, based on on-site observations and the data comprising the application, the Architect certifies to the Owner that to the best of the Architect's knowledge, information and belief the Work has progressed as indicated, the quality of the Work is in accordance with the Contract Documents, and the Contractor is entitled to payment of the AMOUNT CERTIFIED.

AMOUNT CERTIFIED . . . . $ _____

(Attach explanation if amount certified differs from the amount applied. Initial all figures on this Application and on the Continuation Sheet that are changed to conform with the amount certified
AR CHITECT :

By: _____ Date: _____

This Certificate is not negotiable. The AMOUNT CERTIFIED is payable only to the Contractor named herein. Issuance, payment and acceptance of payment are without prejudice to any rights of the Owner or Contractor under this Contract.

THE AMERICAN INSTITUTE OF ARCHITECTS, 1735 NEW YORK AVE. N.W., WASHINGTON, DC 20006-5292

Figure 18.4 Application for Payment and Schedule of Values
Courtesy of Skanska USA Building Inc.

Schedule of Values - General Conditions

Project Name: **Big Manufacturing Facility**
Project Number: **349133**
Architect Project Number: **09048-400**

Project Address: 600 John Ralston Blvd.
Period From: July 1, 2010
Period To: July 31, 2010

Comp No.	Item No.	Description	GMP Budget (for ref. only)	Contract (Buyout) Value	From Previous Application	Total This Period	% This Period	Stored Materials (not in D,E)	Completed and Stored to Date (D+E+F)	% Comp (G/C)	Balance to Finish (C-G)	Retainage (5% of G)
		General Conditions										
	GC-1	On-Site Project Management Staff		$839,038	$60,300	$38,430	4.58%	$0	$98,730	11.77%	$839,038	$4,937
	GC-1A	Project Executives						$0	$0			
	GC-1B	Superintendent						$0	$0			
	GC-1C	Asst. Superintendent						$0	$0			
	GC-1D	MEP Coordinator						$0	$0			
	GC-1E	Senior Project Manager						$0	$0			
	GC-1F	Project Manager						$0	$0			
	GC-1G	Project Engineers						$0	$0			
	GC-1H	Project Support Staff						$0	$0			
	GC-1I	Safety Manager (20%)						$0	$0			
		Field Offices and Office Supplies										
	GC-2	Job Photos, Videos		$9,763	$0		0.00%	$0	$0	0.00%	$9,763	$0
	GC-3	Office Trailer		$24,356	$10,957	$664	2.73%	$0	$11,621	47.71%	$12,735	$581
	GC-4	Office Set up and Demobilization		$8,118	$6,690		0.00%	$0	$6,690	82.41%	$1,428	$335
	GC-5	Office Furniture		$6,711	$0		0.00%	$0	$0	0.00%	$6,711	$0
	GC-6	Office Supplies		$6,711	$6,449	$2,200	32.78%	$0	$8,650	128.89%	($1,939)	$432
	GC-7	Office Machines		$13,910	$0		0.07%	$0	$0	0.00%	$13,910	$0
	GC-8	Postage/Delivery		$7,750	$228	$130	1.68%	$0	$358	4.62%	$7,392	$18
	GC-9	Telephone System		$2,706	$0		0.00%	$0	$0	0.00%	$2,706	$0
	GC-10	As-Builts		$5,000	$5,513	$3,046	60.92%	$0	$8,560	171.19%	($3,560)	$428
		Dump Fees and Hauling		$17,536	$0		0.00%	$0	$0	0.00%	$17,536	$0
		Watchman/Security		$1,938	$0		0.00%	$0	$0	0.00%	$1,938	$0
		Telephone Service		$21,649	$0		0.00%	$0	$0	0.00%	$21,649	$0
		Temp Water (Trailers)		$8,389	$326	$1,030	12.28%	$0	$1,356	16.16%	$7,033	$68
		Temp Electrical (Trailers)		$11,745	$1,130		0.00%	$0	$1,130	9.62%	$10,615	$57
		Janitor Sanitary Supplies		$3,622	$0		0.00%	$0	$0	0.00%	$3,622	$0
	GC-11	Job Signs		$2,706	$2,476	$518	19.14%	$0	$2,994	110.66%	($288)	$150
	GC-12	Safety Supplies		$7,767	$2,940		0.00%	$0	$2,940	37.86%	$4,827	$147
	GC-13	Progress Photos		$1,678	$411	$206	12.25%	$0	$617	36.74%	$1,062	$31
	GC-14	Temp. Toilets		$10,067	$0	$1,031	10.24%	$0	$1,031	10.24%	$9,036	$52
	GC-15	Accounting Processing		$3,875	$0	$467	12.05%	$0	$467	12.05%	$3,408	$23
	GC-15	Misc. Travel Expenses		$15,000	$80	$339	2.26%	$0	$419	2.79%	$14,581	$21
	GC-15	Computer Services		$3,875	$1,682		0.00%	$0	$1,682	43.41%	$2,193	$84
	GC-15	Relocation Expenses		$15,000	$0		#DIV/0	$0	$0	0.00%	$15,000	$0
		General Conditions Sub-Total	$1,048,910	$1,048,910	$99,184	$48,061	4.58%	$0	$147,244	14.04%	$1,000,396	$7,362
		COST OF WORK										

Figure 18.4 (*Continued*)

Schedule of Values - General Conditions

Project Name: Big Manufacturing Facility
Project Number: 349133
Architect Project Number: 09048.400

Project Address: 600 John Ralston Blvd.
Period From: July 1, 2010
Period To: July 31, 2010

A Comp No.	Item No.	B Description	GMP Budget (for ref. only)	C Contract (Buyout) Value	D Work Completed From Previous Application	E Total This Period	% This Period	F Stored Materials (not in D,E)	G Completed and Stored to Date (D+E+F)	% Comp (G/C)	H Balance to Finish (C-G)	I Retainage (5% of G)
		Carlos Paving	$1,530,992	$1,530,992	$618,551	$49,972			$668,524	43.67%	$862,468	$33,426
		Landscape and Irrigation	$100,200	$100,200					$0	0.00%	$100,200	$0
		Becker Concrete	$2,741,757	$2,741,757	$720,431	$775,793			$1,496,224	54.57%	$1,245,533	$74,811
		ABC Masonry	$749,733	$749,733					$0	0.00%	$749,733	$0
		Stone Metals	$817,232	$817,232		$22,067			$22,067	2.70%	$795,166	$1,103
		RWC Millwork	$68,419	$68,419					$0	0.00%	$68,419	$0
		5th Wheel Roofing	$385,290	$385,290					$0	0.00%	$385,290	$0
		Ajax Waterproofing	$172,124	$172,124					$0	0.00%	$172,124	$0
		Arrow Systems	$705,142	$705,142					$0	0.00%	$705,142	$0
		ABC/Triangle	$715,451	$715,451					$0	0.00%	$715,451	$0
		American Tile	$172,150	$172,150					$0	0.00%	$172,150	$0
		XHY Flooring	$33,489	$33,489					$0	0.00%	$33,489	$0
		General Trades	$483,196	$483,196	$6,610				$6,610	1.37%	$476,586	$331
		Dean Painting	$109,537	$109,537					$0	0.00%	$109,537	$0
		Furnishing (Blinds)	$27,335	$27,335					$0	0.00%	$27,335	$0
		Elevators	$438,048	$438,048					$0	0.00%	$438,048	$0
		Fire Protection	$350,000	$350,000		$3,825			$3,825	1.09%	$346,175	$191
		HVAC Mechanical	$1,682,659	$1,682,659	$30,000	$57,450			$87,450	5.20%	$1,595,209	$4,373
		Rice Electric	$1,304,745	$1,304,745	$125,159	$102,692			$227,851	17.46%	$1,076,895	$11,393
		Dumpsters	$145,000	$145,000		$1,441			$1,441	0.99%	$143,559	$72
		Allowance	$280,000	$280,000								

Figure 18.4 (Continued)

Schedule of Values - General Conditions

Project Name: **Big Manufacturing Facility**
Project Number: **349133**
Architect Project Number: **09048.400**

Project Address: 600 John Ralston Blvd.
Period From: July 1, 2010
Period To: July 31, 2010

A	B		C		D	E		F	G		H	I
Comp No.	Item No.	Description	GMP Budget (for ref. only)	Contract (Buyout) Value	Work Completed — From Previous Application	Total This Period	% This Period	Stored Materials (not in D,E)	Completed and Stored to Date (D+E+F)	% Comp (G/C)	Balance to Finish (C-G)	Retainage (5% of G)
		Total Cost of Work	$13,012,499	$13,012,499	$1,500,751	$1,013,240	7.79%	$0	$2,513,991	19.32%		$125,700
		Contingency / Other										
		Construction Contingency	$387,765	$387,765			0.00%	$0	$0	0.00%	$387,765	$0
		General Liability Insurance	$115,595	$115,595	$11,867	$103,728	89.73%	$0	$115,595	100.00%	$0	$5,780
		Subguard	$111,224	$111,224	$111,224	$0	0.00%	$0	$111,224	100.00%	$0	$5,561
		Fee	$733,940	$733,940	$82,461	$55,848	7.61%	$0	$138,309	18.84%	$595,631	$6,915
		Total Contingency/Other	$1,348,524	$1,348,524	$205,552	$159,576	11.83%	$0	$365,128	27.08%	$983,396	$18,256
		TOTAL CONTRACT AMOUNTS	$15,409,933	$15,409,933	$1,805,487	$1,220,877	7.92%	$0	$3,026,364	9.639047%	$12,202,299	$151,318

Total Complete to Date:	$3,026,364
Less Retainage Held:	($151,318)
Less Previous Paid:	($1,707,360)
Total Amount Due:	**$1,167,686**

Figure 18.4 (Continued)

Every 30 days, at the time of the month defined in the contract documents, the contractor will prepare the Application for Payment, which may also be referred to as the Periodic Payment Request. In preparing the Application for Payment, the contractor will determine the percentage complete for each of the activities listed on the Schedule of Values and will perform the mathematical extensions to determine the amounts to be entered in the succeeding columns.

The percentages complete for the activities on the Application for Payment may be determined by measurement, by calculation, by counting, or by estimation. The supervisor is typically directly involved in the making of, or assisting with the making of, these determinations.

Whatever the basis for determination of the percentages complete, the contractor will understand that the percentages will need to be justified to the architect and owner prior to their approval of the Application for Payment. Additionally, the payment provisions of the contract may require that the contractor submit certified payroll records and/or materials invoices or receipts for payment of materials in order to substantiate his earned value in the Request for Payment.

The general contractor will instruct the subcontractors to submit their payment requests in a timely manner for him to analyze, and approve and include in his Application for Payment. Each of the subcontractors' Applications for Payment will have a similar structure to that which the general contractor will submit to the owner. The general contractor will examine and verify each element of each subcontractor's Request for Payment, and when he is satisfied that it is accurate and correct, will include it in his Application for Payment to be submitted to the owner.

The contractor's Application for Payment is submitted to the architect during the time frame indicated in the contract documents. The payment provisions of the contract provide that the architect has a specified number of days (typically 7 days following the receipt of the Application for Payment) to approve the payment request with a Certificate for Payment and to submit it to the owner with his authorization for the owner to make payment to the contractor or to notify the contractor of the reasons for his withholding certification in whole or in part. The contract provides that the owner has a specified number of days (typically 10 days), following his receipt of the architect's Certificate for Payment, to issue payment to the general contractor.

When he receives his payment from the owner, the general contractor will make payment to the subcontractors. All of the provisions regarding payment by the general contractor to each of the subcontractors are defined and described in the Subcontract Agreements. Whether he is in the employ of the general contractor or a subcontractor, the supervisor must fully understand these components of the payment process.

This process will continue every 30 days throughout the project, until the project is complete and the contractor makes his Application for Final Payment. At that point, all of the percentages complete for all of the activities on the project will be at 100% completion. Additionally, at this point, as noted previously, the general contractor and the subcontractors will include in their payment applications request for payment of the retainage that was withheld throughout the project.

RETAINAGE

The payment provisions of the contract documents will typically include a provision that retainage, also referred to as retention, will be withheld by the owner from each payment request submitted by the contractor during the construction of the project. Retainage is defined as a percentage of the amount due to the general contractor every 30 days, which is withheld by the owner from the contractor, pending the contractor's satisfactory completion of all of the requirements set forth in the contract documents.

While retainage amounts vary, the most common provision is that retainage in the amount of 10% of the earned value during each 30 day interval throughout the construction of the project will be withheld from the contractor. Sometimes the owner and architect will provide that the retainage amount will be 10% until the project is 50% complete, whereupon it will be reduced to 5% for the remainder of the project. When the contract provides that retainage will be withheld from the general contractor by the owner, the general contractor will almost always include a provision in all of the subcontract agreements that he will withhold retainage from the subcontractors in like fashion.

The retainage withheld from the general contractor will be usually paid at the time of payment of his Application for Final Payment. Again, in like fashion, the general contractor will typically pay to the subcontractors the retainage that has been withheld from them during the project at the time of payment of their Application for Final Payment.

Because of the importance of the matter of payment to the contractor, and because the supervisor is typically a participant in the application for payment process, it is important for the supervisor to fully understand the workings of the process. Additionally, this underscores the importance of the supervisor having a copy of the contract documents for the projects that he manages, and emphasizes the importance of his reading and understanding the provisions of these documents.

MEETINGS

A number of different kinds of meetings will be held during the performance of a construction project. Supervisors will be participants in many of these meetings, and they will preside at a number of others.

Preconstruction Meetings

Before the project begins, the supervisor will be a participant in handoff meetings and in preconstruction conferences. These important meetings, and the role of the supervisor in the conduct of these meetings, are discussed in Chapter 17. The importance of these meetings is reaffirmed here.

Project Coordination Meetings

During the course of the project, the supervisor will be also be involved in a series of project coordination meetings. These meetings are conducted by the general contractor, usually on a weekly basis. Management staffs of the general contractor and all of the subcontractors typically attend these meetings. Typical agenda items include: project progress updates, schedule updates, scope issues, safety issues, and special matters of concern on the part of any of the participants and/or on the part of the architect or owner.

Supervisors should be prepared to participate in all of these meetings. They should be ready to bring forth any matters they may have for inquiry, or for discussion and resolution. In addition, they should have at the ready any supporting information or documentation as appropriate, to support their inquiry or position. In their participation at these meetings, supervisors will apply the verbal communication, written communication, and documentation skills that have been emphasized throughout this book.

Safety Meetings

The supervisor will often be a participant in safety meetings with his or her company management. Additionally, he or she will frequently be the leader of regular safety meetings with the craft workers on the project. A variety of topics will be included in these meetings, in the interest of informing the workers and to emphasize the concept that safety awareness and safe practices at all times in the conduct of all of the work should be a constant element of the workers' thinking.

Additionally, these meetings can be opportunities for training in the use of personal protective equipment (PPE) or in the use of other safety equipment (e.g., railings, tie-off equipment, body harnesses, etc.) of all kinds. In addition, safety concerns, observations of unsafe practices or situations since the last meeting, or accidents or near misses may be the subjects of these meetings.

Intracompany Meetings

Supervisors should also expect to be involved in a series of intra-company meetings throughout the duration of the project. Typically, they will be meeting with the project manager and/or with company office management personnel (e.g., chief operating officer, company owner) on a regular basis or whenever there is need for an *ad hoc* meeting. Topics that are reviewed at these meetings will typically include: project progress, project schedule and schedule updates, cost reports, payment requests, change orders, quality issues, craft labor issues, materials orders, submittals, tool and equipment issues, safety issues, and any other matters that may be of concern to the supervisor or to the project manager or other company management personnel.

Training Meetings

The supervisor may very well conduct a number of meetings during the course of the project, in order to provide training of various kinds to the craft workers. Instruction in the proper use, care, or maintenance of tools or equipment, as well as training in the application or installation of certain materials or products, training in new work methods, and making trial runs or constructing mockups are examples of training which should be provided on a regular basis. Supervisors should be mindful that the more fully trained the craft workers are, the more productive they can be, and the more safely they can perform the work.

The supervisor should also be aware that there are those who would offer arguments such as the following: "Training is too expensive," or "We do not have time for training because we have a project to build and a project schedule and a project budget to maintain, or "We do not provide training because craft workers may subsequently leave our company, and when they do we will have provided training for our contractor competitors." These shortsighted and outdated arguments, in the many variations they may take, fly in the face of modern management beliefs, as well as extensive research, which indicates that workforce training is a sound investment, and that a better-trained workforce is a more productive workforce and one that works more safely.

Additionally, many believe that training provides an element of positive motivation for the craft workers, adding further to its value as an investment. This topic was further discussed, and this very point was emphasized in Chapter 7, *"Managing the Human Resource."*

CHANGES IN THE WORK

The contract documents for almost all projects contain a provision that the owner reserves the right to make changes in the work after the signing of the contract between the owner and the general contractor. The contract typically provides that all changes requested by the owner will be formalized through the use of change orders. Almost all construction projects contain a number of change orders.

The change order clause of the construction contract provides definition for all of the elements of procedure with regard to the implementation of change orders. It is important for the supervisor to fully understand these provisions as well as his company policy with regard to changes in the work and change orders.

Most commonly, the contract provides that when a change order is to be issued, it will be communicated to the general contractor by the architect, and the change order process will be administered by the architect. Each change order will describe the exact nature and scope of the change to be implemented, as well as the change if any in contract amount and/or contract time which will result from the change.

As noted above, since change orders are a commonplace element of the performance of most construction projects, it is important that the supervisor fully understand the contractual provisions regarding changes in the work. Moreover, the supervisor must be well versed in the procedures which his company utilizes regarding the requirement that a change order be issued before any work is performed which is different from the provisions of the original contract documents.

The supervisor must also develop procedures for handling all aspects of change orders on the job site. He may develop these procedures himself, or they may be prescribed by company policy, or he may receive input and assistance from the project manager.

The supervisor should develop procedures to ensure that any change order that is issued is communicated to all concerned in his company. Depending upon company policy, the supervisor may be involved to a greater or lesser extent in defining the exact scope of the change and in determining the components of direct and indirect cost to be included in the cost of implementing the change, and in determining the effect of the requested change on project duration.

However, the supervisor will most certainly be managing the performance of the work that is defined in the change order when the change order has been approved and authorized. Therefore, it is important that the supervisor know exactly what the scope of the work is, as defined by the change, as well as what the budget is, in terms of dollars and time, for implementing the change.

As noted earlier, change orders occur on almost every construction project. Many projects have numerous change orders issued. The more thoroughly the supervisor understands all aspects of change orders, from the contract provisions that define them through managing the work that the change order authorizes, the more effective he will be.

QUALITY, QUALITY ASSURANCE, AND QUALITY CONTROL

Providing quality work at a fair price is what wins work, both new work and repeat business, for the construction company. Many construction companies have operated by this business philosophy for years. This philosophical approach is especially noteworthy however, in view of the trends toward the increasing usage of negotiated contracts, as well as the increasing use of competitive sealed proposals, where the quality of the work that a contractor performs can be an important criterion in contractor selection. Additionally, owners are frequently today selecting contractors for their bid list or negotiation list who have a sound and well-defined written quality management program, and/or who have a history of producing quality work. Such a reputation and history are of immeasurable value for a construction company.

In like fashion, supervisors whose crews produce quality work and who do so within the cost of the work as defined in the project budget, develop a reputation in for quality management. A company which prides itself on producing quality work will typically place great value upon developing and retaining those supervisors and those craft workers who consistently deliver quality work.

Quality Assurance Programs

A number of contractors have developed written quality assurance policies and plans for their companies. These plans profess the company's philosophy toward quality and the commitment on the part of the company and all of its personnel, management and craft labor alike, to producing quality work in all that the company does. Such plans also typically articulate the control measures which the company employs, in order to assure consistent delivery of quality in its work.

Additionally, many owners have developed quality assurance philosophies and programs of their own. Such owners will typically require certification from contractors who submit proposals for their work or contractors with whom the owners negotiate, that the contractors have read, and understand, and will comply with, the provisions of the owner's quality assurance program.

Frequently today, owners include a requirement in the Instructions to Bidders or in the Contractor Prequalification Statements for their projects, noting for the attention of all of the contractors who would wish to submit a bid, or a competitive sealed proposal, or who would wish to negotiate a contract with the owner for a project the owner has forthcoming, that these contractors will be required to provide documentation along with their proposal, regarding all aspects of their quality assurance program to the owner, for his review and approval. Those contractors who have developed an operational quality assurance and quality management program for their company, who have articulated the elements of a quality control and quality assurance program in a policies document, and who have communicated this philosophy and program throughout their company are, thus, at a decided competitive advantage over those who have not done so.

Definitions of Quality

The consideration and implementation of any quality control program will require definition and understanding of the meaning of the term "quality." The baseline or threshold definition of the quality that the contractor is required to furnish in completing the work on a construction project resides in the provisions of the contract documents for the project. The project specifications will generally contain most of the elements of definition of quality in materials and workmanship that the contractor will be required to deliver in completing the work on the project. The contractor is contractually bound to provide this level of requisite quality in the materials and workmanship, as a minimum.

There are, however, additional levels of quality standards that may be utilized. Sometimes, when a construction company has a quality assurance program in place and is operating by a commitment to provide quality in all of its work, it may become a good business decision, in the interest of providing satisfaction for a client, to provide or to produce a higher level of quality in some instances than that which the specifications set forth. Similarly, at the crew level, the supervisor may observe work that has been or is being performed in such a way that it would (or at the least, could) satisfy project specifications and designer scrutiny but that does not uphold the supervisor's standards for workmanship. The supervisor may well require the work to be redone, on the basis of assuring quality in all that is delivered.

Certainly, decisions such as those noted above must be balanced against the realities of cost control, adherence to project budget, adherence to schedule, and company profitability. However, a great many companies have discovered, and have been highly successful through application of the watchwords "quality does not cost, it pays."

Writings on Quality

Sir Edwards Deming was among the first to espouse a philosophy commending the benefits of consistently delivering high quality in any business venture. He produced the terminology "Total Quality Management," often referred to as TQM. Others, such as Crosby, Juran, and others have continued his work. The writings of these quality management experts are highly recommended for the supervisor's reading.

One of the basic, and yet very effective, principles of Deming's work regarding quality was his recommendation for the application of a cycle, which he referred to as the PDCA Cycle. PDCA stands for Plan, Do, Check, Act. This cycle is illustrated in Figure 18.5.

In Deming's philosophy,

PLAN indicates that quality does not happen of its own accord. Rather, quality must be specifically planned for as an outcome;

DO refers to implementing the plan, mindful that a quality outcome is expected.

CHECK refers to constantly assessing the effectiveness of the plan and of the result being produced, in order to assess the effectiveness of the planning and to determine whether a new plan is indicated in order to achieve the desired result.

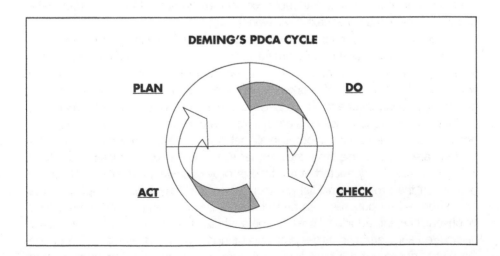

Figure 18.5 Deming's PDCA Cycle

ACT means to carry out the plan in its entirety if it is producing the desired outcome or, alternately, to move toward the formulation of a new plan to replace the first. This portion of the phrase also includes the action of moving toward the formulation of a new plan to achieve the desired quality in the next step or activity in the work.

Deming recommended the application of these four steps in a continuous cycle until the activity or project has been completed and a high-quality outcome assured. These four cyclic steps provide excellent guidance for supervisors as they seek to produce high quality in all that they do.

A Practical Strategy

One of the methods a supervisor can use in order to ensure delivery of a high-quality outcome is to conduct regular examinations of the work as it is being produced. A technique that has been found valuable, is to "walk the project" in the evening following the completion of the workday by the craft workers. The supervisor can make a critical examination of the work that has been installed on that day and can apply his own standards of quality or the company's quality assurance guidelines in order to determine whether the work is of acceptable quality. If any component of the work is not of acceptable quality, the supervisor can make plans for immediate correction in the following day's work assignments, either for his own crews of craft workers or, if the supervisor is working for the general contractor, for his own craft workers as well as for those of the subcontractors on the project.

In another sense, the supervisor can examine the work at this time with a "punchlist consciousness" in mind. If the work is not, or might not be, compliant with contract document requirements and, thus, could possibly appear on the punchlist at the time of project closeout, the supervisor can make note of this, and can make plans for corrections in the next day's work.

These practical approaches have the advantage of detecting errors and work of unacceptable quality early, and providing for their correction at the best possible time. While it may be unpleasant for the supervisor to issue a directive for remedial work, as well as for the craft workers to receive such a directive, and while such action will certainly incur additional cost, the benefits of this approach are manifold.

First, this approach ensures a consistent awareness of, and an unceasing emphasis upon, high-quality outcomes in all that is produced. Additionally, when corrections are made at this time, they are far less onerous and far less costly than when they must be corrected at the time of project closeout, after they have appeared on the project punchlist. Further, when this approach is employed, the end-of-the-project punchlist can be reduced in size and scope, and the resulting inconvenience and additional cost can be minimized. Additionally, when errors are discovered and corrected as the project is being constructed, as the architect and the owner observe the construction in progress, they can see an ongoing and tangible commitment to quality as an outcome on the part of the supervisor and his crew. This, in turn, only enhances progress toward the supervisor's fulfillment of the fifth project objective, which has been defined as leaving a satisfied client.

SUMMARY

As supervisors manage the work on a construction project, certainly they have numerous responsibilities, and certainly a great deal is expected of them. In this chapter, with the information provided regarding documentation, reports of various kinds, meetings, revenue flow, and quality assurance, supervisors should come to realize that there are powerful tools and techniques at their disposal to help them to be more effective and more successful in all that they do, and to help them assure that they are able to consistently accomplish the five project objectives for every project.

Learning Activities

1. Learn more about Deming, and Total Quality Management (TQM), and some of the other principles he espoused.

   Get a copy of the book, *Why Things Go Wrong*, by Gary Fellers. You may be able to get a copy from your local library, or you can buy a copy for about $20.

   This book will explain, in quick, easy-to-read fashion, many of the principles which Deming believed in and espoused.

   Another good book for you to read is, *The Deming Management Method*, by Mary Walton.

   These books will add greatly to your learning, and may well encourage you to read more about Deming and the processes of quality management.

2. On the project where you are now supervising, look for opportunities to implement the "**P**lan, **D**o, **C**heck, **A**ct Cycle (PDCA), and see for yourself how effective it can be.

3. Initiate the practice of following the guidance in this chapter for being more effective at meetings you attend: make an agenda; know what the issues are, that you need to get resolved or addressed at the meeting; never leave a meeting without asking something; never leave a meeting without saying something.

   Demonstrate for yourself how measures such as these can enhance your effectiveness at meetings of all kinds.

CHAPTER **19**

CLOSEOUT OF FIELD OPERATIONS

INTRODUCTION

As the conclusion of the project draws near, the supervisor will be involved in a series of very important activities that are referred to collectively as "project closeout." These activities involve assuring that all contract requirements are fulfilled, preparing to hand over the completed facility to the owner, demobilizing and moving off the site, and conducting a final review of the project and its management. It is important for the supervisor to know that the things that a supervisor does and does not do, as well as the things he or she says or does not say at this critically important time in the project, will have a large bearing upon the success of the project and, therefore, upon the success of the supervisor.

To gain perspective, the supervisor does well to consider the overall project environment at this time. The management team, as well as all of the craft workers on the project, those working for the general contractor and all of the subcontractors alike, have been involved on this project for a considerable time. Most of the time, by this stage of the project, they are ready to see the project completed and are ready to move on.

The owner, meanwhile, has been involved with the project for an even longer time than the contractors and their craft workers and management teams. He is also ready to see the work finished, and is anxious to take possession of the completed facility. There is typically a natural excitement on the part of the owner at this time. All of the planning, design, and construction are about to culminate in the completion of his project. He can see the project nearing completion, and he is eager to be finished, and to move in and take possession. At the same time, the owner realizes that when the project is closed out, the contractor and subcontractors will be leaving. The owner wishes to ensure that all of the contract requirements have been fulfilled and that he is receiving everything that the contract provides.

It is imperative for the supervisor, and all of the management team, to realize the significance of this time. As noted earlier, one of the objectives for every project is leaving a satisfied client at the conclusion of the project. This is the time for the supervisor to focus her management attention on ensuring that the expectations of the owner are fulfilled, in order to ensure that he is satisfied with the outcome. This is also the time to ensure that all of the contract requirements have been fulfilled, in order that the supervisor's construction company can receive final payment and can move to the successful termination of the contract.

PROJECT CLEANUP

Project cleanup is an important consideration throughout the time of construction of all construction projects. Safety, the morale of the workforce, and the reputation of the contractor all are directly affected by the cleanliness and orderliness of the project.

This consideration is especially important at the time of project closeout. Typically, the architect and the owner will be visiting the project much more frequently as the project nears completion than at any other time. Additionally, it is not uncommon for owners to visit the site frequently and to bring members of their company staff to view the project as it approaches completion. Having a clean and orderly project at this time is very important for making a good impression and for helping ensure a satisfied client.

Ironically, it is at the time of project closeout that craft workers are installing finish and trim items. Usually, these items are delivered to the job site in boxes and crates that are filled with protective wrappings, and packing materials, and foam padding, and other forms of protection. This debris is a safety hazard and is unsightly, especially if it is allowed to accumulate, or even to remain on the floor of the newly constructed facility. The supervisor does well to devote an extra measure of care and attention to the matter of timely cleanup at this very important time in the life of the project.

THE COMMISSIONING PROCESS

The contract documents will define a series of actions that the contractor must complete as part of the project closeout process, called commissioning. While the exact steps to be performed in commissioning will vary somewhat with different projects, as defined in the contract documents for each project, some of the common steps are outlined and discussed below.

The primary point of emphasis for the supervisor is that all of the closeout and commissioning actions to be performed are defined in the project contract documents and, therefore, should be included in the planning of the work for the project in order that the project can be completed on time and with a minimum of difficulty. The supervisor and/or the project manager should make a list of all of the commissioning actions that are specified for the project, should include them in the planning process, and should make a checklist so as to follow through on each of them to ensure their completion.

Tests and Certifications

A variety of tests that are to be conducted on various components of the project may be specified in the contract documents. Typically, tests are specified for verifying the proper operation and/or output of equipment that has been installed in the course of constructing the project.

Commonly, for example, specifications will require tests for startup amperage draw, as well as operational amperage draw, of the blowers and compressors that are part of the heating, ventilating, and air conditioning system in a building. Likewise, tests may be required to verify cfm (cubic feet per minute) of air delivered by blowers in the mechanical system, as well as temperature measurements taken at registers to determine the temperature of air delivered during the heating and cooling cycles of the mechanical equipment.

Similarly, fire alarm and notification systems, intrusion alarm and notification systems, electronic surveillance systems, elevators, escalators, moving sidewalks, emergency lighting systems, electronic lock systems, emergency generators, and so forth will have testing and certification procedures specified. Additionally, mechanical locks, and the accompanying keying systems (grand master and submaster keys, duplicate keys, key control cabinets, etc.) will commonly have procedures specified for verifying proper operation, for delivery of the correct type and number of keys, for the provision for key security, and so on.

Further, in the construction of a manufacturing facility or process plant, a battery of tests may be specified for the various types of equipment and machinery that are operational components of the systems in the facility. In like fashion, all of the required tests and commissioning procedures will be defined and described in all of their detail in the contract documents. The supervisor should know what these required tests and certifications are, and should have them recorded in a log so that all elements of the procedures that are specified can be properly scheduled and managed.

As has been noted previously, the supervisor must be aware of all of the testing and certification procedures that are required for the entire facility. They are set forth in the specifications and other contract documents for the project. They must become part of the long-term and short-term planning for the work to be completed.

Additionally, the supervisor should realize that many of these tests and certifications will require certified technicians and/or calibrated and certified instrumentation for their performance. Careful planning is required to ensure that the requisite personnel and equipment are available at the time when the tests and certifications need to be performed.

Additionally, project requirements often specify that certain witnesses (owner, architect, mechanical engineer, process engineers, owner's facilities management and maintenance personnel, etc.) must be present to observe the performance of the tests and to monitor and certify the results, and to ensure that all contract requirements have been fulfilled. Such a requirement, of course, adds yet another planning and scheduling dimension for the supervisor. In addition, requirements often stipulate that the witnesses must sign a certificate or a form to verify their observance of the tests, and to certify the results. This means that the supervisor will need to ensure that the proper forms are present when the tests and certifications are conducted, to ascertain that the proper signatures are obtained, and to ensure that the documentation is filed as a matter of record.

Owner's Manuals, Parts Lists, Equipment Warranties, and Spare Parts

Project specifications typically require that the contractor provide the owner with owner's manuals, parts lists, operating instructions, maintenance instructions, and equipment warranties for all equipment installed or furnished during the course of construction. Frequently, the specifications require that some of these documents be completed by the contractor with model numbers and serial numbers of the installed equipment, as well as the date of installation, date of operationalizing, and so forth. Often, the contractor will organize these documents into folders, binders, or notebooks for the owner's convenience.

For equipment that the contractor has installed, it becomes a duty for the project management team to deliver all of the documents as well as all of the accessories and spare parts that are specified. For equipment that has been furnished or installed by subcontractors or sub-subcontractors, the management team must ensure that these items are provided and that they are complete and in the proper format, prior to closing out the subcontract agreements.

Similarly, the project requirements will often stipulate that spare parts and accessories of various kinds be provided by the contractor. Filters for air handlers, blower belts, cleaning kits, and special lubricating oils are typical examples, but there are numerous others that may be specified in similar fashion. The items that are to be provided must be identified from the specifications, and should be listed on a log and on a checklist, so that they can be included in the steps to be completed at project closeout.

Owner Training

Project specifications commonly require that the contractor conduct, or coordinate the conduct of, training in the operation and maintenance of various components of the project for the owner, the owner's maintenance staff, or the owner's facilities management personnel. Frequently, this training must be conducted by factory-trained personnel, by certified technicians, or by engineers.

It is imperative that the supervisor extract from the project specifications, a list of all of these training requirements, and that they be placed on a log and checklist for planning and completion as components of project closeout. It is not uncommon that many of these training sessions require scheduling long in advance of the time they are to be conducted, again emphasizing the need for careful long-term and short-term planning on the part of the supervisor.

DEVELOPING FINAL DOCUMENTATION

Another important element of the project closeout process is ensuring that all necessary project documentation is completed and finalized. While the components of final documentation to be completed may vary somewhat from one project to another, discussion in this chapter will include some elements of final documentation that are almost always required.

Change Orders Finalized

Prior to submitting his final Application for Payment to the owner (or in the case of a subcontractor, prior to submitting his final Application for Payment to the general contractor), the contractor or subcontractor should check to be certain that all change orders that were issued during the course of constructing the project have been completed and settled.

The supervisor can assist in this regard by providing information to the project manager and/or to the company office regarding the status of changes in the work. As discussed in Chapter 18, the supervisor will have firsthand knowledge with respect to all changes issued in the course of constructing the project inasmuch as he and his crew will have been the ones who implemented the changes on the site. Additionally, the supervisor may have provided assistance in the pricing and/or the administration of the change order and, thus, will be familiar with all of the details of the change order.

The supervisor should be able to verify for the project manager and/or the company office that all change order work is complete. All change orders can then be tracked through the contractor's system in order to ensure that payment has been received for all of the changes, and that all change orders have been finalized and closed out.

As-Built Drawings

The contract documents for a project typically require that, when the project is complete, the contractor is to provide a set of accurate *as-built drawings* to the owner, which reflect all changes from the original drawings and specifications and represent all aspects of the project as they were actually constructed. The owner typically requires that these as-built drawings be submitted to the architect for approval and requires further that they be furnished in a format (line drawings, marked copy of original drawings, CADD files, etc.) suitable to the architect and the owner.

The owner needs these drawings as an accurate portrayal of all that was actually constructed. Additionally, these drawings are necessary for safety in maintaining the building and for use when modifications or renovations are to be made in the future. In addition, copies of the as-built drawings are kept in a suitable location by the owner, so as to be readily available for use by emergency personnel who may be called to the building.

The contractor, and therefore the supervisor, has a contractual duty, as well as a moral duty, to ensure that an accurate and complete set of as-built drawings is provided to the owner at the conclusion of the project. The contractual duty is, as noted above, stipulated in the contract documents. This contractual duty is accompanied by a moral imperative that the as-built drawings be complete and correct, and that they accurately represent all that is present in the facility. Those who perform work on the facility in the future, as well as emergency responders who may be called to the facility, are literally entrusting their safety to the accuracy and completeness of what is represented on the as-built drawings.

The best way for the supervisor to ensure delivery of an accurate and complete set of as-built drawings in a format suitable to the owner is to make note of this requirement early in his planning for the work on the project, in order to ensure that as-builts are produced and maintained *beginning on the first day of the project.* If as-built drawings have been properly developed and maintained from the onset of construction operations, and have been developed and maintained in the format specified by the owner throughout the performance of the project, then at the time of project closeout only a final updating and final verification of these drawings will be necessary before they are handed over to the architect and the owner. It is also worthwhile for the supervisor to consider that the task of producing and maintaining as-built drawings is one which could be delegated to an employee, accompanied by proper training as may be necessary, as well as by proper supervision and follow-up on the part of the supervisor.

It is also important to realize that if some of the work on the project is subcontracted, all of the subcontractors' work that is in any way different from what is shown on the original contract documents must be reflected on the as-built drawings as well. The general contractor may elect to have his staff maintain as-builts to reflect the subcontractors' work, or more typically, he may require each subcontractor to develop and maintain as-builts for their portion of the work in the proper format as specified by the owner, and then to provide them to the general contractor at the time of project closeout for inclusion in the set that will be provided to the owner. In either case, it is very important that diligence be exercised, from the beginning of the project to the end, in order to ensure that accurate and complete as-built drawings are developed for the entire project.

The Punchlist Process

As the completion of the project draws near, a date and time will be established for the architect to conduct the final inspection of the project as a precedent to final payment to the general contractor. This final inspection, and the elements that accompany it, is commonly referred to as the punchlist process, or as "punching out" the project, or as "making the punch." The punchlist can be defined as a listing, made by the architect on behalf of the owner during this final inspection of the project, of items of work remaining to be done, unsatisfactory materials or workmanship that must be remedied, and errors that must be corrected before the project will be accepted by the owner from the contractor.

At the time scheduled for making the punchlist, the architect and often the owner will be present, as well as representatives of the general contractor (project manager, superintendent, supervisor) and representatives of the key subcontractors on the project. A tour of the site will be conducted on a room-by-room basis, and a careful and detailed examination of all aspects of the construction will be made by the architect on behalf of the owner. When an error or variance from contract requirements is discovered, a written notation will be made. The list of all of these notations constitutes the punchlist.

At the conclusion of the inspection tour, a copy of the punchlist will be handed over to the general contractor. The list of items on the punchlist becomes his list of "action items" in terms of matters in need of remedy in order to make the project acceptable to the owner and, in turn, to warrant final payment by the owner to the contractor. The general contractor will provide copies of the punchlist to each of the subcontractors, in order that each of them can address the correction of those items of work that pertain to their scope of work.

As punchlist items are corrected, the general contractor will typically inspect them to ensure that errors or omissions that were noted have been brought fully into compliance with contract requirements, and if so will check those items off the punchlist. The supervisor may be called upon to assist with this process. When all of the punchlist items have been completed, the general contractor will notify the architect, who will again visit the site and will verify for himself that satisfactory remedy has been made for everything that was noted on the punchlist, and that the work is now acceptable.

The objective of the punchlist process is to ensure that the owner is receiving all that the contract documents call for and to ensure that all of the work and its quality are in compliance with the requirements set forth in the contract documents. When this has been ascertained, the architect will issue a document called the Certificate of Substantial Completion. This document provides certification on the part of the architect and the owner that the contractor has substantially fulfilled all of the requirements of the contract documents.

It is important for supervisors to understand the punchlist process, because they will have a key role in completing the punchlist requirements and in bringing the project to substantial completion. Punchlist work must be very carefully planned and performed, in order to ensure that the finished work complies with contract requirements. Additionally, great care must be exercised, so as to ensure that other work on the project is not damaged while the punchlist items are being corrected.

The work of completing punchlist items can frequently be tedious and demanding, accompanied by the frustration inherent in needing to make corrections, and in the desire to get the project completed. The supervisor will need to utilize all of his human relations, communication, and motivation skills, in order to see that punchlist items are properly completed and that all contract requirements are fulfilled.

Additionally it should be noted that punchlist work is often very expensive, because of the difficulty of access, and because of the need to protect all other work. The supervisor who has been mindful of maintaining quality control in the work from the very beginning of the project, and who has performed the work with "punchlist consciousness" throughout the duration of the project, as has been recommended in previous chapters, will find the actual punchlist process much less stressful, and much less costly.

CLOSING OUT SUBCONTRACTORS AND SUPPLIERS

Subcontractors

The supervisor for the general contractor may very well be providing assistance with project closeout activities for the various subcontractors who have performed work on the project. This will include ascertaining that the subcontractors' as-built drawings are completed and up to date, accurate, and in a form acceptable to the owner. If the supervisor has maintained a practice of requiring each subcontractor to maintain and update as- built drawings from the beginning of the project and continuing throughout the life of the project, as discussed in a previous section, this effort will prove much less formidable at the time of project closeout.

Additionally, when the punchlist has been made, the general contractor's supervisor will help ensure that all of the subcontractors' punchlist items are completed properly and in a timely manner. Careful coordination of the work of the several trades is required. Of course, the supervisor for each of the several subcontractors will have the primary responsibility for ensuring that all of the subcontract items for their company's scope of work are properly completed in accordance with the requirements of the contract documents and the subcontract agreement.

Protection of existing work is an important consideration for all who are working on the project at this time. In addition, as noted previously, extra care and attention to cleanup is another important element of consideration at this time for everyone on the project.

When punchlist items have been completed by the subcontractors, and when their final touch-up and cleanup activities are completed, the subcontractors will be making their final Application for Payment to the general contractor. The subcontractors will be requesting the last of their earned value, in accord with their

Schedule of Values, as well as the retainage that has been withheld throughout the project.

This in turn, will typically require their completion of a Waiver of Lien Form, to ensure that they attest that they have been paid for all expenses that they incurred in conjunction with their work on this project. This will preclude the possibility of their filing a lien or a claim against the property, the owner, or the general contractor.

Suppliers

As a part of project closeout functions, the supervisor will need to ensure that all accounts with all materials suppliers are up to date and paid in full. Additionally, any materials that can be returned for credit should be sent back to the supplier, and credit invoices should be processed and included in the project documentation files.

FINAL REQUEST FOR PAYMENT

As punchlist items are completed, and as the contractor's final project touch-up and cleanup are completed, the supervisor will assist with preparation and submittal of the contractor's Final Request for Payment. This request will include all of the remaining earned value items as indicated in the Schedule of Values. Additionally, this payment request will include a request for payment of all retainage that has been withheld by the owner from the contractor throughout the duration of the project. This final payment will convey the balance of the contract amount to the general contractor. The general contractor's submittal of the Request for Final Payment and the owner's payment of the amount due are often linked, by the terms of the contract documents, to the issuance of the Certificate of Occupancy for a new building and to the issuance of the Certificate of Substantial Completion for the building. These two documents will be discussed in the sections that follow.

THE CERTIFICATE OF OCCUPANCY

When buildings are constructed or remodeled, city governments frequently require that a Certificate of Occupancy, often referred to as a "CO," be issued by the city, before the completed building can be occupied by the owner of the facility. The Certificate of Occupancy provides certification by city government that the structure and the site are compliant with the building code and zoning ordinance statutes of the city. To obtain this certificate, the contractor notifies the city that he is ready for a "CO inspection."

The city government then sends building code compliance inspectors and zoning ordinance officials to the site. These officials will make a thorough examination of the building in all of its aspects and will also make a complete inspection of the site in all of its aspects, to ensure compliance with the city's building code and zoning ordinance requirements. When all of the code and zoning requirements have been fulfilled, the city issues the Certificate of Occupancy document, authorizing the owner to take possession of and to occupy the newly completed building.

The owner and the architect typically include a clause in the contract documents for a building construction project that final payment will not be forthcoming, and that the Certificate of Substantial Completion will not be issued, until the city has issued a Certificate of Occupancy. The supervisor should understand these requirements, since he will typically be involved in scheduling the CO inspection and in coordinating the site visit by the city officials. The Certificate of Occupancy itself is, obviously, a very important component of the documentation for the project. Therefore, a copy of this document will be retained in the project documentation files.

CERTIFICATE OF SUBSTANTIAL COMPLETION

The Certificate of Substantial Completion is a document issued by the architect, through the owner, which certifies that the general contractor (prime contractor) has substantially fulfilled all of the requirements set forth in the contract documents for the project. This document certifies that the contractor is relieved of any further responsibilities in the performance of contract requirements, with the exception of the warranty provisions as set forth in the contract.

The issuance of the Certificate of Substantial completion will occur only when:

- All of the punchlist items have been completed to the satisfaction of the architect and a final inspection has been conducted to verify this.
- A satisfactory set of as-built drawings has been submitted and is approved by the architect and the owner.
- The Certificate of Occupancy has been issued by city government.
- The prime contractor's Final Request for Payment has been submitted and approved by the architect and owner.
- The prime contractor has signed a Waiver of Liens, certifying that he has paid any and all debts and expenses associated with the construction of the project.
- There is consent of the contractor's surety or bonding company.

When the Certificate of Substantial Completion and Final Payment are issued by the owner to the prime contractor, the prime contractor will immediately make final payment to all of the subcontractors. The general contractor and subcontractors will immediately move all of their facilities, equipment, and remaining materials off of the site. Frequently, the owner will conduct a grand opening or ribbon-cutting event for the new facility.

The supervisor should expect to be involved, directly or indirectly, with all of these activities. The project is not complete until all of these events have taken place. Therefore, the supervisor needs to understand them, and to make them a part of his planning for the work.

PROJECT REVIEW

After the project has been completed, and after the contractor has moved off of the site, one very important function remains to be conducted with regard to this project before the contractor's full focus and attention shift to the performance of the next project. This important step is referred to as project review. Many managers believe that project reviews are among their most valuable project management and company management tools.

Project review involves assembling all of the project documentation and then assembling the key people who were participants in the project for a final management review of the project in all of its aspects. The documentation will typically include the Job Log, the cost reports, the change order file, and the schedule and schedule revisions, as well as any other documentation that the management team believes may be relevant. The people who participate in the project review will typically include a representative from company management (president or chief operating officer), the estimator, the scheduler, the project manager, the superintendent, and the supervisor.

The intent of the project review is to reflect upon the project and to thoughtfully analyze all aspects of its performance in order to determine what lessons can be learned and stored for future reference. The expectation is that lessons learned, as well as practices that were employed on the project, can be brought forth, considered, discussed, and written down, and then can be included in the company's institutional memory, or incorporated into the company's policies and procedures, the estimating or scheduling processes, or the day-to-day management processes.

When the project review meeting is held, a facilitator should be present to conduct the meeting and to keep it on track. Additionally, a person should be designated to take notes and to distribute the notes from the meeting to everyone on the management team following the meeting.

The meeting should involve open discussion and should be conducted in a cordial and professional manner. It should be emphasized that the purpose of the meeting is not to find fault, nor to assess blame for things that did not go well or for things that may not have turned out as planned in the construction of the project. Where the project experienced cost or schedule overruns, or safety difficulties, for example, a root cause analysis should be performed by continuously asking "Why?" until there is no more "Why?" Once the root cause of the problem or difficulty has been identified, the solution, or management steps to be taken to avoid a recurrence, can be identified by the members of the team who are present at the meeting and written down. These thoughts and notes then become part of the record for the project. Moreover, they become part of the institutional memory for the company, so that this information can be accessed and used to management advantage in the future.

It is very important to realize, and very important to include as a part of the project review, as many as possible of the successes that were experienced on the project. Management must ensure that problems and difficulties are not the only matters to be examined and recorded. Any and all successes should also be identified, celebrated, and written down, such as completing an activity or set of activities at significantly lower costs than indicated in the project budget, finding special savings on materials purchases, completing activities in less time than indicated on the project schedule, discovering especially productive ways to utilize equipment, achieving better-than-expected productivity rates, developing new work methods, and so forth.

At the conclusion of the project review meeting, the notes from the meeting should be transcribed and sent to all who participated in the project review session. Additionally, and importantly, the notes from this meeting, and from all similar meetings for all of the projects that the company performs, should be bound and should be retained by company management as the institutional record of the company's projects. Many companies label these manuals "Best Practices" or "Project Review Summaries," or something similar.

When managers, including supervisors, periodically take the time to read and reflect upon these project reviews, a wealth of useful information emerges. This information is often extremely valuable in the management of future projects. There are many managers who believe very strongly that these project review sessions, and the "Best Practices" manuals that result, are among the most valuable resources at a supervisor's disposal. Supervisors should make it a regular practice to carefully read these project review summaries on a regular recurring basis for projects that they and others have managed and to avail themselves of the best practices and lessons learned that they contain. When they do so, it is a virtual certainty that they will discover in these project reviews a wealth of useful information that will help them to be more effective and more successful in the management of their construction projects.

SUMMARY

The time of closeout of field operations on a project is certainly a time filled with important matters that must be attended by the supervisor. As has been emphasized throughout this book, planning and attention to detail are keys to success. Additionally, because of the nature of this time on the project, supervisors will make extensive use in a special way of their leadership and motivational skills, as well as their verbal and written communications skills.

This chapter has provided some valuable information, as well as some useful tools and techniques, which can assist in making the time of project closeout as productive and as pleasant as possible. From being mindful of the special importance of this time, through understanding the elements of the commissioning process and final documentation, and the Certificate of Occupancy and the Certificate of Substantial Completion, through recognizing the importance of the final project review, the supervisor must provide his care and management attention at this time. The summary guidance that is reaffirmed for the supervisor at this time is to remember that what is done and not done, and what is said and not said, at this vitally important time of the project will have a lasting effect upon the success of the project and, therefore, upon the reputation and the success of the supervisor.

Learning Activities

1. Implement punchlist consciousness.

   On the project you are now supervising, initiate a practice of managing the project through the time of its closeout with the "punchlist consciousness" discussed in this and other chapters.

   Take care of managing for a quality outcome while the work is in progress.

   Demonstrate for yourself the principle that usually the best and most economical time to assure a quality outcome, is while the work is in progress, and certainly before project closeout and the preparation of the project punchlist.

   After this project is complete perform a personal assessment to determine for yourself whether you think this mindset provided tangible results in the management of the project, including the project closeout.

2. Implement Project Review.

   If your company does not already have a policy of conducting project reviews following the completion of its projects, talk to your project manager or company executive.

   Share with them the "Project Review" section of this chapter.

   Ask whether they think it would be possible to implement this process in your company. Assure your superiors that many companies and their management have found tremendous value in this practice.

CHAPTER **20**

THE SUPERVISOR'S CONTINUING DEVELOPMENT

INTRODUCTION

Construction supervision is dynamic. It is constantly changing. Each job is different. The industry is changing. Companies are changing as they adapt to social and economic changes, as well as to changes in business practices. Our understanding of the processes of construction and how best to manage them is changing. Therefore, it is critical that the construction supervisor changes, as well, throughout her career.

This final chapter begins by considering some of the ways in which the industry is changing and how those changes will affect the supervisor. It then suggests how both the character and the role of the supervisor may change. Finally, the chapter concludes by considering how supervisors can not only keep up with the changes but also stay ahead of change, enabling them to embrace change as a positive and creative way of advancing both the job and their professional career.

CHANGE IN THE CONSTRUCTION INDUSTRY

Observations of the construction industry demonstrate that, over time, the industry has changed and that it is continuing to change. There are a number of drivers that are imposing change on the industry. Among them are:

- Technology
- Environmental change
- The economy
- Demographics and the construction workforce
- Understanding of construction processes
- Perceptions of the roles of management and supervision

An understanding of how these drivers are changing the industry will help the supervisor better understand how supervision and the role of the supervisor are evolving.

Technology

Technology is a very broad category that includes construction materials and installed equipment, construction tools and equipment, and a variety of electronic technologies that are revolutionizing the way business is conducted.

Changes in construction materials and installed equipment are probably the easiest to visualize. As a new product is introduced, designers learn how to incorporate it into the design of the construction project. For the contractor, the new product shows up in the construction documents. Material suppliers and distributors train contractors in how the new products are to be handled and incorporated into the project. Estimators need to understand how to price out the cost of the new items, both in terms of item cost and in terms of the labor and equipment required to install them. Supervisors must understand the appropriate ways to incorporate the new products into the project and how this will affect scheduling and resource management. New materials and installed equipment have been a part of the industry for a long time, and the industry has learned how to incorporate them with little disturbance.

Whereas new materials and installed equipment are introduced to the industry through the design process, new construction tools and equipment are introduced by contractors, who are made aware of new products by manufacturers and vendors. The path to introduce new tools and equipment often leads through the supervisor, who is looking for ways to improve field operations. Occasionally, a totally new item will come on the market, but most changes in tools and equipment come through upgrades to existing tools and equipment. An example of a significant improvement in existing tools is the replacement of power cords for electrical hand tools with powerful, small and light weight rechargeable battery packs. What appears to be a small change has made a significant impact on both production and safety.

It is up to the supervisor to stay abreast of new refinements to construction tools and equipment and to determine what can make a valuable contribution to company operations and what is not worth the time, cost, and effort to implement.

Changes in electronic technology are having a much more profound effect on how construction is done. Some changes are positive and some are negative.

One class of electronic technology is communications technology, which has evolved dramatically in recent years. Now, virtually everyone has a cell phone and most cell phones have broad capabilities, including texting, Internet access and the ability to take pictures. Numerous applications are available for advanced telephones, and new applications are constantly being developed. For most craft workers, cell phones are a distraction and their use must be limited on the job site. For the supervisor, advance communication technologies can prove to be powerful tools. However, if these are improperly used, they can be more disruptive than useful.

Better communication is generally considered an improvement, but there is also a negative side to the broad implementation of this technology. The advent of touch-to-talk technology, which made instantaneous communication available throughout a company, provides an example of how technology can be useful if properly used but how it is counterproductive if not used with constraint. Pre-planning, a fundamental supervisory responsibility, has suffered because planners know that if they forget something, they can immediately summon what they forgot in the pre-plan. Spending time in thorough pre-planning is perceived as becoming less important because of the technology. The result is deterioration in both the quality and the quantity of planning. Touch-to-talk technology also impacts the person on the other end of the communication. A person with a problem contacts another who may, or may not have the answer, and in so doing disrupts the work of the second individual. The conclusion with regard to new communications technologies is that the use must be regulated to take advantage of the new opportunities, but to guard against disruptions that can deteriorate performance and safety.

Computer technology is evolving to provide increased power in smaller, more robust units at continually diminishing cost, so that on site computers are becoming practical for field use. Not only are they becoming practical, but because of changes in the form of construction documents to digital format, job site computers are becoming a requirement. Computer-based electronic communications can be very effective, not only for soliciting timely responses but for documenting discussions. The supervisor must seek ways to creatively use job site computers while regulating the use to eliminate detrimental applications.

One of the greatest technology advances just evolving is Building Information Modeling (BIM), which will eventually incorporate all information for a construction project in a digital format within a single, unified database. As a very complex and powerful emerging technology, the impact of BIM on the job of the supervisor has yet to be clearly defined.

Environmental Change

Environmental change is driving the construction industry to incorporate more environmentally appropriate construction materials and equipment into energy efficient systems. These changes are coming about through the incorporation of new materials and installed equipment into the project design.

However, another very important aspect of environmentally appropriate construction is the use of environmentally sensitive construction tools, equipment, methods, and techniques. This is a realm in which the supervisor plays a significant role. One focus of environmentally sensitive construction is minimizing pollutants. For example, the supervisor must be well aware of how to handle water runoff from the site to ensure that polluting materials are not washed off the site. Another example is the supervisor's involvement in appropriately handling the many toxic chemicals that are used on site and responsibly disposing of any waste materials or residue from such materials.

Another focus in environmentally sensitive construction is on the use and maintenance of construction equipment. The most energy efficient equipment should be selected for a job. Fuels and lubricants for construction equipment should be nonpolluting and, if possible, made from renewable resources. Lubricants must be properly handled and properly disposed of.

The Economy

The economy has many impacts on construction work. A robust economy drives construction schedules, demanding that the execution time for projects be continually reduced. A robust economy also creates significant construction work, which tends to increase the number of contractors and raise the competition such that prices need to be driven down. A faltering economy means more contractors will be going after less work, again increasing competition and driving prices down. A fundamental principal of the construction industry is that more complex projects are expected to be completed in less time at less cost with increased safety and quality.

The global nature of the economy is having a significant impact on local projects. It is true that, unlike manufacturing, construction projects cannot be built outside the country and imported. However, the global economy impacts construction in many other ways. Although an entire building or industrial complex cannot be built overseas and imported, many of the materials and pieces of installed equipment that are incorporated into a construction project are being manufactured overseas and imported. With the increasing trend in prefabrication (see below) prefabricated assemblies can be manufactured overseas and imported. The result is to move construction toward a process of assembly rather than traditional building in place. In addition, construction equipment and temporary structures, such as forming systems, are increasingly coming from outside the country.

Finally, economic booms in other countries can tie up copious amounts of construction materials, causing supervisors to have to cope with local material shortages. One impact of material shortages is to cause disturbances to both cost and schedule. Another impact is to require supervisors to seek alternative sources of materials to draw from should the long lines of transportation break down. Finally, shortages of materials often result in the substitution of lesser-quality materials or even counterfeit items that do not meet specifications.

Demographics and the Construction Workforce

The nature of the construction workforce is changing dramatically. Generational changes that affect all of society, also affect construction workers. What motivates people, attitudes toward work and employers, perceptions of time, and standards of quality are examples of areas where worker perceptions differ, depending upon various generational categories.

Cultural diversity is also changing throughout the industry. In addition to differences noted above, this brings diversity in language, religion, and social customs.

The construction worker has traditionally been personified as white, Anglo, and male, although minority groups have always been represented. As minority groups increase and the proportion of white, Anglo, males decreases, the workforce can no longer be thought of as monolithic. It is very diverse in many ways, including gender, race, and language.

The supervisor must become proficient in leadership skills in the context of a diverse environment. Directing labor with a heavy hand is neither acceptable within the current culture nor is it effective. Although the supervisor takes ultimate responsibility for the efficiency of field operations, the effective supervisor will engage workers in planning and designing the construction processes, taking advantage of diversity in the workforce to develop more effective solutions.

Understanding of Construction Processes

Construction processes continue to evolve. Traditional "stick building" is giving way to prefabrication of assemblies, and installation of these assemblies. Prefabrication has many advantages, including improvements in productivity, safety and quality. Installation of assemblies tends to be less technically challenging than more traditional construction methods, allowing successful incorporation of a larger proportion of lower skilled workers on the site. The supervisor must learn how to effectively employ lower skilled workers in mixed crews consisting of a few traditionally skilled craft workers and various degrees of lesser skilled workers.

Recent developments in the application of lean production methods to construction, as described in Chapter 15 are providing a much better understanding of the complex processes involved in construction. As supervisors understand the processes better, they can manage them more effectively. They can also find ways to improve the processes leading to opportunities to execute projects at lower cost, more quickly, with higher quality and fewer accidents.

The project delivery system, or the way construction projects are packaged has a profound impact on the supervisor's role. In the traditional design-bid-build approach, the supervisor has a completed design prior to starting the project. Few questions arise during the project, except those resulting from errors and omissions in the construction documents. However, at the same time, there is little latitude for innovations in the design by the supervisor. The supervisor's innovation is focused on finding the most efficient way to turn the design drawings and specifications into reality.

In addition to design-bid-build, there are many other ways to package a project, each with its own set of opportunities and pitfalls. Some of the most popular among these project delivery systems are design-build, agency construction management, and construction management at risk. New project delivery systems are being developed as creative ways are sought to better package projects, and also as technology advances. BIM technology has led to the development of integrated project delivery, which is in the early stages of development so that its impact on field supervision is not yet well understood.

Contractors are also finding ways to expand their involvement in projects such as participating in the financing, taking an ownership position in the project, or making long term maintenance and operation a part of the contract.

Supervisors rarely get the opportunity to choose, or even influence the choice of how the project is packaged, so it behooves supervisors to recognize that when a different project delivery system is used, field operations will be affected. It is important for the supervisor to learn as much as possible about the project delivery system for the specific project to which they are assigned and how best to supervise within the context of that specific project delivery system.

Perceptions of the Roles of Management and Supervision

The perception of the role of the supervisor is evolving. Years ago, supervisors were often known as "pushers." This term implies a role of forcing production from workers. Changes in what motivates workers, the way contractual relationships are laid out, and our perception of the treatment of workers have essentially eliminated the role of pusher. Supervisors are now considered leaders, motivators, and planners. Improvements in production come through innovations rather than pushing workers to work harder.

Relationships between management and supervision are also changing. In some companies, the project manager takes a strong role in running the project. In other companies, the project manager has many diverse responsibilities outside the project, such as managing other projects and estimating new projects, so that on a specific job, more responsibility is delegated to the supervisor.

Many general contractors are moving away from self-performing work to become brokers, who subcontract all the work to specialty contractors. The work of the supervisor for the general contractor on that type of project is coordination rather than detailed assignment-level planning. The preparation of such supervisors is moving away from gaining experience through craft training and hands-on work toward preparation through an educational or academic background. Supervisors on projects where all the work is performed by specialty contractors need a different skill set than the traditional supervisor involved on a day-to-day basis in work activities.

The result of these drivers is a construction industry that is constantly changing. In order to keep up with the changing industry, the supervisor must constantly update his or her own skill set, and be a life-time learner.

RECRUITING AND PREPARING CONSTRUCTION SUPERVISORS

As the construction industry changes, demands on supervisors are also changing, so the source of new supervisors and the preparation of supervisors is evolving. Construction supervisors have traditionally come out of the trades. Highly qualified journeymen are identified because they display well-developed craft skills, have a strong work ethic, and get along well with others. They also tend to have the interest of the company at heart, can communicate well, and demonstrate leadership capabilities. Their supervisor will bring them to the attention of management and suggest that their next assignment might be in a supervisory role.

Traditionally, little if any training has been provided for supervisors. It is expected that since the craft worker worked under a supervisor, they understand what a supervisor does. New supervisors may get a very brief overview of company requirements for supervision, such as company standard documentation and communication requirements for the home office, but little more. The new supervisor is dropped into the new supervisory position and given the opportunity to learn on the job. For some this works, for others, it soon becomes apparent that, though this person is a skilled craft worker, they are not able to make the transition into supervision and they are soon reassigned back to craft work. Others may like the idea of being a supervisor, but soon find that the expectations are too much and the rewards are too small to support their desire to be a supervisor. Without proper training, the dropout rate for new supervisors is unacceptably high.

As shortages continue to plague the construction workforce, shortages in well-qualified, highly skilled supervisors are even more acute. With the changes summarized in the previous section, the job of the supervisor is becoming more complex and demanding. The traditional means to obtain supervisors are no longer adequate to meet the demands. Thus, new methods are needed to recruit and prepare supervisors.

Certainly, the traditional approach of bring supervisors up from the craft ranks is still important. However, such promising young supervisors must be nurtured both to retain them as supervisors and to enable them to perform optimally as a supervisor. At the same time, the increasing demand for new, more highly skilled supervisors is driving the development of new means to recruit, develop, and retain qualified supervisors.

Recruitment can be improved by employers developing their own supervisory recruitment programs within the company. The position of supervisor needs to be well defined and then recognized as important and rewarded accordingly. Monetary rewards are limited, in many instances by contract, and in others by economics that do not allow supervisors to be paid significantly higher than craft workers while maintaining company competitiveness. Human resources experts can provide many ideas for non-monetary compensation for supervisors, such as recognition in company publications and at company get-togethers and small rewards such as gift certificates, badges, and other acknowledgments.

Many companies are starting recruitment at a much younger level by scrutinizing the ranks of apprentices to find those with strong supervisory potential. The companies then work out professional development plans for them that will lead them into supervision at an early stage of their career.

In construction companies engaged primarily in management of construction projects, rather than in self performing work, supervisors are being recruited from two-year, four-year, or even graduate academic programs. From these programs, construction companies are seeking graduates who would like a field, rather than an office orientation. They then work out a professional development plan, covering several years to transition the graduate from academic learning to a practical knowledge of the industry and a specific knowledge of the company.

Supervisory training is becoming much more readily available through trade associations and other educational organizations. Broad training programs are available that address many different topics important to the supervisor. However, no training program covers all topics at a comprehensive level. A broad training program can provide significant learning opportunities to both less experienced and more experienced supervisors at all levels from foreman to general superintendent. It can then be supplemented in specific areas of need either by the company or by the supervisor through any of a wide variety of shorter, more focused training programs. Examples of more focused training areas include construction safety, written and oral communication, and leadership. Many of these focused training areas are not construction-specific, but can be readily adapted to construction by the supervisor.

Rather than starting with a broad-based supervisory training program, supervisors can work out, with the company or on their own, a long-term learning experience comprising a number of specific training programs that will build strength and provide skills in specific areas. Many of these programs will be construction-oriented and perhaps even focused in a specific trade area. However, many supervisory skills, such as negotiating, communicating one on one, or technical writing, can be obtained from generic training programs with the specific focus desired.

STAYING AHEAD OF THE CURVE

As the role of the supervisor changes, the knowledge and skills of the supervisor must also change. This last section suggests ways in which the supervisor can evolve as the role of the supervisor changes. Four means will be reviewed that can enable the supervisor to keep up with change: networking, mentoring, training, and education.

Networking

Networking: The exchange of information or services among individuals, groups, or institutions; specifically: the cultivation of productive relationships for employment or business.

Merriam-Webster Online Dictionary

Networking is important to the supervisor because it is a powerful way in which to expand one's career. It provides an opportunity to learn in an informal environment. One can learn from other experienced supervisors about "tricks of the trade" that they have developed to facilitate their jobs or to avoid problems. Supervisors can also learn about problems experienced on other jobs that make them more aware of potential problems that may occur on their jobs. The supervisor also learns about the effects of problems and potential solutions that might be used on the supervisor's own job. It is far easier to learn from the experience of others rather than to have to encounter problems firsthand on one's own job and then try to rush to solve the problem without knowledge of the effects and potential solutions.

Networking creates contacts that can be used when difficult situations or problems arise on the job. It is important to be able to give a trusted colleague a call to discuss a situation that has just arisen that could have significant ramifications on your job. It is also helpful when dealing with difficult people to be able to call another supervisor to discuss how best to deal with that person.

The process of networking is not complex. It begins with meeting a broad cross-section of people. These people will have common interests but do not necessarily have to be supervisors or even employed within the construction industry. Supervisors from other industries deal with many of the same challenges with which construction supervisors deal. The supervisor can learn a great deal from people who are in other businesses and professions. For example, a human resources specialist in any type of business environment can help the supervisor work more effectively with various aspects of managing people. A lawyer can provide insight into contracts even if they are not specifically construction contracts. An accountant can offer an understanding of cash flow without specifically focusing on construction.

From among the broad pool of acquaintances, specific people are selected to get to know better. They will be perceived as someone who might be of help in some way, who has common interests and professional goals, and who shares a similar business ethic. It is very important that networking partners be able to share mutual respect. Also networking partners should be sought for whom the supervisor might be able to provide some value. The focus in networking is mutual, not unilateral.

Networking should take place both within and outside the company. Within the company, the supervisor should seek opportunities to get together with other supervisors. These could be supervisors at about the same level, but it is also valuable to interact with supervisors with more experience and also with those with less experience. Networking with less experienced supervisors is valuable because these young professionals often ask questions that the more experienced supervisor may have forgotten, leading the more experienced supervisor to look at problems from a different perspective than the one he or she has become accustomed to. Networking with experienced craft workers can also be beneficial. Craft workers often provide a perspective of the job different from that of the supervisor. Because they are actively involved in the work on a daily basis, they tend to see incipient problems that the supervisor has not yet seen, and they might have solutions that the supervisor has overlooked.

To develop a network within the company, get to know as many people on each job as possible and retain these acquaintances over the long term. Retain relationships developed on previous jobs with respected journeymen. Consider company events, whether business events or social events, as opportunities to expand the network. It might be possible to visit other company jobs, thus meeting new company employees, maintaining relationships previously developed, and also seeing another project.

Outside the company, supervisors should maintain contacts with craft workers they have met earlier in their career who have left the company. Participation in professional meetings where other supervisors with common interests are likely to participate also provides the opportunity to expand the network. As the supervisor moves up to higher levels, it might be appropriate to join a professional organization or become involved in a trade organization that is focused on the work of the supervisor. The Project Management Institute and the American Institute of Constructors are two professional organizations that might prove valuable. There are many trade associations that cater to specific specialty areas within the construction industry. These associations are constantly looking for volunteers to participate on committees focusing on areas of interest to the supervisor, such as workforce development and safety.

Mentoring

Mentor: A trusted counselor or guide, a tutor or coach.

Merriam-Webster Online Dictionary

Mentoring is a special type of networking in which the mentor is essentially a knowledgeable, often influential, individual who takes an interest in, and advises, another person concerning that person's career.

Mentor relationships tend not to be as mutual as networking relationships; however, the mentor does receive value. By mentoring younger professionals, the mentor can maintain contact with the ideas and culture of the newer generation of supervisors and can often be introduced to new technologies that longer-term professionals tend not to get involved with. Mentoring relationships certainly are of benefit to the company in preparing younger professionals for positions of higher responsibility. Mentoring relations also can be very fulfilling to the mentor who desires to give back to the profession that has served them well.

Training

Craft workers, supervisors and managers at all levels must participate in training from time to time. Training provides many benefits. It is essential to renew skills and knowledge learned earlier, especially if those skills or that knowledge have not been recently used. If one moves into a new position, or is preparing to move into a new position, training enables one to gain new knowledge and acquire new skills required for the new position. It has already been noted that the role of supervisors evolves as the industry changes, so training is required for supervisors to keep up with innovations in the industry and changes in supervision. Training is so important that most professional organizations have minimum requirements for training and continuing education to retain membership or certification.

Training is accessible through many sources. A great variety of training opportunities in the traditional classroom are available through many trade and professional organizations. More sophisticated companies will bring training programs, often tailored specifically for the company, into the company. Training is also available through many companies specializing in industry training. Training in the form of continuing education is also available through local colleges and universities.

Newer, nontraditional training opportunities have become abundantly available. Asynchronous, online training programs that can be accessed at any time and in any location with Internet access abound. There are also many live interactive training programs available online.

In assessing training programs, supervisors should look for a program that specifically addresses the area of learning sought. They should ensure, as much as possible, that the training is being provided by qualified facilitators and sponsored by recognized organizations. With training available in various formats, offered on a variety of schedules, supervisors should select the format and scheduling that meets their learning style and that is most convenient within the context of professional and personal commitments. Online learning has the advantage of convenience and flexibility. However, it requires a high level of self-discipline and the ability to learn independently. A live group setting is rigid in terms of scheduling, but it creates opportunities beyond just learning the topic, such as networking and the ability to ask questions and enter into discussions. Well-designed training works well within various formats. Supervisors need to select the format that works best for them.

Some training programs provide recognition at the end of the training experience. A certificate at the end of the program may recognize that the supervisor did participate in the learning experience. Certification is different from a certificate of participation or a certificate of completion. Certification certifies that the participant gained specific knowledge and/or skills. Certification requires testing throughout and/or at the end of the experience, demonstrating that specific learning objectives were met.

Education

The terms education and training are often used interchangeably, but they are significantly different. Training is focused on developing skills and capabilities. Education is more broadly focused on learning about the topic of interest, typically within a broad context. Certainly there is overlap because a participant in an educational program should gain skills and a participant in a training program learns a good deal about the topic. However, the objectives of training are different from those of education.

It is also important that participants in either training or educational programs understand the difference between a teaching and a learning environment. In a teaching environment, often used for young children in school, the teacher is presumed to have knowledge and skills that he or she wants to impart to the student, and it is the teacher's responsibility to make this happen. In a learning environment, which works far better for adult learners, the learning experience is under the leadership of a facilitator, not a teacher. The focus is on the learner rather than on the facilitator, and it is the learners' responsibility to determine what they want from the program and then to shape the program to meet their learning objectives. The learning experience can be shaped by the learner through interaction, by asking questions, and by entering into discussion. Supervisors should approach their learning opportunities (whether education or training) as active participants who take responsibility for their own learning.

Educational programs tend to lead to academic degrees. These are typically associated with an educational institution and the program of study is generally spread over a significant amount of time, such as 2-years, 4-years, or longer for graduate degrees.

Numerous construction trade associations and construction professional organizations can be found through a computer search on such websites as http://www.constructionweblinks.com/. This will provide access to organizations through which a broad selection of training is available as well as significant opportunities for networking.

SUMMARY

In this chapter, the following key points have been presented:

- The role of the supervisor is constantly changing.
- Changing supervisory roles reflect industry changes.
- Industry change is the result of a number of drivers.
- Changing supervisory roles also reflect changes in culture and in business practices.
- New supervisors need to be recruited from a variety of sources
- Development of new supervisors makes them more effective and stems the rapid loss of supervisors.

- To maintain and enhance value as a supervisor, a long-term professional development plan should be implemented.
- The supervisor's professional development should include networking, mentoring, training, and education.

Learning Activities

1. Developing a Personal Professional Development Plan

   The objective of this activity is to write a personal professional development plan tailored to the individual. Start by briefly defining your current job and the next job you would like. (For example, I am a foreman and would like to be a general foreman on my next assignment). Next, define the supervisory skills that you feel could be improved, both to help you in your present job and to better prepare you for the next step. Finally, identify at least one opportunity to develop a networking (or mentoring) relationship that can build strength in an identified area, and identify at least one training program that you would like to participate in to develop a new skill or strengthen an existing skill. If you would like to develop a more comprehensive plan, repeat the process identifying various networking relationships and training programs. To enhance the value of this plan, make it a continuous process by reevaluating the plan from time to time to see if you are improving your supervisory skills and whether the plan needs to be modified, expanded, or realigned based upon where you are at the time of reevaluation.

2. Developing a Training Agenda

   The objective of this activity is to plan your own training program to strengthen your supervisory skills. Start by doing a self-analysis, identifying and prioritizing supervisory skills you would like to gain or to strengthen. Pick one of these skills and do research to find available training programs focusing on this skill. One source of information might be training opportunities offered by local chapters of trade and professional organizations. Another source might be local academic or training institutions. A third source might involve a web search of training in the skill area you have chosen. Once training opportunities have been identified, evaluate the value of an opportunity to you and select the best value opportunity. You might ask such questions as: Does this focus specifically on my area of interest? Does the provider have appropriate credentials? Does the delivery format fit my best learning style? Is this program accessible in terms of timing and location? Will the company sponsor me in this learning endeavor, or do I have the means to pay for it? To develop a more comprehensive training agenda, repeat the process for various skills and then develop a potential schedule to obtain the training.

APPENDIX 1

CONSENSUSDOCS 750

STANDARD FORM OF AGREEMENT BETWEEN CONTRACTOR AND SUBCONTRACTOR

This document was developed through a collaborative effort of entities representing a wide cross-section of the construction industry. The organizations endorsing this document believe it represents a fair and reasonable consensus among the collaborating parties of allocation of risk and responsibilities in an effort to appropriately balance the critical interests and concerns of all project participants.

These endorsing organizations recognize and understand that users of this document must review and adapt this document to meet their particular needs, the specific requirements of the project, and applicable laws. Users are encouraged to consult legal, insurance and surety advisors before modifying or completing this document. Further information on this document and the perspectives of endorsing organizations is available in the ConsensusDOCS Guidebook.

TABLE OF ARTICLES

1. AGREEMENT
2. SCOPE OF WORK
3. SUBCONTRACTOR'S RESPONSIBILITIES
4. CONTRACTOR'S RESPONSIBILITIES
5. PROGRESS SCHEDULE

6. SUBCONTRACT AMOUNT

7. CHANGES IN THE SUBCONTRACT WORK

8. PAYMENT

9. INDEMNITY, INSURANCE AND WAIVER OF SUBROGATION

10. CONTRACTOR'S RIGHT TO PERFORM SUBCONTRACTOR'S RESPONSI-
BILITIES AND TERMINATION OF AGREEMENT

11. DISPUTE RESOLUTION

12. MISCELLANEOUS PROVISIONS

13. EXISTING SUBCONTRACT DOCUMENTS

This Agreement has important legal and insurance consequences. Consulta-
tions with an attorney and with insurance and surety consultants are encouraged
with respect to its completion or modification. Notes indicate where information is
to be inserted to complete this Agreement.

ARTICLE 1: AGREEMENT

This Agreement is made this _____ Day of
_____ in the year _____, by and between the
CONTRACTOR

and the SUBCONTRACTOR

for services in connection with the SUBCONTRACT WORK

for the following PROJECT

whose OWNER is

The ARCHITECT/ENGINEER for the Project is

Notice to the Parties shall be given at the above addresses.

ARTICLE 2: SCOPE OF WORK

2.1 SUBCONTRACT WORK   The Contractor contracts with the Subcontractor as
an independent contractor to provide all labor, materials, equipment and services
necessary or incidental to complete the work for the project described in Article
1 and as may be set forth in further detail in Exhibit A, in accordance with, and
reasonably inferable from, that which is indicated in the Subcontract Documents,
and consistent with the Progress Schedule, as may change from time to time. The
Subcontractor shall perform the Subcontract Work under the general direction of
the Contractor and in accordance with the Subcontract Documents.

2.2 CONTRACTOR'S WORK   The Contractor's Work is the construction and
services required of the Contractor to fulfill its obligations pursuant to its agreement
with the Owner (the Work). The Subcontract Work is a portion of the Contractor's
Work.

2.2.1 The Contractor and the Subcontractor shall perform their obligations with integrity, ensuring at a minimum that:

2.2.1.1 Conflicts of interest shall be avoided or disclosed promptly to the other Party; and

2.2.1.2 The Contractor and the Subcontractor warrant that they have not and shall not pay nor receive any contingent fees or gratuities to or from the other Party, including their agents, officers and employees, Subcontractors or others for whom they may be liable, to secure preferential treatment.

2.3 SUBCONTRACT DOCUMENTS   The Subcontract Documents include this Agreement, the Owner-Contractor agreement, special conditions, general conditions, specifications, drawings, addenda, Subcontract Change Orders, approved submittals, amendments and any pending and exercised alternates. The Contractor shall provide to the Subcontractor, prior to the execution of this Agreement, copies of the existing Subcontract Documents to which the Subcontractor will be bound. The Subcontractor similarly shall provide copies of applicable portions of the Subcontract Documents to its proposed subcontractors and suppliers. Nothing shall prohibit the Subcontractor from obtaining copies of the Subcontract Documents from the Contractor at any time after the Subcontract Agreement is executed. The Subcontract Documents existing at the time of the execution of this Agreement are listed in Article 13.

2.3.1 ELECTRONIC DOCUMENTS   If the Owner requires that the Owner, Architect/Engineer, Contractor and Subcontractors exchange documents and data in electronic or digital form, prior to any such exchange, the Owner, Architect/Engineer and Contractor shall agree in ConsensusDOCS 200.2 or a written protocol governing all exchanges, which, at a minimum, shall specify: (1) the definition of documents and data to be accepted in electronic or digital form or to be transmitted electronically or digitally; (2) management and coordination responsibilities; (3) necessary equipment, software and services; (4) acceptable formats, transmission methods and verification procedures; (5) methods for maintaining version control; (6) privacy and security requirements; and (7) storage and retrieval requirements. The Subcontractor shall provide whatever input is needed to assist the Contractor in developing the protocol and shall be bound by the requirements of the written protocol. Except as otherwise agreed to by the Parties in writing, the Parties shall each bear their own costs as identified in the protocol. In the absence of a written protocol, use of documents and data in electronic or digital form shall be at the sole risk of the recipient.

2.4 CONFLICTS   In the event of a conflict between this Agreement and the other Subcontract Documents, this Agreement shall govern.

2.5 EXTENT OF AGREEMENT   Nothing in this Agreement shall be construed to create a contractual relationship between persons or entities other than the Contractor and Subcontractor. This Agreement is solely for the benefit of the Parties, represents the entire and integrated agreement between the Parties, and supersedes all prior negotiations, representations, or agreements, either written or oral.

2.6 DEFINITIONS

2.6.1 Wherever the term Progress Schedule is used in this Agreement, it shall be read as Project Schedule when that term is used in the Subcontract Documents.

2.6.2 Whenever the term Change Order is used in this Agreement, it shall be read as Change Document when that term is used in the Subcontract Documents.

2.6.3 Unless otherwise indicated, the term Day shall mean calendar day.

ARTICLE 3: SUBCONTRACTOR'S RESPONSIBILITIES

3.1 OBLIGATIONS   The Contractor and Subcontractor are hereby mutually bound by the terms of this Agreement. To the extent the terms of the Owner-Contractor agreement apply to the Subcontract Work, then the Contractor hereby assumes toward the Subcontractor all the obligations, rights, duties, and redress that the Owner under the prime contract assumes toward the Contractor. In an identical way, the Subcontractor hereby assumes toward the Contractor all the same obligations, rights, duties, and redress that the Contractor assumes toward the Owner and Architect/Engineer under the prime contract. In the event of an inconsistency among the documents, the specific terms of this Agreement shall govern.

3.2 RESPONSIBILITIES   The Subcontractor agrees to furnish its diligent efforts and judgment in the performance of the Subcontract Work and to cooperate with the Contractor so that the Contractor may fulfill its obligations to the Owner. The Subcontractor shall furnish all of the labor, materials, equipment, and services, including but not limited to, competent supervision, shop drawings, samples, tools, and scaffolding as are necessary for the proper performance of the Subcontract Work. The Subcontractor shall provide the Contractor a list of its proposed subcontractors and suppliers, and be responsible for taking field dimensions, providing tests, obtaining required permits related to the Subcontract Work and affidavits, ordering of materials and all other actions as required to meet the Progress Schedule.

3.3 INCONSISTENCIES AND OMISSIONS   The Subcontractor shall make a careful analysis and comparison of the drawings, specifications, other Subcontract Documents and information furnished by the Owner relative to the Subcontract Work. Such analysis and comparison shall be solely for the purpose of facilitating the Subcontract Work and not for the discovery of errors, inconsistencies or omissions in the Subcontract Documents nor for ascertaining if the Subcontract Documents are in accordance with applicable laws, statutes, ordinances, building codes, rules or regulations. Should the Subcontractor discover any errors, inconsistencies or omissions in the Subcontract Documents, the Subcontractor shall report such discoveries to the Contractor in writing within three (3) Days. Upon receipt of notice, the Contractor shall instruct the Subcontractor as to the measures to be taken, and the Subcontractor shall comply with the Contractor's instructions. If the Subcontractor performs work knowing it to be contrary to any applicable laws, statutes, ordinances, building codes, rules or regulations without notice to the Contractor and advance approval by appropriate authorities, including the Contractor, the Subcontractor shall assume appropriate responsibility for such work and shall bear all associated costs, charges, fees and expenses necessarily incurred to remedy the violation. Nothing in this paragraph shall relieve the Subcontractor of responsibility for its own errors, inconsistencies and omissions.

3.4 SITE VISITATION   Prior to performing any portion of the Subcontract Work, the Subcontractor shall conduct a visual inspection of the Project site to become generally familiar with local conditions and to correlate site observations with the Subcontract Documents. If the Subcontractor discovers any discrepancies between its site observations and the Subcontract Documents, such discrepancies shall be promptly reported to the Contractor.

3.5 INCREASED COSTS OR TIME   The Subcontractor may assert a Claim as provided in Article 7 if Contractor's clarifications or instructions in responses to requests for information are believed to require additional time or cost. If the Subcontractor fails to perform the reviews and comparisons required in Paragraphs 3.3 and 3.4, above, to the extent the Contractor is held liable to the Owner because of the Subcontractor's failure, the Subcontractor shall pay the costs and damages to the Contractor that would have been avoided if the Subcontractor had performed those obligations.

3.6 COMMUNICATIONS   Unless otherwise provided in the Subcontract Documents and except for emergencies, Subcontractor shall direct all communications related to the Project to the Contractor.

3.7 SUBMITTALS

3.7.1 The Subcontractor promptly shall submit for approval to the Contractor all shop drawings, samples, product data, manufacturers' literature and similar submittals required by the Subcontract Documents. Submittals shall be submitted in electronic form if required in accordance with Subparagraph 2.3.1. The Subcontractor shall be responsible to the Contractor for the accuracy and conformity of its submittals to the Subcontract Documents. The Subcontractor shall prepare and deliver its submittals to the Contractor in a manner consistent with the Progress Schedule and in such time and sequence so as not to delay the Contractor or others in the performance of the Work. The approval of any Subcontractor submittal shall not be deemed to authorize deviations, substitutions or changes in the requirements of the Subcontract Documents unless express written approval is obtained from the Contractor and Owner authorizing such deviation, substitution or change. Such approval shall be promptly memorialized in a Subcontract Change Order with in seven (7) Days following approval by the Contractor and, if applicable, provide for an adjustment in the Subcontract Amount or Subcontract Time. In the event that the Subcontract Documents do not contain submittal requirements pertaining to the Subcontract Work, the Subcontractor agrees upon request to submit in a timely fashion to the Contractor for approval any shop drawings, samples, product data, manufacturers' literature or similar submittals as may reasonably be required by the Contractor, Owner or Architect/Engineer.

3.7.2 The Contractor, Owner, and Architect/Engineer are entitled to rely on the adequacy, accuracy and completeness of any professional certifications required by the Subcontract Documents concerning the performance criteria of systems, equipment or materials, including all relevant calculations and any governing performance requirements.

3.8 DESIGN DELEGATION

3.8.1 If the Subcontract Documents (1) specifically require the Subcontractor to procure design services and (2) specify all design and performance criteria, the Subcontractor shall provide those design services necessary to satisfactorily complete the Subcontract Work. Design services provided by the Subcontractor shall be procured from licensed design professionals retained by the Subcontractor as permitted by the law of the place where the Project is located (the Designer). The Designer's signature and seal shall appear on all drawings, calculations, specifications, certifications, Shop Drawings and other submittals prepared by the Designer. Shop Drawings and other submittals related to the Subcontract Work designed or certified by the Designer, if prepared by others, shall bear the Subcontractor's and the Designer's written approvals when submitted to the Contractor. The Contractor shall be entitled to rely upon the adequacy, accuracy and completeness of the services, certifications or approvals performed by the Designer.

3.8.2 If the Designer is an independent professional, the design services shall be procured pursuant to a separate agreement between the Subcontractor and the Designer. The Subcontractor- Designer agreement shall not provide for any limitation of liability, except to the extent that consequential damages are waived pursuant to Subparagraph 5.4.1, or exclusion from participation in the multiparty proceedings requirement of Paragraph 11.4. The Designer(s) is (are) _____, The Subcontractor shall notify the Contractor in writing if it intends to change the Designer. The Subcontractor shall be responsible for conformance of its design with the information given and the design concept expressed in the Subcontract Documents. The Subcontractor shall not be responsible for the adequacy of the performance or design criteria required by the Subcontract Documents.

3.8.3 The Subcontractor shall not be required to provide design services in violation of any applicable law.

3.9 TEMPORARY SERVICES   The Subcontractor's and Contractor's respective responsibilities for temporary services are set forth in Exhibit _____.

3.10 COORDINATION   The Subcontractor shall:

3.10.1 cooperate with the Contractor and all others whose work may interface with the Subcontract Work;

3.10.2 specifically note and immediately advise the Contractor of any such interface with the Subcontract Work; and

3.10.3 participate in the preparation of coordination drawings and work schedules in areas of congestion.

3.11 SUBCONTRACTOR'S REPRESENTATIVE   The Subcontractor shall designate a person, subject to Contractor's approval, who shall be the Subcontractor's authorized representative. This representative shall be the only person to whom the Contractor shall issue instructions, orders or directions, except in an emergency. The Subcontractor's representative is _____, who is agreed to by the Contractor.

3.12 TESTS AND INSPECTIONS   The Subcontractor shall schedule all required tests, approvals and inspections of the Subcontract Work at appropriate times so as not to delay the progress of the work. The Subcontractor shall give proper written notice to all required Parties of such tests, approvals and inspections. Except as otherwise provided in the Subcontract Documents the Subcontractor shall bear all expenses associated with tests, inspections and approvals required of the Subcontractor by the Subcontract Documents which, unless otherwise agreed to, shall be conducted by an independent testing laboratory or entity approved by the Contractor and Owner. Required certificates of testing, approval or inspection shall, unless otherwise required by the Subcontract Documents, be secured by the Subcontractor and promptly delivered to the Contractor.

3.13 CLEANUP

3.13.1 The Subcontractor shall at all times during its performance of the Subcontract Work keep the Work site clean and free from debris resulting from the Subcontract Work. Prior to discontinuing the Subcontract Work in an area, the Subcontractor shall clean the area and remove all its rubbish and its construction equipment, tools, machinery, waste and surplus materials. Subcontractor shall make provisions to minimize and confine dust and debris resulting from its construction activities. The Subcontractor shall not be held responsible for unclean conditions caused by others.

3.13.2 If the Subcontractor fails to commence compliance with cleanup duties within two (2) business Days after written notification from the Contractor of noncompliance, the Contractor may implement appropriate cleanup measures without further notice and the cost thereof shall be deducted from any amounts due or to become due the Subcontractor in the next payment period.

3.14 SAFETY

3.14.1 The Subcontractor is required to perform the Subcontract Work in a safe and reasonable manner. The Subcontractor shall seek to avoid injury, loss or damage to persons or property by taking reasonable steps to protect:

3.14.1.1 Employees and other persons at the site;

3.14.1.2 Materials and equipment stored at the site or at off-site locations for use in performance of the Subcontract Work; and

3.14.1.3 All property and structures located at the site and adjacent to work areas, whether or not said property or structures are part of the Project or involved in the Work.

3.14.2 The Subcontractor shall give all required notices and comply with all applicable rules, regulations, orders and other lawful requirements established to prevent injury, loss or damage to persons or property.

3.14.3 The Subcontractor shall implement appropriate safety measures pertaining to the Subcontract Work and the Project, including establishing safety rules, posting appropriate warnings and notices, erecting safety barriers, and establishing proper notice procedures to protect persons and property at the site and adjacent to the site from injury, loss or damage.

3.14.4 The Subcontractor shall exercise extreme care in carrying out any of the Subcontract Work which involves explosive or other dangerous methods of construction or hazardous procedures, materials or equipment. The Subcontractor shall use properly qualified individuals or entities to carry out the Subcontract Work in a safe and reasonable manner so as to reduce the risk of bodily injury or property damage.

3.14.5 Damage or loss not insured under property insurance and to the extent caused by the negligent acts or omissions of the Subcontractor, or anyone for whose acts the Subcontractor may be liable, shall be promptly remedied by the Subcontractor. Damage or loss to the extent caused by the negligent acts or omissions of the Contractor, or anyone for whose acts the Contractor may be liable, shall be promptly remedied by the Contractor.

3.14.6 The Subcontractor is required to designate an individual at the site in the employ of the Subcontractor who shall act as the Subcontractor's designated safety representative with a duty to prevent accidents. Unless otherwise identified by the Subcontractor in writing to the Contractor, the designated safety representative shall be the Subcontractor's project superintendent. Such safety representative shall attend site safety meetings as requested by the Contractor.

3.14.7 The Subcontractor has an affirmative duty not to overload the structures or conditions at the site and shall take reasonable steps not to load any part of the structures, or site so as to give rise to an unsafe condition or create an unreasonable risk of bodily injury or property damage. The Subcontractor shall have the right to request, in writing, from the Contractor loading information concerning the structures at the site.

3.14.8 The Subcontractor shall give prompt written notice to the Contractor of any accident involving bodily injury requiring a physician's care, any property damage exceeding Five Hundred Dollars ($500.00) in value, or any failure that could have resulted in serious bodily injury, whether or not such an injury was sustained.

3.14.9 Prevention of accidents at the site is the responsibility of the Contractor, Subcontractor, and all other subcontractors, persons and entities at the site. Establishment of a safety program by the Contractor shall not relieve the Subcontractor or other Parties of their safety responsibilities. The Subcontractor shall establish its own safety program implementing safety measures, policies and standards conforming to those required or recommended by governmental and quasi-governmental authorities having jurisdiction and by the Contractor and Owner, including, but not limited to, requirements imposed by the Subcontract Documents. The Subcontractor shall comply with the reasonable recommendations of insurance companies having an interest in the Project, and shall stop any part of the Subcontract Work which the Contractor deems unsafe until corrective measures satisfactory to the Contractor shall have been taken. The Contractor's failure to stop the Subcontractor's unsafe practices shall not relieve the Subcontractor of the responsibility therefor. The Subcontractor shall notify the Contractor immediately following a reportable incident under applicable rules, regulations, orders and other lawful requirements, and promptly confirm the notice in writing. A detailed written report shall be furnished if requested by the Contractor. To the fullest extent permitted by law, each Party to this Agreement shall indemnify the other party from and against fines or penalties imposed as a result of safety violations, but only to the extent that such fines or penalties are caused by its failure to comply with applicable safety requirements.

This indemnification obligation does not extend to additional or increased fines that result from repeated or willful violations not caused by the Subcontractor's failure to comply with applicable rules, regulations, orders and other lawful requirements.

3.15 PROTECTION OF THE WORK   The Subcontractor shall take necessary precautions to properly protect the Subcontract Work and the work of others from damage caused by the Subcontractor's operations. Should the Subcontractor cause damage to the Work or property of the Owner, the Contractor or others, the Subcontractor shall promptly remedy such damage to the satisfaction of the Contractor, or the Contractor may, after forty-eight (48) hours written notice to the Subcontractor, remedy the damage and deduct its cost from any amounts due or to become due the Subcontractor, unless such costs are recovered under applicable property insurance.

3.16 PERMITS, FEES, LICENSES AND TAXES   The Subcontractor shall give timely notices to authorities pertaining to the Subcontract Work, and shall be responsible for all permits, fees, licenses, assessments, inspections, testing and taxes necessary to complete the Subcontract Work in accordance with the Subcontract Documents. To the extent reimbursement is obtained by the Contractor from the Owner under the Owner-Contractor agreement, the Subcontractor shall be compensated for additional costs resulting from taxes enacted after the date of this Agreement.

3.17 ASSIGNMENT OF SUBCONTRACT WORK   The Subcontractor shall neither assign the whole nor any part of the Subcontract Work without prior written approval of the Contractor.

3.18 HAZARDOUS MATERIALS   To the extent that the Contractor has rights or obligations under the Owner-Contractor agreement or by law regarding hazardous materials as defined by the Subcontract Document within the scope of the Subcontract Work, the Subcontractor shall have the same rights or obligations.

3.19 MATERIAL SAFETY DATA (MSD) SHEETS   The Subcontractor shall submit to the Contractor all Material Safety Data Sheets required by law for materials or substances necessary for the performance of the Subcontract Work. MSD sheets obtained by the Contractor from other subcontractors or sources shall be made available to the Subcontractor by the Contractor.

3.20 LAYOUT RESPONSIBILITY AND LEVELS   The Contractor shall establish principal axis lines of the building and site, and benchmarks. The Subcontractor shall lay out and be strictly responsible for the accuracy of the Subcontract Work and for any loss or damage to the Contractor or others by reason of the Subcontractor's failure to lay out or perform Subcontract Work correctly. The Subcontractor shall exercise prudence so that the actual final conditions and details shall result in alignment of finish surfaces.

3.21 WARRANTIES   The Subcontractor warrants that all materials and equipment shall be new unless otherwise specified, of good quality, in conformance with the Subcontract Documents, and free from defective workmanship and materials. The Subcontractor further warrants that the Work shall be free from material defects not intrinsic in the design or materials required in the Subcontract Documents. The Subcontractor's warranty does include remedies for defects or damages caused by normal wear and tear during normal usage, use for a purpose for which the Project was not intended, improper or insufficient maintenance, modifications performed by Others, or abuse. The Subcontractor's warranties shall commence on the date of Substantial Completion of the Work or a designated portion.

3.22 UNCOVERING/CORRECTION OF SUBCONTRACT WORK

3.22.1 UNCOVERING OF SUBCONTRACT WORK

3.22.1.1 If required in writing by the Contractor, the Subcontractor must uncover any portion of the subcontract Work which has been covered by the Subcontractor in violation of the Subcontract documents or contrary to a directive issued to the Subcontractor by the Contractor. Upon receipt of a written directive from the Contractor, the Subcontractor shall uncover such work for the Contractor's or Owner's inspection and restore the uncovered Subcontract Work to its original condition at the Subcontractor's time and expense.

3.22.1.2 The Contractor may direct the Subcontractor to uncover portions of the Subcontract Work for inspection by the Owner or Contractor at any time. The Subcontractor is required to uncover such work whether or not the Contractor or Owner had requested to inspect the Subcontract Work prior to it being covered. Except as provided in Subparagraph 3.22.1.1, this Agreement shall be adjusted by change order for the cost and time of uncovering and restoring any work which is uncovered for inspection and proves to be installed in accordance with the Subcontract Documents, provided the Contractor had not previously instructed the Subcontractor to leave the work uncovered. If the Subcontractor uncovers work pursuant to a directive issued by the Contractor, and such work upon inspection does not comply with the Subcontract Documents, the Subcontractor shall be responsible for all costs and time of uncovering, correcting and restoring the work so as to make it conform to the Subcontract Documents. If the Contractor or some other entity for which the Subcontractor is not responsible caused the nonconforming condition, the Contractor shall be required to adjust this Agreement by change order for all such costs and time.

3.22.2 CORRECTION OF WORK

3.22.2.1 If the Architect/Engineer or Contractor rejects the Subcontract Work or the Subcontract Work is not in conformance with the Subcontract Documents, the Subcontractor shall promptly correct the Subcontract Work whether it had been fabricated, installed or completed. The Subcontractor shall be responsible for the costs of correcting such Subcontract Work, any additional testing, inspections, and compensation for services and expenses of the Architect/Engineer and Contractor made necessary by the defective Subcontract Work.

3.22.2.2 In addition to the Subcontractor's obligations under Paragraph 3.21, the Subcontractor agrees to promptly correct, after receipt of a written notice from the Contractor, all Subcontract Work performed under this Agreement which proves to be defective in workmanship or materials within a period of one year from the date of Substantial Completion of the Subcontract Work or for a longer period of time as may be required by specific warranties in the Subcontract Documents. Substantial Completion of the Subcontract Work, or of a designated portion, occurs on the date when construction is sufficiently complete in accordance with the Subcontract Documents so that the Owner can occupy or utilize the Project, or a designated portion, for the use for which it is intended. If, during the one-year period, the Contractor fails to provide the Subcontractor with prompt written notice of the discovery of defective or nonconforming Subcontract Work, the Contractor shall neither have the right to require the Subcontractor to correct such Subcontract Work nor the right to make claim for breach of warranty. If the Subcontractor fails to correct defective or nonconforming Subcontract Work within a reasonable time after receipt of notice from the Contractor, the Contractor may correct such Subcontract Work pursuant to Subparagraph 10.1.1.

3.22.3 The Subcontractor's correction of Subcontract Work pursuant to this Paragraph 3.22 shall not extend the one-year period for the correction of Subcontract Work, but if Subcontract Work is first performed after Substantial Completion, the one-year period for corrections shall be extended by the time period after Substantial Completion and the performance of that portion of Subcontract Work. The Subcontractor's obligation to correct Subcontract Work within one year as described in this Paragraph 3.22 does not limit the enforcement of Subcontractor's other obligations with regard to the Agreement and the Subcontract Documents.

3.22.4 If the Subcontractor's correction or removal of Subcontract Work destroys or damages completed or partially completed work of the Owner, the Contractor or any separate contractors or subcontractors, the Subcontractor shall be responsible for the reasonable cost of correcting such destroyed or damaged property.

3.22.5 If portions of Subcontract Work which do not conform with the requirements of the Subcontract Documents are neither corrected by the Subcontractor nor accepted by the Contractor, the Subcontractor shall remove such Subcontract Work from the Project site if so directed by the Contractor.

3.23 MATERIALS OR EQUIPMENT FURNISHED BY OTHERS   In the event the scope of the Subcontract Work includes installation of materials or equipment furnished by others, it shall be the responsibility of the Subcontractor to exercise proper care in receiving, handling, storing and installing such items, unless otherwise provided in the Subcontract Documents. The Subcontractor shall examine the items provided and report to the Contractor in writing any items it may discover that do not conform to requirements of the Subcontract Documents. The Subcontractor shall not proceed to install non-conforming items without further instructions from the Contractor. Loss or damage due to acts or omissions of the Subcontractor shall, upon two (2) business Days written notice to the Subcontractor be deducted from any amounts due or to become due the Subcontractor.

3.24 SUBSTITUTIONS   No substitutions shall be made in the Subcontract Work unless permitted in the Subcontract Documents, and only upon the Subcontractor first receiving all approvals required under the Subcontract Documents for substitutions.

3.25 USE OF CONTRACTOR'S EQUIPMENT   The Subcontractor, its agents, employees, subcontractors or suppliers shall use the Contractor's equipment only with the express written permission of the Contractor's designated representative and in accordance with the Contractor's terms and conditions for such use. If the Subcontractor or any of its agents, employees, subcontractors or suppliers utilize any of the Contractor's equipment, including machinery, tools, scaffolding, hoists, lifts or similar items owned, leased or under the control of the Contractor, the Subcontractor shall indemnify and be liable to the Contractor as provided in Article 9 for any loss or damage (including bodily injury or death) which may arise from such use, except to the extent that such loss or damage is caused by the negligence of the Contractor's employees operating the Contractor's equipment.

3.26 WORK FOR OTHERS   Until final completion of the Subcontract Work, the Subcontractor agrees not to perform any work directly for the Owner or any tenants, or deal directly with the Owner's representatives in connection with the Subcontract Work, unless otherwise approved in writing by the Contractor.

3.27 SYSTEMS AND EQUIPMENT STARTUP   With the assistance of the Owner's maintenance personnel and the Contractor, the Subcontractor shall direct the check-out and operation of systems and equipment for readiness, and assist in their initial startup and the testing of the Subcontract Work.

3.28 COMPLIANCE WITH LAWS   The Subcontractor agrees to be bound by, and at its own costs comply with, all federal, state and local laws, ordinances and regulations (the Laws) applicable to the Subcontract Work, including but not limited to, equal employment opportunity, minority business enterprise, women's business enterprise, disadvantaged business enterprise, safety and all other Laws with which the Contractor must comply. The Subcontractor shall be liable to the Contractor and the Owner for all loss, cost and expense attributable to any acts of commission or omission by the Subcontractor, its employees and agents resulting from the failure to comply with Laws, including, but not limited to, any fines, penalties or corrective measures, except as provided in Subparagraph 3.14.9.

3.29 CONFIDENTIALITY   To the extent the Owner-Contractor agreement provides for the confidentiality of any of the Owner's proprietary or otherwise confidential information disclosed in connection with the performance of this Agreement, the Subcontractor is equally bound by the Owner's confidentiality requirements.

3.30 ROYALTIES, PATENTS AND COPYRIGHTS   The Subcontractor shall pay all royalties and license fees which may be due on the inclusion of any patented or copyrighted materials, methods or systems selected by the Subcontractor and incorporated in the Subcontract Work. The Subcontractor shall defend, indemnify and hold the Contractor and Owner harmless from all suits or claims for infringement of any patent rights or copyrights arising out of such selection. The Subcontractor shall be liable for all loss, including all costs, expenses, and attorneys' fees, but shall not be responsible for such defense or loss when a particular design, process or product of a particular manufacturer or manufacturers is required by the Subcontract Documents. However, if the Subcontractor has reason to believe that a particular design, process or product required by the Subcontract Documents is an infringement of a patent, the Subcontractor shall promptly furnish such information to the Contractor or be responsible to the contractor and Owner for any loss sustained as a result.

3.31 LABOR RELATIONS   (Insert here any conditions, obligations or requirements relative to labor relations and their effect on the project. Legal counsel is recommended.)

ARTICLE 4: CONTRACTOR'S RESPONSIBILITIES

4.1 CONTRACTOR'S REPRESENTATIVE   The Contractor shall designate a person who shall be the Contractor's authorized representative. The Contractor's representative shall be the only person the Subcontractor shall look to for instructions, orders or directions, except in an emergency. The Contractor's Representative is

_____

4.2 OWNER'S ABILITY TO PAY

4.2.1 The Subcontractor shall have the right upon request to receive from the Contractor such information as the Contractor has obtained relative to the Owner's financial ability to pay for the Work, including any subsequent material variation in such information. The Contractor, however, does not warrant the accuracy or completeness of the information provided by the Owner.

4.2.2 If the Subcontractor does not receive the information referenced in Subparagraph 4.2.1 with regard to the Owner's ability to pay for the Work as required by the Contract Documents, the Subcontractor may request the information from the Owner or the Owner's lender.

4.3 CONTRACTOR APPLICATION FOR PAYMENT   Upon request, the Contractor shall give the Subcontractor a copy of the most current Contractor application for payment reflecting the amounts approved or paid by the Owner for the Subcontract Work performed to date.

4.4 INFORMATION OR SERVICES   The Subcontractor is entitled to request through the Contractor any information or services relevant to the performance of the Subcontract Work which is under the Owner's control. The Subcontractor also is entitled to request through the Contractor any information necessary to give notice of or enforce mechanics lien rights and, where applicable, stop notices. This information shall include the Owner's interest in the real property on which the Project is located and the recorded legal title. To the extent the Contractor receives such information and services, the Contractor shall provide them to the Subcontractor. The Contractor, however, does not warrant the accuracy or completeness of the information provided by the Owner. To the extent the Owner provides any warranty of Owner provided information, the Contractor agrees to permit the Subcontractor to prosecute a claim in the name of the Contractor for the use and benefit of the Subcontractor, pursuant to Subparagraph 5.3.2.

4.5 STORAGE AREAS   The Contractor shall allocate adequate storage areas, if available, for the Subcontractor's materials and equipment during the course of the Subcontract Work. Unless otherwise agreed upon, the Contractor shall reimburse the Subcontractor for the additional costs of having to relocate such storage areas at the direction of the Contractor.

4.6 TIMELY COMMUNICATIONS   The Contractor shall transmit to the Subcontractor, with reasonable promptness, all submittals, transmittals, and written approvals relative to the Subcontract Work. Unless otherwise specified in the Subcontract Documents, communications by and with the Subcontractor's subcontractors, materialmen and suppliers shall be through the Subcontractor.

4.7 USE OF SUBCONTRACTOR'S EQUIPMENT   The Contractor, its agents, employees or suppliers shall use the Subcontractor's equipment only with the express written permission of the Subcontractor's designated representative and in accordance with the Subcontractor's terms and conditions for such use. If the Contractor or any of its agents, employees or suppliers utilize any of the Subcontractor's equipment, including machinery, tools, scaffolding, hoists, lifts or similar items owned, leased or under the control of the Subcontractor, the Contractor shall indemnify and be liable to the Subcontractor as provided in Article 9 for any loss or damage (including bodily injury or death) which may arise from such use, except to the extent that such loss or damage is caused by the negligence of the Subcontractor's employees operating the Subcontractor's equipment.

ARTICLE 5: PROGRESS SCHEDULE

5.1 TIME IS OF THE ESSENCE   Time is of the essence for both Parties. They mutually agree to see to the performance of their respective obligations so that the entire Project may be completed in accordance with the Subcontract Documents and particularly the Progress Schedule as set forth in Exhibit _____.

5.2 SCHEDULE OBLIGATIONS   The Subcontractor shall provide the Contractor with any scheduling information proposed by the Subcontractor for the Subcontract Work. In consultation with the Subcontractor, the Contractor shall prepare the schedule for performance of the Work (the Progress Schedule) and shall revise and update such schedule, as necessary, as the Work progresses. Both the Contractor and the Subcontractor shall be bound by the Progress Schedule. The Progress Schedule and all subsequent changes and additional details shall be submitted to the Subcontractor promptly and reasonably in advance of the required performance. The Contractor shall have the right to determine and, if necessary, change the time, order and priority in which the various portions of the Work shall be performed and all other matters relative to the Subcontract Work. To the extent such changes increase Subcontractor's time and costs, the Subcontract Amount and Subcontract Time shall be equitably adjusted.

5.3 DELAYS AND EXTENSIONS OF TIME

5.3.1 OWNER CAUSED DELAY   Subject to Subparagraph 5.3.2, if the commencement or progress of the Subcontract Work is delayed without the fault or responsibility of the Subcontractor, the time for the Subcontract Work shall be extended by Subcontract Change Order and the Subcontract Price equitably adjusted to the extent obtained by the Contractor under the Subcontract Documents, and the Progress Schedule shall be revised accordingly.

5.3.2 CLAIMS RELATING TO OWNER   The Subcontractor agrees to initiate all claims for which the Owner is or may be liable in the manner and within the time limits provided in the Subcontract Documents for like claims by the Contractor upon the Owner and in sufficient time for the Contractor to initiate such claims against the Owner in accordance with the Subcontract Documents. At the Subcontractor's request and expense to the extent agreed upon in writing, the Contractor agrees to permit the Subcontractor to prosecute a claim in the name of the Contractor for the use and benefit of the Subcontractor in the manner provided in the Subcontract Documents for like claims by the Contractor upon the Owner.

5.3.3 CONTRACTOR CAUSED DELAY   Nothing in this Article shall preclude the Subcontractor's recovery of delay damages caused by the Contractor to the extent not otherwise precluded by this Agreement.

5.3.4 CLAIMS RELATING TO CONTRACTOR   The Subcontractor shall give the Contractor written notice of all claims not included in Subparagraph 5.3.2 within fourteen (14) Days of the Subcontractor's knowledge of the facts giving rise to the event for which claim is made. Thereafter, the Subcontractor shall submit written documentation of its claim, including appropriate supporting documentation, within twenty-one (21) Days after giving notice, unless the Parties agree upon a longer period of time. The Contractor shall respond in writing denying or approving, in whole or in part the Subcontractor's claim no later than fourteen (14) Days after receipt of the Subcontractor's documentation of claim. All unresolved claims, disputes and other matters in question between the Contractor and the Subcontractor not relating to claims included in Subparagraph 5.3.2 shall be resolved in the manner provided in Article 11.

5.4 LIMITED MUTUAL WAIVER OF CONSEQUENTIAL DAMAGES

5.4.1 Except for damages provided for by the Subcontract Documents as liquidated damages and excluding losses covered by insurance required by the Subcontract Documents, the Contractor and Subcontractor waive claims against each other for consequential damages arising out of or relating to this Agreement, to the same extent the Owner-Contractor agreement furnished to the Subcontractor in accordance with Paragraph 2.3 provides for a mutual waiver of consequential damages by the Owner and Contractor, including to the extent provided in the Owner-Contractor agreement, damages for loss of business, loss of financing, principal office overhead and expenses, loss of profits not related to this Project, loss of bonding capacity, loss of reputation, or insolvency. Similarly, the Subcontractor shall obtain in another agreement from its Sub-Subcontractors mutual waivers of consequential damages that correspond to the Subcontractor's waiver of consequential damages herein. To the extent applicable, this mutual waiver applies to consequential damages due to termination by the Contractor or the Owner in accordance with this Agreement or the Owner-Contractor agreement. The provisions of this Article shall also apply to and survive termination of this Agreement.

5.5 LIQUIDATED DAMAGES

5.5.1 If the Subcontract Documents furnished to the Subcontractor in accordance with Paragraph 2.3 provide for liquidated damages or other damages for delay beyond the completion date set forth in the Subcontract Documents that are not specifically addressed as a liquidated damage item in this Agreement, and such damages are assessed, the Contractor may assess a share of the damages against the Subcontractor in proportion to the Subcontractor's share of the responsibility for the damages. However, the amount of such assessment shall not exceed the amount assessed against the Contractor. This Paragraph shall not limit the

Subcontractor's liability to the Contractor for the Contractor's actual damages caused by the Subcontractor.

5.5.2 To the extent the Owner-Contractor Agreement provides for a mutual waiver of consequential damages by the Owner and the Contractor, damages for which the Contractor is liable to the Owner including those related to Paragraph 9.1. are not consequential damages for the purpose of this waiver. Similarly, to the extent the Subcontractor-Sub-Subcontractor agreement provides for a mutual waiver of consequential damages by the Owner and the Contractor, damages for which the Subcontractor is liable to lower-tiered parties due to the fault of the Owner or Contractor are not consequential damages for the purpose of this waiver.

ARTICLE 6: SUBCONTRACT AMOUNT

As full compensation for performance of this Agreement, Contractor agrees to pay Subcontractor in current funds for the satisfactory performance of the Subcontract Work subject to all applicable provisions of the Subcontract:

(a) the fixed-price of _____ Dollars ($_____) subject to additions and deductions as provided for in the Subcontract Documents; or (b) alternates and unit prices in accordance with the attached schedule of Alternates and Unit Prices and estimated quantities, which is incorporated by reference and identified as Exhibit _____; or (c) time and material rates and prices in accordance with the attached Schedule of Labor and Material Costs which is incorporated by reference and identified as Exhibit _____. The fixed-price, unit prices or time and material rates and prices are referred to as the Subcontract Amount.

ARTICLE 7: CHANGES IN THE SUBCONTRACT WORK

7.1 SUBCONTRACT CHANGE ORDERS   When the Contractor orders in writing, the Subcontractor, without nullifying this Agreement, shall make any and all changes in the Subcontract Work which are within the general scope of this Agreement. Any adjustment in the Subcontract Amount or Subcontract Time shall be authorized by a Subcontract Change Order. No adjustments shall be made for any changes performed by the Subcontractor that have not been ordered by the Contractor. A Subcontract Change Order is a written instrument prepared by the Contractor and signed by the Subcontractor stating their agreement upon the change in the Subcontract Work.

7.2 CONSTRUCTION CHANGE DIRECTIVES   To the extent that the Subcontract Documents provide for Construction Change Directives in the absence of agreement on the terms of a Subcontract Change Order, the Subcontractor shall promptly comply with the Construction Change Directive and be entitled to apply for interim payment if the Subcontract Documents so provide.

7.3 UNKNOWN CONDITIONS   If in the performance of the Subcontract Work the Subcontractor finds latent, concealed or subsurface physical conditions which differ materially from those indicated in the Subcontract Documents or unknown physical conditions of an unusual nature, which differ materially from those ordinarily found to exist, and not generally recognized as inherent in the kind of work provided for in this Agreement, the Subcontract Amount or the Progress Schedule shall be equitably adjusted by a Subcontract Change Order within a reasonable time after the conditions are first observed. The adjustment which the Subcontractor may receive shall be limited to the adjustment the Contractor receives from the Owner on behalf of the Subcontractor, or as otherwise provided under Subparagraph 5.3.2.

7.4 ADJUSTMENTS IN SUBCONTRACT AMOUNT   If a Subcontract Change Order requires an adjustment in the Subcontract Amount, the adjustment shall be established by one of the following methods:

7.4.1 mutual acceptance of an itemized lump sum;

7.4.2 unit prices as indicated in the Subcontract Documents or as subsequently agreed to by the Parties; or

7.4.3 costs determined in a manner acceptable to the Parties and a mutually acceptable fixed or percentage fee; or

7.4.4 another method provided in the Subcontract Documents.

7.5 SUBSTANTIATION OF ADJUSTMENT   If the Subcontractor does not respond promptly or disputes the method of adjustment, the method and the adjustment shall be determined by the Contractor on the basis of reasonable expenditures and savings of those performing the Work attributable to the change, including, in the case of an increase in the Subcontract Amount, an allowance for overhead and profit of the percentage provided in Paragraph 7.6, or if none is provided as mutually agreed upon by the Parties. The Subcontractor may contest the reasonableness of any adjustment determined by the Contractor. The Subcontractor shall maintain for the Contractor's review and approval an appropriately itemized and substantiated accounting of the following items attributable to the Subcontract Change Order:

7.5.1 labor costs, including Social Security, health, welfare, retirement and other fringe benefits as normally required, and state workers' compensation insurance;

7.5.2 costs of materials, supplies and equipment, whether incorporated in the Subcontract Work or consumed, including transportation costs;

7.5.3 costs of renting machinery and equipment other than hand tools;

7.5.4 costs of bond and insurance premiums, permit fees and taxes attributable to the change; and

7.5.5 costs of additional supervision and field office personnel services necessitated by the change.

7.6 Adjustments shall be based on net change in Subcontractor's reasonable cost of performing the changed Subcontract Work plus, in case of a net increase in cost, an agreed upon sum for overhead and profit not to exceed _____ percent (_____ %).

7.7 NO OBLIGATION TO PERFORM   The Subcontractor shall not perform changes in the Subcontract Work until a Subcontract Change Order has been executed or written instructions have been issued in accordance with Paragraphs 7.2 and 7.9.

7.8 EMERGENCIES   In an emergency affecting the safety of persons or property, the Subcontractor shall act, at its discretion, to prevent threatened damage, injury or loss. Any change in the Subcontract Amount or the Progress Schedule on account of emergency work shall be determined as provided in this Article.

7.9 INCIDENTAL CHANGES   The Contractor may direct the Subcontractor to perform incidental changes in the Subcontract Work which do not involve adjustments in the Subcontract Amount or Subcontract Time. Incidental changes shall be consistent with the scope and intent of the Subcontract Documents. The Contractor shall initiate an incidental change in the Subcontract Work by issuing a written order to the Subcontractor. Such written notice shall be carried out promptly and are binding on the Parties.

ARTICLE 8: PAYMENT

8.1 SCHEDULE OF VALUES   As a condition to payment, the Subcontractor shall provide a schedule of values satisfactory to the Contractor not more than fifteen (15) Days from the date of execution of this Agreement.

8.2 PROGRESS PAYMENTS

8.2.1 APPLICATIONS   The Subcontractor's applications for payment shall be itemized and supported by substantiating data as required by the Subcontract Documents. If the Subcontractor is obligated to provide design services pursuant to Paragraph 3.8, Subcontractor's applications for payment shall show the Designer's fee and expenses as a separate cost item. The Subcontractor's application shall be notarized if required and if allowed under the Subcontract Documents may include properly authorized Subcontract Construction Change Directives. The Subcontractor's progress payment application for the Subcontract Work performed in the preceding payment period shall be submitted for approval of the Contractor in accordance with the schedule of values if required and Subparagraphs 8.2.2, 8.2.3, and 8.2.4. The Contractor shall incorporate the approved amount of the Subcontractor's progress payment application into the Contractor's payment application to the Owner for the same period and submit it to the Owner in a timely fashion. The Contractor shall immediately notify the Subcontractor of any changes in the amount requested on behalf of the Subcontractor.

8.2.2 RETAINAGE   The rate of retainage shall be _____ percent (_____ %), which is equal to the percentage retained from the Contractor's payment by the Owner for the Subcontract Work. If the Subcontract Work is satisfactory and the Subcontract Documents provide for reduction of retainage at a specified percentage of completion, the Subcontractor's retainage shall also be reduced when the Subcontract Work has attained the same percentage of completion and the Contractor's retainage for the Subcontract Work has been so reduced by the Owner.

8.2.3 TIME OF APPLICATION   The Subcontractor shall submit progress payment applications to the Contractor no later than the _____ Day of each payment period for the Subcontract Work performed up to and including the _____ Day of the payment period indicating work completed and, to the extent allowed under Subparagraph 8.2.4, materials suitably stored during the preceding payment period.

8.2.4 STORED MATERIALS   Unless otherwise provided in the Subcontract Documents, applications for payment may include materials and equipment not yet incorporated in the Subcontract Work but delivered to and suitably stored on-site or off-site including applicable insurance, storage and costs incurred transporting the materials to an off-site storage facility. Approval of payment applications for such stored items on or off the site shall be conditioned upon submission by the Subcontractor of bills of sale and required insurance or such other procedures satisfactory to the Owner and Contractor to establish the Owner's title to such materials and equipment, or otherwise to protect the Owner's and Contractor's interest including transportation to the site.

8.2.5 TIME OF PAYMENT   Progress payments to the Subcontractor for satisfactory performance of the Subcontract Work shall be made no later than seven (7) Days after receipt by the Contractor of payment from the Owner for the Subcontract Work. If payment from the Owner for such Subcontract Work is not received by the Contractor, through no fault of the Subcontractor, the Contractor will make payment to the Subcontractor within a reasonable time for the Subcontract Work satisfactorily performed.

8.2.6 PAYMENT DELAY   If the Contractor has received payment from the Owner and if for any reason not the fault of the Subcontractor, the Subcontractor does not receive a progress payment from the Contractor within seven (7) Days after the date such payment is due, as defined in Subparagraph 8.2.5, or, if the Contractor has failed to pay the Subcontractor within a reasonable time for the Subcontract Work satisfactorily performed, the Subcontractor, upon giving seven (7) Days' written notice to the Contractor, and without prejudice to and in addition to any other legal remedies, may stop work until payment of the full amount owing to the Subcontractor has been received. The Subcontract Amount and Time shall be adjusted by the amount of the Subcontractor's reasonable and verified cost of shutdown, delay, and startup, which shall be effected by an appropriate Subcontractor Change Order.

8.2.7 PAYMENTS WITHHELD   The Contractor may reject a Subcontractor payment application in whole or in part or withhold amounts from a previously approved Subcontractor payment application, as may reasonably be necessary to protect the Contractor from loss or damage for which the Contractor may be liable and without incurring an obligation for late payment interest based upon:

8.2.7.1 the Subcontractor's repeated failure to perform the Subcontract Work as required by this Agreement;

8.2.7.2 loss or damage arising out of or relating to this Agreement and caused by the Subcontractor to the Owner, Contractor or others to whom the Contractor may be liable;

8.2.7.3 the Subcontractor's failure to properly pay for labor, materials, equipment or supplies furnished in connection with the Subcontract Work;

8.2.7.4 rejected, nonconforming or defective Subcontract Work which has not been corrected in a timely fashion;

8.2.7.5 reasonable evidence of delay in performance of the Subcontract Work such that the Work will not be completed within the Subcontract Time, and that the unpaid balance of the Subcontract Amount is not sufficient to offset the liquidated damages or actual damages that may be sustained by the Contractor as a result of the anticipated delay caused by the Subcontractor;

8.2.7.6 reasonable evidence demonstrating that the unpaid balance of the Sub-contract Amount is insufficient to cover the cost to complete the Subcontract Work;

8.2.7.7 third party claims involving the Subcontractor or reasonable evidence demonstrating that third party claims are likely to be filed unless and until the Sub-contractor furnishes the Contractor with adequate security in the form of a surety bond, letter of credit or other collateral or commitment which are sufficient to dis-charge such claims if established. No later than seven (7) Days after receipt of an application for payment, the Contractor shall give written notice to the Sub-contractor, at the time of disapproving or nullifying all or part of an application for payment, stating its specific reasons for such disapproval or nullification, and the remedial actions to be taken by the Subcontractor in order to receive payment. When the above reasons for disapproving or nullifying an application for payment are removed, payment will be promptly made for the amount previously withheld.

8.3 FINAL PAYMENT

8.3.1 APPLICATION   Upon acceptance of the Subcontract Work by the Owner and the Contractor and receipt from the Subcontractor of evidence of fulfillment of the Subcontractor's obligations in accordance with the Subcontract Documents and Subparagraph 8.3.2, the Contractor shall incorporate the subcontractor's ap-plication for final payment into the Contractor's next application for payment to the Owner without delay, or notify the Subcontractor if there is a delay and the reasons therefor.

8.3.2 REQUIREMENTS   Before the Contractor shall be required to incorporate the Subcontractor's application for final payment into the Contractor's next applica-tion for payment, the Subcontractor shall submit to the Contractor:

8.3.2.1 an affidavit that all payrolls, bills for materials and equipment, and other indebtedness connected with the Subcontract Work for which the Owner or its property or the Contractor or the Contractor's surety might in any way be liable, have been paid or otherwise satisfied;

8.3.2.2 consent of surety to final payment, if required;

8.3.2.3 satisfaction of required closeout procedures;

8.3.2.4 other data, if required by the Contractor or Owner, such as receipts, releases, and waivers of liens to the extent and in such form as may be required by the Subcontract Documents;

8.3.2.5 written warranties, equipment manuals, startup and testing required in Paragraph 3.28; and

8.3.2.6 as-built drawings if required by the Subcontract Documents.

8.3.3 TIME OF PAYMENT   Final payment of the balance due of the Subcontract Amount shall be made to the Subcontractor within seven (7) Days after receipt by the Contractor of final payment from the Owner for such Subcontract Work.

8.3.4 FINAL PAYMENT DELAY   If the Owner or its designated agent does not issue a certificate for final payment or the Contractor does not receive such payment for any cause which is not the fault of the Subcontractor, the Contractor shall promptly inform the Subcontractor in writing. The Contractor shall also diligently pursue, with the assistance of the Subcontractor, the prompt release by the Owner of the final payment due for the Subcontract Work. At the Subcontractor's request and expense, to the extent agreed upon in writing, the Contractor shall institute reasonable legal remedies to mitigate the damages and pursue payment of the Subcontractor's final payment including interest. If final payment from the Owner for such Subcontract Work is not received by the Contractor, through no fault of the Subcontractor, the Contractor will make payment to the Subcontractor within a reasonable time.

8.3.5 WAIVER OF CLAIMS   Final payment shall constitute a waiver of all claims by the Subcontractor relating to the Subcontract Work, but shall in no way relieve the Subcontractor of liability for the obligations assumed under Paragraphs 3.21 and 3.22, or for faulty or defective work or services discovered after final payment, nor relieve the Contractor for claims made in writing by the Subcontractor as required by the Subcontract Documents prior to its application for final payment as unsettled at the time of such payment.

8.4 LATE PAYMENT INTEREST   Progress payments or final payment due and unpaid under this Agreement, as defined in Subparagraphs 8.2.5, 8.3.3 and 8.3.4, shall bear interest from the date payment is due at the prevailing Statutory rate at the place of the Project. However, if the Owner fails to timely pay the Contractor as required under the Owner-Contractor agreement through no fault or neglect of the Contractor, and the Contractor fails to timely pay the Subcontractor as a result of such nonpayment, the Contractor's obligation to pay the Subcontractor interest on corresponding payments due and unpaid under this Agreement shall be extinguished by the Contractor promptly paying to the Subcontractor the Subcontractor's proportionate share of the interest, if any, received by the Contractor from the Owner on such late payments.

8.5 CONTINUING OBLIGATIONS   Provided the Contractor is making payments on or has made payments to the Subcontractor in accordance with the terms of this Agreement, the Subcontractor shall reimburse the Contractor for any costs and expenses for any claim, obligation or lien asserted before or after final payment is made that arises from the performance of the Subcontract Work. The Subcontractor shall reimburse the Contractor for costs and expenses including attorneys' fees and costs and expenses incurred by the Contractor in satisfying, discharging or defending against any such claims, obligation or lien including any action brought or judgment recovered. In the event that any applicable law, statute, regulation or bond requires the Subcontractor to take any action prior to the expiration of the reasonable time for payment referenced in Subparagraph 8.2.5 in order to preserve or protect the Subcontractor's rights, if any, with respect to mechanic's lien or bond claims, then the Subcontractor may take that action prior to the expiration of the reasonable time for payment and such action will not create the reimbursement obligation recited above nor be in violation of this Agreement or considered premature for purposes of preserving and protecting the Subcontractor's rights.

8.6 PAYMENT USE RESTRICTION   Payments received by the Subcontractor shall be used to satisfy the indebtedness owed by the Subcontractor to any person furnishing labor or materials, or both, for use in performing the Subcontract Work through the most current period applicable to progress payments received from the Contractor before it is used for any other purpose. In the same manner, payments received by the Contractor from the Owner for the Subcontract Work shall be dedicated to payment to the Subcontractor. This provision shall bear on this Agreement only, and is not for the benefit of third parties. Moreover, it shall not be construed by the Parties to this Agreement or third parties to require that dedicated sums of money or payments be deposited in separate accounts, or that there be other restrictions on commingling of funds. Neither shall these mutual covenants be construed to create any fiduciary duty on the Subcontractor or Contractor, nor create any tort cause of action or liability for breach of trust, punitive damages, or other equitable remedy or liability for alleged breach.

8.7 PAYMENT USE VERIFICATION   If the Contractor has reason to believe that the Subcontractor is not complying with the payment terms of this Agreement, the Contractor shall have the right to contact the Subcontractor's subcontractors and suppliers to ascertain whether they are being paid by the Subcontractor in accordance with this Agreement.

8.8 PARTIAL LIEN WAIVERS AND AFFIDAVITS   As a prerequisite for payments, the Subcontractor shall provide, in a form satisfactory to the Owner and Contractor, partial lien or claim waivers in the amount of the application for payment and affidavits covering its subcontractors and suppliers for completed Subcontract Work. Such waivers may be conditional upon payment. In no event shall Contractor require the Subcontractor to provide an unconditional waiver of lien or claim, either partial or final, prior to receiving payment or in an amount in excess of what it has been paid.

8.9 SUBCONTRACTOR PAYMENT FAILURE   Upon payment by the Contractor, the Subcontractor shall promptly pay its subcontractors and suppliers the amounts to which they are entitled. In the event the Contractor has reason to believe that labor, material or other obligations incurred in the performance of the Subcontract Work are not being paid, the Contractor may give written notice of a potential claim or lien to the Subcontractor and may take any steps deemed necessary to assure that progress payments are utilized to pay such obligations, including but not limited to the issuance of joint checks. If upon receipt of notice, the Subcontractor does not (a) supply evidence to the satisfaction of the Contractor that the moneys owing have been paid; or (b) post a bond indemnifying the Owner, the Contractor, the Contractor's surety, if any, and the premises from a claim or lien, the Contractor shall have the right to withhold from any payments due or to become due to the Subcontractor a reasonable amount to protect the Contractor from any and all loss, damage or expense including attorneys' fees that may arise out of or relate to any such claim or lien.

8.10 SUBCONTRACTOR ASSIGNMENT OF PAYMENTS   The Subcontractor shall not assign any moneys due or to become due under this Agreement, without the written consent of the Contractor, unless the assignment is intended to create a new security interest within the scope of Article 9 of the Uniform Commercial Code. Should the Subcontractor assign all or any part of any moneys due or to become due under this Agreement to create a new security interest or for any other purpose, the instrument of assignment shall contain a clause to the effect that the assignee's right in and to any money due or to become due to the Subcontractor shall be subject to the claims of all persons, firms and corporations for services rendered or materials supplied for the performance of the Subcontract Work.

8.11 PAYMENT NOT ACCEPTANCE   Payment to the Subcontractor does not constitute or imply acceptance of any portion of the Subcontract Work.

ARTICLE 9: INDEMNITY, INSURANCE AND WAIVER OF SUBROGATION

9.1 INDEMNITY

9.1.1 INDEMNITY   To the fullest extent permitted by law, the Subcontractor shall indemnify and hold harmless the Contractor, Architect/Engineer, the Owner and their agents, consultants and employees (the Indemnitees) from all claims for bodily injury and property damage other than to the Work itself that may arise from the performance of the Subcontract Work, including reasonable attorneys' fees, costs and expenses, that arise from the performance of the Work, but only to the extent caused by the negligent acts or omissions of the Subcontractor, the Subcontractor's Sub-Subcontractors or anyone employed directly or indirectly by any of them or by anyone for whose acts any of them may be liable. The Subcontractor shall be entitled to reimbursement of any defense cost paid above Subcontractor's percentage of liability for the underlying claim to the extent attributable to the negligent acts or omissions of the Indemnitees.

9.1.2 NO LIMITATION ON LIABILITY   In any and all claims against the Indemnitees by any employee of the Subcontractor, anyone directly or indirectly employed by the Subcontractor or anyone for whose acts the Subcontractor may be liable, the indemnification obligation shall not be limited in any way by any limitation on the amount or type of damages, compensation or benefits payable by or for the Subcontractor under workers' compensation acts, disability benefit acts or other employee benefit acts.

9.2 INSURANCE

9.2.1 SUBCONTRACTOR'S INSURANCE   Before commencing the Subcontract Work, and as a condition of payment, the Subcontractor shall purchase and maintain insurance that will protect it from the claims arising out of its operations under this Agreement, whether the operations are by the Subcontractor, or any of its consultants or subcontractors or anyone directly or indirectly employed by any of them, or by anyone for whose acts any of them may be liable.

9.2.2 MINIMUM LIMITS OF LIABILITY   The Subcontractor shall procure and maintain with insurance companies licensed in a the jurisdiction in which the Project is located and acceptable to the Contractor, which acceptance shall not be unreasonably withheld, at least the limits of liability as set forth in Exhibit _____.

9.2.3 PROFESSIONAL LIABILITY INSURANCE

9.2.3.1 PROFESSIONAL LIABILITY INSURANCE   The Subcontractor shall require the Designer(s) to maintain Professional Liability Insurance with a company reasonably satisfactory to the Contractor, including contractual liability insurance against the liability assumed in Paragraph 3.8, and including coverage for any professional liability caused by any of the Designer's(s') consultants. Said insurance shall have specific minimum limits as set forth below:

Limit of $_____ per claim. General Aggregate of $_____ for the subcontract services rendered. The Professional Liability Insurance shall contain prior acts coverage sufficient to cover all subcontract services rendered by the Designer. Said insurance shall be continued in effect with an extended period of _____ years following final payment to the Designer. Such insurance shall have a maximum deductible amount of $_____ per occurrence. The deductible shall be paid by the Subcontractor or Designer.

9.2.3.2 The Subcontractor shall require the Designer to furnish to the Subcontractor and Contractor, before the Designer commences its services, a copy of its professional liability policy evidencing the coverages required in this Paragraph. No policy shall be cancelled or modified without thirty (30) Days' prior written notice to the Subcontractor and Contractor.

9.2.4 NUMBER OF POLICIES   Commercial General Liability Insurance and other liability insurance may be arranged under a single policy for the full limits required or by a combination of underlying policies with the balance provided by an Excess or Umbrella Liability Policy.

9.2.5 CANCELLATION, RENEWAL AND MODIFICATION   The Subcontractor shall maintain in effect all insurance coverages required under this Agreement at the Subcontractor's sole expense and with insurance companies acceptable to the Contractor, which acceptance shall not be unreasonably withheld. The policies shall contain a provision that coverage will not be cancelled or not renewed until at least thirty (30) Days' prior written notice has been given to the Contractor. Certificates of insurance showing required coverage to be in force pursuant to Subparagraph 9.2.2 shall be filed with the Contractor prior to commencement of the Subcontract Work. In the event the Subcontractor fails to obtain or maintain any insurance coverage required under this Agreement, the Contractor may purchase such coverage as desired for the Contractor's benefit and charge the expense to the Subcontractor, or terminate this Agreement.

9.2.6 CONTINUATION OF COVERAGE   The Subcontractor shall continue to carry Completed Operations Liability Insurance for at least one year after either ninety (90) Days following Substantial Completion of the Work or final payment to the Contractor, whichever is earlier. Prior to commencement of the Work, Subcontractor shall furnish the Contractor with certificates evidencing the required coverages.

9.2.7 PROPERTY INSURANCE

9.2.7.1 Upon written request of the Subcontractor, the Contractor shall provide the Subcontractor with a copy of the Builder's Risk Policy of insurance or any other property or equipment insurance in force for the Project and procured by the Owner or Contractor. The Contractor shall advise the Subcontractor if a Builder's Risk Policy of insurance is not in force.

9.2.7.2 If the Owner or Contractor has not purchased property insurance reasonably satisfactory to the Subcontractor, the Subcontractor may procure such insurance as will protect the interests of the Subcontractor, its subcontractors and their subcontractors in the Subcontract Work. The cost of this insurance shall be charged to the Contractor in a Change Order.

9.2.7.3 If not covered under the Builder's Risk Policy of insurance or any other property or equipment insurance required by the Subcontract Documents, the Subcontractor shall procure and maintain at the Subcontractor's own expense property and equipment insurance for the Subcontract Work including portions of the Subcontract Work stored off the site or in transit, when such portions of the Subcontract Work are to be included in an application for payment under Article 8.

9.2.8 WAIVER OF SUBROGATION

9.2.8.1 The Contractor and Subcontractor waive all rights against each other, the Owner and the Architect/Engineer, and any of their respective consultants, subcontractors, and sub-subcontractors, agents and employees, for damages caused by perils to the extent covered by the proceeds of the insurance provided in Subparagraph 9.2.7, except such rights as they may have to the insurance proceeds. The Subcontractor shall require similar waivers from its subcontractors.

9.2.9 ENDORSEMENT   If the policies of insurance referred to in this Article require an endorsement to provide for continued coverage where there is a waiver of subrogation, the owners of such policies will cause them to be so endorsed.

9.2.10 CONTRACTOR'S LIABILITY INSURANCE   The Contractor shall obtain and maintain its own liability insurance for protection against claims arising out of the performance of this Agreement, including without limitation, loss of use and claims, losses and expenses arising out of the Contractor's errors or omissions.

9.2.11 ADDITIONAL LIABILITY COVERAGE   Contractor _____ shall/_____ shall not (indicate one) require Subcontractor to purchase and maintain liability coverage, primary to Contractor's coverage under Subparagraph 9.2.10.

9.2.11.1 If required by Subparagraph 9.2.11, the additional liability coverage required of the Subcontractor shall be:

[Designate Required Coverage(s)]

————.1 ADDITIONAL INSURED.   Contractor shall be named as an additional insured on Subcontractor's Commercial General Liability Insurance specified, for operations and completed operations, but only with respect to liability for bodily injury, property damage or personal and advertising injury to the extent caused by the negligent acts or omissions of Subcontractor, or those acting on Subcontractor's behalf, in the performance of Subcontract Work for Contractor at the Project site.

———— .2 OCP. Subcontractor shall provide an Owners' and Contractors' Protective Liability Insurance ("OCP") policy with limits equal to the limits on Commercial General Liability Insurance specified, or limits as otherwise required by Contractor. Any documented additional cost in the form of a surcharge associated with procuring the additional liability coverage in accordance with this Subparagraph shall be paid by the Contractor directly or the costs may be reimbursed by Contractor to Subcontractor by increasing the Subcontract Amount to correspond to the actual cost required to purchase and maintain the additional liability coverage. Prior to commencement of the Subcontract Work, Subcontractor shall obtain and furnish to the Contractor a certificate evidencing that the additional liability coverages have been procured.

9.2.12 RISK OF LOSS Except to the extent a loss is covered by applicable insurance, risk of loss or damage to the Subcontract Work shall be upon the Subcontractor until the Date of Substantial Completion, unless otherwise agreed to by the Parties.

9.3 BONDS

9.3.1 The Subcontractor _____shall/_____ shall not furnish to the Contractor, as the named Obligee, appropriate surety bonds to secure the faithful performance of the Subcontract Work and to satisfy all Subcontractor payment obligations related to Subcontract Work. Such bonds shall be issued by a surety admitted in the State in which the Project is located and shall be acceptable to the Contractor. Contractor's acceptance shall not be withheld without reasonable cause.

9.3.2 If a performance or payment bond, or both, is required of the Subcontractor under this Agreement, the bonds shall be in a form and by a surety acceptable to the Contractor, and in the full amount of the Subcontract Amount, unless otherwise specified. Contractor's acceptance shall not be withheld without reasonable cause.

9.3.3 The Subcontractor shall be reimbursed, without retainage, for the cost of any required performance or payment bonds simultaneously with the first progress payment. The reimbursement amount for the Subcontractor bonds shall be _____ percent (_____%) of the Subcontract Amount, which sum is included in the Subcontract Amount. If acceptable to the Contractor, the Subcontractor may in lieu of retainage, furnish a retention bond or other security interest, acceptable to the Contractor, to be held by the Contractor.

9.3.4 In the event the Subcontractor shall fail to promptly provide any required bonds, the Contractor may terminate this Agreement and enter into a subcontract for the balance of the Subcontract Work with another subcontractor. All Contractor costs and expenses incurred by the Contractor as a result of said termination shall be paid by the Subcontractor.

9.3.5 PAYMENT BOND REVIEW   The Contractor _____has/_____ has not provided the Owner a payment bond. The Contractor's payment bond for the Project, if any, shall be made available by the Contractor for review and copying by the Subcontractor.

ARTICLE 10:CONTRACTOR'S RIGHT TO PERFORM SUBCONTRACTOR'S RESPONSIBILITIES AND TERMINATION OF AGREEMENT

10.1 FAILURE OF PERFORMANCE

10.1.1 NOTICE TO CURE   If the Subcontractor refuses or fails to supply enough properly qualified workers, proper materials, or maintain the Progress Schedule, or fails to make prompt payment to its workers, subcontractors or suppliers, or disregards laws, ordinances, rules, regulations or orders of any public authority having jurisdiction, or otherwise is guilty of a material breach of a provision of this Agreement, the Subcontractor shall be deemed in default of this Agreement. If the Subcontractor fails within three (3) business Days after written notification to commence and continue satisfactory correction of the default with diligence and promptness, then the Contractor without prejudice to any other rights or remedies, shall have the right to any or all of the following remedies:

10.1.1.1 supply workers, materials, equipment and facilities as the Contractor deems necessary for the completion of the Subcontract Work or any part which the Subcontractor has failed to complete or perform after written notification, and charge the cost, including reasonable overhead, profit, attorneys' fees, costs and expenses to the Subcontractor;

10.1.1.2 contract with one or more additional contractors to perform such part of the Subcontract Work as the Contractor determines will provide the most expeditious completion of the Work, and charge the cost to the Subcontractor as provided under Clause 10.1.1.1; or

10.1.1.3 withhold any payments due or to become due the Subcontractor pending corrective action in amounts sufficient to cover losses and compel performance to the extent required by and to the satisfaction of the Contractor. In the event of an emergency affecting the safety of persons or property, the Contractor may proceed as above without notice, but the Contractor shall give the Subcontractor notice promptly after the fact as a precondition of cost recovery.

10.1.2 TERMINATION BY CONTRACTOR   If the Subcontractor fails to commence and satisfactorily continue correction of a default within three (3) business Days after written notification issued under Subparagraph 10.1.1, then the Contractor may, in lieu of or in addition to the remedies provided for in Subparagraph 10.1.1, issue a second written notification, to the Subcontractor and its surety, if any. Such notice shall state that if the Subcontractor fails to commence and continue correction of a default within seven (7) Days of the written notification, the Agreement will be deemed terminated. A written notice of termination shall be issued by the Contractor to the Subcontractor at the time the Subcontractor is terminated. The Contractor may furnish those materials, equipment or employ such workers or subcontractors as the Contractor deems necessary to maintain the orderly progress of the Work. All costs incurred by the Contractor in performing the Subcontract Work, including reasonable overhead, profit and attorneys' fees, costs and expenses, shall be deducted from any moneys due or to become due the Subcontractor. The Subcontractor shall be liable for the payment of any amount by which such expense may exceed the unpaid balance of the Subcontract Amount. At the Subcontractor's request, the Contractor shall provide a detailed accounting of the costs to finish the Subcontract Work.

10.1.3 USE OF SUBCONTRACTOR'S EQUIPMENT   If the Contractor performs work under this Article, either directly or through other subcontractors, the Contractor or other subcontractors shall have the right to take and use any materials, implements, equipment, appliances or tools furnished by, or belonging to the Subcontractor and located at the Project site for the purpose of completing any remaining Subcontract Work. Immediately upon completion of the Subcontract Work, any remaining materials, implements, equipment, appliances or tools not consumed or incorporated in performance of the Subcontract Work, and furnished by, belonging to, or delivered to the Project by or on behalf of the Subcontractor, shall be returned to the Subcontractor in substantially the same condition as when they were taken, normal wear and tear excepted.

10.2. BANKRUPTCY

10.2.1 TERMINATION ABSENT CURE   If the Subcontractor files a petition under the Bankruptcy Code, this Agreement shall terminate if the Subcontractor or the Subcontractor's trustee rejects the Agreement or, if there has been a default, the Subcontractor is unable to give adequate assurance that the Subcontractor will perform as required by this Agreement or otherwise is unable to comply with the requirements for assuming this Agreement under the applicable provisions of the Bankruptcy Code.

10.2.2 INTERIM REMEDIES   If the Subcontractor is not performing in accordance with the Progress Schedule at the time a petition in bankruptcy is filed, or at any subsequent time, the Contractor, while awaiting the decision of the Subcontractor or its trustee to reject or to assume this Agreement and provide adequate assurance of its ability to perform, may avail itself of such remedies under this Article as are reasonably necessary to maintain the Progress Schedule. The Contractor may offset against any sums due or to become due the Subcontractor all costs incurred in pursuing any of the remedies provided including, but not limited to, reasonable overhead, profit and attorneys' fees. The Subcontractor shall be liable for the payment of any amount by which costs incurred may exceed the unpaid balance of the Subcontract Amount.

10.3 SUSPENSION BY OWNER FOR CONVENIENCE   Should the Owner suspend the Work or any part which includes the Subcontract Work for the convenience of the Owner and such suspension is not due to any act or omission of the Contractor, or any other person or entity for whose acts or omissions the Contractor may be liable, the Contractor shall notify the Subcontractor in writing and upon receiving notification the Subcontractor shall immediately suspend the Subcontract Work. To the extent provided for under the Owner-Contractor Agreement and to the extent the Contractor recovers such on the Subcontractor's behalf, the Contract Price and the Contract Time shall be equitably adjusted by Change Order for the cost and delay resulting from any such suspension. The Contractor agrees to cooperate with the Subcontractor, at the Subcontractor's expense, in the prosecution of any Subcontractor claim arising out of an Owner suspension and to permit the Subcontractor to prosecute the claim, in the name of the Contractor, for the use and benefit of the Subcontractor.

10.4 TERMINATION BY OWNER   Should the Owner terminate its contract with the Contractor or any part which includes the Subcontract Work, the Contractor shall notify the Subcontractor in writing within three (3) business Days of the termination and upon written notification, this Agreement shall be terminated and the Subcontractor shall immediately stop the Subcontract Work, follow all of Contractor's instructions, and mitigate all costs. In the event of Owner termination, the Contractor's liability to the Subcontractor shall be limited to the extent of the

Contractor's recovery on the Subcontractor's behalf under the Subcontract Documents, except as otherwise provided in this Agreement. The Contractor agrees to cooperate with the Subcontractor, at the Subcontractor's expense, in the prosecution of any Subcontractor claim arising out of the Owner termination and to permit the Subcontractor to prosecute the claim, in the name of the Contractor, for the use and benefit of the Subcontractor, or assign the claim to the Subcontractor. In the event Owner terminates Contractor for cause, through no fault of the Subcontractor, Subcontractor shall be entitled to recover from the Contractor its reasonable costs arising from the termination of this Agreement, including overhead and profit on Work not performed.

10.5 CONTINGENT ASSIGNMENT OF THIS AGREEMENT   The Contractor's contingent assignment of this Agreement to the Owner, as provided in the Owner-Contractor agreement, is effective when the Owner has terminated the Owner-Contractor agreement for cause and has accepted the assignment by notifying the Subcontractor in writing. This contingent assignment is subject to the prior rights of a surety that may be obligated under the Contractor's bond, if any. Subcontractor consents to such assignment and agrees to be bound to the assignee by the terms of this Agreement, provided that the assignee fulfills the obligations of the Contractor.

10.6 SUSPENSION BY CONTRACTOR   The Contractor may order the Subcontractor in writing to suspend all or any part of the Subcontract Work for such period of time as may be determined to be appropriate for the convenience of the Contractor. Phased Work or interruptions of the Subcontract Work for short periods of time shall not be considered a suspension. The Subcontractor, after receipt of the Contractor's order, shall notify the Contractor in writing in sufficient time to permit the Contractor to provide timely notice to the Owner in accordance with the Owner-Contractor agreement of the effect of such order upon the Subcontract Work. The Subcontract Amount or Subcontract Time shall be adjusted by Subcontract Change Order for any increase in the time or cost of performance of this Agreement caused by such suspension. No claim under this Paragraph shall be allowed for any costs incurred more than fourteen (14) Days prior to the Subcontractor's notice to the Contractor. Neither the Subcontract Amount nor the Progress Schedule shall be adjusted for any suspension, to the extent that performance would have been suspended, due in whole or in part to the fault or negligence of the Subcontractor or by a cause for which Subcontractor would have been responsible. The Subcontract Amount shall not be adjusted for any suspension to the extent that performance would have been suspended by a cause for which the Subcontractor would have been entitled only to a time extension under this Agreement.

10.7 WRONGFUL EXERCISE   If the Contractor wrongfully exercises any option under this Article, the Contractor shall be liable to the Subcontractor solely for the reasonable value of Subcontract Work performed by the Subcontractor prior to the Contractor's wrongful action, including reasonable overhead and profit on the Subcontract Work performed, less prior payments made, together with reasonable overhead and profit on the Subcontract Work not executed, and other reasonable costs incurred by reason of such action.

10.8 TERMINATION BY SUBCONTRACTOR   If the Subcontract Work has been stopped for thirty (30) Days because the Subcontractor has not received progress payments or has been abandoned or suspended for an unreasonable period of time not due to the fault or neglect of the Subcontractor, then the Subcontractor may terminate this Agreement upon giving the Contractor seven (7) Days' written notice. Upon such termination, Subcontractor shall be entitled to recover from the Contractor payment for all Subcontract Work satisfactorily performed but not yet paid for, including reasonable overhead, profit and attorneys' fees, costs and expenses. However, if the Owner has not paid the Contractor for the satisfactory performance of the Subcontract Work through no fault or neglect of the Contractor, and the Subcontractor terminates this Agreement under this Article because it has not received corresponding progress payments, the Subcontractor shall be entitled to recover from the Contractor, within a reasonable period of time following termination, payment for all Work executed and for any proven loss, cost or expense in connection with the Work, including all demobilization costs plus reasonable overhead and profit on Work not performed. The Contractor's liability for any other damages claimed by the Subcontractor under such circumstances shall be extinguished by the Contractor pursuing said damages and claims against the Owner, on the Subcontractor's behalf, in the manner provided for in subparagraphs 10.3 and 10.4 of this Agreement.

ARTICLE 11: DISPUTE RESOLUTION

11.1 WORK CONTINUATION AND PAYMENT   Unless otherwise agreed in writing, the Subcontractor shall continue the Subcontract Work and maintain the Progress Schedule during any dispute mitigation or resolution proceedings. If the Subcontractor continues to perform, the Contractor shall continue to make payments in accordance with this Agreement.

11.2 NO LIMITATION OF RIGHTS OR REMEDIES   Nothing in this Article shall limit any rights or remedies not expressly waived by the Subcontractor which the Subcontractor may have under lien laws or payment bonds.

11.3 MULTIPARTY PROCEEDING   The Parties agree that all parties necessary to resolve a claim shall be parties to the same dispute resolution proceeding. To the extent disputes between the Contractor and Subcontractor involve in whole or in part disputes between the Contractor and the Owner, disputes between the Subcontractor and the Contractor shall be decided by the same tribunal and in the same forum as disputes between the Contractor and the Owner.

11.4 DISPUTES BETWEEN CONTRACTOR AND SUBCONTRACTOR   In the event that the provisions for resolution of disputes between the Contractor and the Owner contained in the Subcontract Documents do not permit consolidation or joinder with disputes of third parties, such as the Subcontractor, or if such dispute is only between the Contractor and Subcontractor, then the Parties shall submit the dispute to the dispute resolution procedures set forth in Paragraph 11.5.

11.5 CONTRACTOR-SUBCONTRACTOR DISPUTE RESOLUTION

11.5.1 DIRECT DISCUSSIONS   If the Parties cannot reach resolution on a matter relating to or arising out of the Agreement, the Parties shall endeavor to reach resolution through good faith direct discussions between the Parties' representatives, who shall possess the necessary authority to resolve such matter and who shall record the date of first discussions. If the Parties' representatives are not able to resolve such matter within seven (7) Days, the Parties' representatives shall immediately inform senior executives of the Parties in writing that resolution was not affected. Upon receipt of such notice, the senior executives of the Parties shall meet within seven (7) Days to endeavor to reach resolution. If the matter remains unresolved after fifteen (15) Days from the date of first discussion, the Parties shall submit such matter to the dispute resolution procedures selected in Article 11.

11.5.2 MEDIATION   If direct discussions pursuant to Subparagraph 11.6.1 do not result in resolution of the matter, the Parties shall endeavor to resolve the matter by mediation through the current Construction Industry Mediation Rules of the American Arbitration Association, or the Parties may mutually agree to select another set of mediation rules. The administration of the mediation shall be as mutually agreed by the Parties. The mediation shall be convened within thirty (30) working Days of the matter first being discussed and shall conclude within forty-five (45) working Days of the matter being first discussed. Either Party may terminate the mediation at any time after the first session, but the decision to terminate shall be delivered in person by the terminating Party to the non-terminating Party and to the mediator. The costs of the mediation shall by shared equally by the Parties.

11.5.3 BINDING DISPUTE RESOLUTION   If the matter is unresolved after submission of the matter to a mitigation procedure or mediation, the Parties shall submit the matter to the binding dispute resolution procedure selected herein: (Designate only one) _____ Arbitration using the current Construction Industry Arbitration Rules of the American Arbitration Association or the Parties may mutually agree to select another set of arbitration rules. The administration of the arbitration shall be as mutually agreed by the Parties. _____ Litigation in either the state or federal court having jurisdiction of the matter in the location of the Project.

11.6 COST OF DISPUTE RESOLUTION   The costs of any binding dispute resolution procedure shall be borne by the non-prevailing Party, as determined by the adjudicator of the dispute.

11.7 VENUE   The venue for any binding dispute resolution proceeding shall be the location of the Project unless the Parties agree on a mutually convenient location.

ARTICLE 12: MISCELLANEOUS PROVISIONS

12.1 GOVERNING LAW   This Agreement shall be governed by the law in effect at the location of the Project.

12.2 SEVERABILITY   The partial or complete invalidity of any one or more provisions of this Agreement shall not affect the validity or continuing force and effect of any other provision.

12.3 NO WAIVER OF PERFORMANCE   The failure of either Party to insist, in any one or more instances, upon the performance of any of the terms, covenants or conditions of this Agreement, or to exercise any of its rights, shall not be construed as a waiver or relinquishment of term, covenant, condition or right with respect to further performance.

12.4 TITLES   The titles given to the Articles and Paragraphs of this Agreement are for ease of reference only and shall not be relied upon or cited for any other purpose.

12.5 OTHER PROVISIONS AND DOCUMENTS   Other provisions and documents applicable to the Subcontract Work are set forth in Exhibit _____.

12.6 JOINT DRAFTING   The Parties expressly agree that this Agreement was jointly drafted, and that they both had opportunity to negotiate its terms and to obtain the assistance of counsel in reviewing its terms prior to execution. Therefore, this Agreement shall be construed neither against nor in favor of either Party, but shall be construed in a neutral manner.

ARTICLE 13: EXISTING SUBCONTRACT DOCUMENTS

13.1 INTERPRETATION OF SUBCONTRACT DOCUMENTS

13.1.1 The drawings and specifications are complementary. If Work is shown only on one but not on the other, the Subcontractor shall perform the Subcontract Work as though fully described on both consistent with the Subcontract Documents and reasonably inferable from them as being necessary to produce the indicated results.

13.1.2 In case of conflicts between the drawings and specifications, the specifications shall govern. In any case of omissions or errors in figures, drawings or specifications, the Subcontractor shall immediately submit the matter to the Contractor for clarification by the Owner. The Owner's clarifications are final and binding on all Parties, subject to an equitable adjustment in Subcontract Time or Price pursuant to Articles 5 and 6 or dispute resolution in accordance with Article 11.

13.1.3 Where figures are given, they shall be preferred to scaled dimensions.

13.1.4 Any terms that have well-known technical or trade meanings, unless otherwise specifically defined in this Agreement, shall be interpreted in accordance with their well-known meanings.

13.1.5 In case of any inconsistency, conflict or ambiguity among the Subcontract Documents, the documents shall govern in the following order: (a) Change Orders and written amendments to this Agreement; (b) this Agreement; (c) subject to Subparagraph 13.1.2 the drawings (large scale governing over small scale), specifications and addenda issued prior to the execution of this Agreement; (d) approved submittals; (e) information furnished by the Owner pursuant to Paragraph 4.5; (f) other documents listed in this Agreement. Among categories of documents having the same order of precedence, the term or provision that includes the latest date shall control. Information identified in one Contract Document and not identified in another shall not be considered a conflict or inconsistency. As defined in Paragraph 2.3, the following Exhibits are a part of this Agreement.

EXHIBIT _____ The Subcontract Work, _____ pages.

EXHIBIT _____ The Drawings, Specifications, General and other conditions, addenda and other information. (Attach a complete listing by title, date and number of pages.)

EXHIBIT _____ Progress Schedule, _____ pages.

EXHIBIT _____ Alternates and Unit Prices, include dates when alternates and unit prices no longer apply, _____ pages.

EXHIBIT _____ Temporary Services, stating specific responsibilities of the Subcontractor, and Contractor _____ pages.

EXHIBIT _____ Temporary Services, stating specific responsibilities of the Subcontractor, _____ pages.

EXHIBIT _____ Insurance Provisions, _____ pages.

EXHIBIT _____ Other Provisions and Documents, _____ pages.

This Agreement is entered into as of the date entered in Article 1.

CONTRACTOR _____

BY: ....

PRINT NAME: _____

PRINT TITLE: _____

ATTEST ....

SUBCONTRACTOR: _____

BY: ....

PRINT NAME: _____

PRINT TITLE: _____

ATTEST ....

APPENDIX 2

SIMPLE FILING SYSTEM

The outline below provides an example of what a simple fining system might look like in a construction field office.

1. General Files
   a. Correspondence
   b. Transmittal sheets
   c. Weekly project reports
   d. Job percentage completion reports
   e. Billing records
   f. Minutes of job meetings
   g. Rental records
2. Supplier files - A section for each supplier, with folders for
   a. Submittals
   b. Correspondence
   c. Documentation
3. Change order files
   a. Change requests
   b. Field authorizations
   c. Numerically ordered change order quotations
   d. Approved change orders

4. Material File
   a. Purchase orders completed - separate by type
   b. Purchase orders outstanding - separate by type
   c. Requisitions completed
   d. Requisitions outstanding
   e. Back orders to be followed up
   f. Material release forms
   g. Receiving forms
   h. Tool and materials transfer reports

5. Labor File
   a. Labor reports
   b. Weekly / Monthly reports
   c. Productivity reports
   d. Accident reports
   e. Minutes of safety meetings

6. Claims File
   a. Forms
   b. Copies of delays correspondence
   c. Copies of instructions
   d. Copies of productivity reports
   e. Everything else that might be relevant to a claim

INDEX